Ideology and Evolution in Nineteenth Century Britain

Written over several decades and collected together for the first time, these richly detailed contextual studies by a leading historian of science examine the diverse ways in which cultural values and political and professional considerations impinged upon the construction, acceptance and applications of nineteenth-century evolutionary theory. They include a number of interrelated analyses of the highly politicised roles of embryos and monsters in pre- and post- Darwinian evolutionary theorizing, including Darwin's; several studies of the intersection of Darwinian science and its practitioners with issues of gender, race and sexuality, featuring a pioneering contextual analysis of Darwin's theory of sexual selection; and explorations of responses to Darwinian science by notable Victorian women intellectuals, including the crusading anti-feminist and ardent Darwinian, Eliza Lynn Linton, the feminist and leading anti-vivisectionist Frances Power Cobbe, and Annie Besant, the bible-bashing, birth-control advocate who confronted Darwin's opposition to contraception at the notorious Knowlton Trial.

Evelleen Richards is Honorary Professor in the School of History and Philosophy of Science, University of Sydney. She has published widely on the social analysis of contemporary scientific and medical controversies, with particular reference to the sociology of clinical trials and on the contextual history of evolutionary biology. Her most recent book, *Darwin and the Making of Sexual Selection* (2017), won the 2018 Suzanne J. Levinson Prize, awarded biennially by the History of Science Society for the "best book in the history of the life sciences and natural history".

www.routledge.com/history/series/VARIORUMCS

VARIORUM COLLECTED STUDIES

Ideology and Evolution in Nineteenth
Century Britain

Evelleen Richards

Ideology and Evolution in Nineteenth Century Britain

Embryos, Monsters, and Racial and Gendered Others in the Making of Evolutionary Theory and Culture

Routledge
Taylor & Francis Group

LONDON AND NEW YORK

First published 2021
by Routledge
2 Park Square, Milton Park, Abingdon, Oxon OX14 4RN

and by Routledge
605 Third Avenue, New York, NY 10017

First issued in paperback 2022

Routledge is an imprint of the Taylor & Francis Group, an informa business

British Library Cataloguing-in-Publication Data
A catalogue record for this book is available from the British Library

Library of Congress Cataloging-in-Publication Data
A catalog record has been requested for this book

ISBN 13: 978-0-367-50182-2 (pbk)
ISBN 13: 978-1-138-60771-2 (hbk)
ISBN 13: 978-0-429-46704-2 (ebk)

DOI: 10.4324/9780429467042

Typeset in Times New Roman
by Apex CoVantage, LLC

VARIORUM COLLECTED STUDIES SERIES CS1089

For my daughters, Morgan, Caitlin and Kirstie,
who grew up alongside these essays

CONTENTS

ix

CONTENTS

FIGURES

ACKNOWLEDGMENTS

Grateful acknowledgement is made to the following institutions and publishers for their permission to reproduce the articles included in this volume: Chapter 1, Cambridge University Press www.cambridge.org/gb/academic/subjects/literature/english-literature-1700-1830/romanticism-and-sciences?format=PB&isbn=9780521356855; Chapter 2, Cambridge University Press; Chapter 3, Springer Nature www.springer.com/gp/book/9789027714770; Chapter 4, University of Chicago Press; Chapter 5, Springer Nature; Chapter 6, Cambridge University Press http://services.cambridge.org/hk/academic/subjects/history/history-science-and-technology/history-humanity-and-evolution-essays-john-c-greene?format=HB&isbn=9780521335119; Chapter 7, University of Chicago Press www.press.uchicago.edu/ucp/books/book/chicago/V/bo3642614.html.

For their supply of and permission to use illustrative images, I thank the British Library; the Wellcome Library; the Hunterian Museum of the Royal College of Surgeons, London; and the University of Sydney Library, Rare Books and Special Collections. I am very grateful to those libraries and archives that allowed access to unpublished manuscript material in their possession; notably the Darwin Papers in Cambridge University Library; the Bradlaugh Papers in the Bishopsgate Institute; the Huxley Papers in Archives, Imperial College London; the Richard Owen Papers in the Natural History Museum, London; and the Ethnological Society of London minutes and related material, Archives, Royal Anthropological Institute, London. My warmest thanks go to my daughter Morgan Richards who Photoshopped the illustrations.

INTRODUCTION

This volume brings together a collection of interrelated essays in the history of evolutionary biology. They were written over several decades, dating from the early 1980s, when I adopted the then emerging contextual historiography of evolutionary biology as a refreshing and necessary corrective to the positivist historiography that dominated evolutionary studies. These essays insist on the interpretation of evolutionary theory as embedded in an ideological context that is shaped by professional and institutional power plays and the larger issues of the day. They laid the groundwork for my major contextual study of the genesis and reception of Charles Darwin's concept of sexual selection (E. Richards 2017). They are of interest in their own right, as offering detailed analyses of theory construction that open up conventional narratives of "discovery" to competing players and alternative interpretations, giving them context and contingency, and through their foregrounding of racial and gender issues in subsequent ideological battles over the definition and control of Darwinism.

The collection is divided into two parts, which reflect my personal development as an historian of evolutionary biology in response to larger shifts in the historiography and sociology of science and gender studies and to the on-going emergence of the archival and other resources that made this possible. Part I, comprising four chapters, examines the ways in which embryos and monsters were inserted in various constellations of political, institutional, and professional power relations in the construction of a number of early evolution theories, including Darwin's. Part II consists of three published essays and one previously unpublished (based on two interrelated papers delivered at international conferences in the history of science), which concern the intersection of Darwinian science and its practitioners and critics with issues of race, gender, and sexuality.

Large, indeed ideological, disagreements have led to differences over the historical interpretation of "Darwinism" and evolutionary biology. This volume does not rise above these disputes but engages with them. The studies it contains were written against the canonical "precursor" history, which retold partisan narratives of heroic conflicts and victories, picking out tell-tale prefigurations of incipient Darwinism, marginalising "losers", and rigidly excluding "externalist" considerations – social ideology, institutions, class, race, and gender – from the progressive

1

march of "pure" science. In the 1970s, when I began my research, the history of evolutionary biology was dominated by philosophers and scientist/historians who rigorously policed their collective claims that Darwinism had ousted religion and superstition from science, that Darwinism itself was objective and value free, and that so-called Social Darwinism was an abuse of pure science. These carefully maintained distinctions were just beginning to be eroded by specialist historians, the most iconoclastic being the American Marxist historian Robert M. Young, then based in Cambridge. Young argued forcefully that Darwinism was "social from the start", that we could not readily separate evolutionary biology from ideology, then or now. He insisted on the historiographic necessity of relating theory content to context and coined the term "contextualist" for the kind of history he was advocating.[1]

Young's radical claims provoked strong reaction, but they were a breath of fresh air to younger historians like myself and very influential, even on those who continued to consider themselves internalist in orientation. Alongside and reinforcing the historiographic turn to contextualism came the post-Kuhnian social constructivist turn in science studies. The "new" sociology of scientific knowledge (SSK) undercut the standard view of the special epistemological status accorded to science. It promoted the interpretation of scientific knowledge claims as not directly given by nature but as the products of social processes and negotiations that mediate scientists' accounts of the natural world. Accordingly, science is not made by heroic individuals, but it is a social process, dependent on networks of power, professional affiliations, audiences, and institutions.

Yet another significant contribution to the reconfiguration of evolutionary history was the introjection of critiques of Darwin and Darwinism by second-wave feminists then engaged in the fiercely contested sociobiology wars of the 1970s. Although generally disregarded by evolutionary historians and structured within the standard view of science, so standing apart from SSK, these early feminist critiques brought to the fore the perceived "bias" of Darwinian conceptions of female abilities, sexuality, and evolution. They provoked my earliest attempt at an integration of contextual historiography with SSK and feminist analyses in the 1983 paper, "Darwin and the Descent of Woman", which leads the second part of this collection. I shall come back to this early paper, which is best considered in relation to the papers that follow it in Part II.

It is useful to indicate how the articles in the first grouping (Part I) connect to my doctoral thesis at the University of New South Wales (1976). This was a history of the intersection of embryology and evolution theory in Britain up to the publication of the *Origin of Species* in 1859. Anyone working in this field today would go to the indispensable websites and other electronic resources that facilitate long-distance research, allowing researchers to have at hand the contents of whole archives and the plethora of secondary sources that comprise the so-called

---

1 Young's pioneering studies are collected in Young 1985.

Darwin industry. However, Australia in the 1970s was a very long way from what was then a still nascent and highly centralised business. The opportunities to engage with archival, explicational, and contextual possibilities were restricted to what was available in Australian libraries (which, wonderfully, did hold excellent collections of early Victorian periodicals and medical and zoology text books), spiced with some laboriously acquired material sourced from international archival collections. My discovery of Young's inspirational studies came too late to make much impact on my essentially internalist analysis, though I made some gestures towards the cultural "milieu".

My central concern was with tracing British "pre-Darwinian" deployments of the transcendental "law of development" that governed both individual development (ontogeny) and the history of life on earth. The concept that the embryo, as it develops, successively repeats or recapitulates the adult forms of "lower" organisms (or, in the Estonian Karl Ernst von Baer's version, their diverging embryonic forms) was closely associated with German romantic thought and French transcendentalism. The romantic gestation of nature was sometimes given an overt evolutionary interpretation, notably by the French anatomist, Étienne Geoffroy St Hilaire. It had a certain empirical validation in the similarities discerned between embryonic forms and the forms of lower adult organisms – such as the transient branchial arches of the human embryo and the gills of a fish. In accordance with this linking of phylogeny with embryogenesis, "monstrosities" (malformed human foetuses), which seemingly resembled non-human animals, were interpreted as "arrests of development". This latter interpretation, a commonplace of early teratology (the study of abnormalities of development), particularly in French transcendental versions promoted by Geoffroy and his son Isidore, was a specific source of transformist or evolutionary speculation: if arrest of development resulted in monstrosity, then prolongation or deviation of development might lead to the origin of a new, viable species.

This was the basis of the "development hypothesis" of the anonymous evolutionary potboiler, *Vestiges of the Natural History of Creation* (1844), secretly authored by Robert Chambers, the prominent Edinburgh publisher. *Vestiges*, although it had undoubtedly put evolution "in the air", was viewed largely as a mere amateur production, notorious product of "pseudo-science", a deviation from the critical path to "Darwinism" that had obscured rather than advanced the acceptance of evolution by serious British scientists.[2] Other British theorists and anatomists whose work could be identified as connected to one degree or another with early attempts to articulate a connection between individual development and the evolution of life on earth included Erasmus Darwin (grandfather of Charles) and the Edinburgh-based or educated Robert Edmond Grant, Robert Knox and Richard Owen. Grant and Knox were virtually unknown to historians of

---

2 Lovejoy and Hodge were among the earliest to challenge this orthodoxy; Lovejoy 1959; Hodge 1972.

3

evolutionary biology, while Owen, though recognised as a major re-interpreter of romantic morphological concepts (through his adaptations of the vertebrate archetype, unity of plan, homologies, repetition of parts, and the principle of polarity to the dominant Cuvierian functionalism), was far better known as the scheming, devious, dishonourable anti-Darwinian of the evolutionary disputes of the 1860s. Owen suffered such a bad press at the hands of the triumphant Darwinians of the late nineteenth century that it was difficult to separate the expert morphologist, well-informed embryologist, and putative evolutionist from the mythology that had accreted around this designated arch-enemy. More than this, historians, with few exceptions, toed the partisan line established by the leading Darwinian, Thomas Henry Huxley, of the vacuity and nullity of romantic science. Huxley's influential denigration of its "metaphorical mystifications" was not only designed to ridicule and undermine Owen – his institutional and professional rival – but also to promote the much vaunted "objectivity" and ideological purity of Huxley's Darwinian programme of scientific naturalism.

This Huxleyan historiographical heritage was alive and well until far into the twentieth century. Romanticism, particularly in its suspect German form of *Naturphilosophie*, being identified with mad Teutonic speculation and wild-eyed poets, was the adversary of all that science stood for. Any major scientific figure suspected of the taint of romanticism underwent a vigorous laundering and fumigation. The accepted purification ritual was either to dismiss the subject's romanticism as a fleeting youthful fling from which he emerged the wiser (Darwin and Huxley), and/or to attest his devotion to observation. Where necessary, romanticism might be sidestepped by invoking neoplatonic influence. Those figures (such as Owen) whose romanticism was blatantly constitutive of their science were readily dismissed as losers, and their losing status sheeted home to their romanticism. Even by the 1970s, when the inevitable reaction against the cavalier wholesale dismissal of romanticism had taken shape, there was a dearth of literature in English by those few historians who dealt sympathetically or knowledgeably with *Naturphilosophie* (notably Temkin 1950, 1963; Gode-von Aesch [1941] 1966; Mason 1953, Ch. 29).[3]

When it came to British romantic biology, the trickle of information virtually dried up. Apart from the obvious sources of Owen's writings and the sparse secondary literature on Owen (notably, Roy Macleod's prize winning study, 1965), one was forced back onto literary sources for clues in relation to notable figures like Coleridge or Thomas Carlyle or seized on the few hints in the history of science literature. Taken together, these pointed to the centrality of the Edinburgh

---

3 Timothy Lenoir's ground-breaking analysis of nineteenth-century German biology that differentiated among early German approaches to biology was yet to appear, although Lenoir disappointed by rejecting any formative influence of Romantic *Naturphilosophie* on the leading German morphologists of the early nineteenth century; Lenoir 1982. From 1992 on, Robert Richards has been the single most influential and prolific historian of romantic biology, focusing particularly on the relations of embryology and evolutionary thought (R. Richards 1992, 2002, 2008).

medical schools (including extramural ones, such as Knox's anatomy school) and the strong continental ties of the dissenting Scottish universities in the dissemination of German and French romantic thought in British biology. A key text was the much maligned *Vestiges* and its sources, which strongly implicated Chambers' Edinburgh milieu. This was substantiated and further developed through analyses of the published writings of Erasmus Darwin, Grant, Knox and Owen, and a survey of contemporary anatomy and physiology texts, including available translations of German texts.

The literature of the prevailing natural theology, particularly the popular writings of the devout Scottish stonemason Hugh Miller, also had to be factored into the mix. Here Peter Bowler's early study of palaeontology and the idea of progressive evolution offered valuable hints (Bowler 1976). It is arguable that it was largely through its assimilation by the orthodox followers of William Paley into natural theology that the romantic gestation of nature was so thoroughly absorbed into mid-nineteenth-century British biology. "Progressionism" (or "developmentalism", as it was sometimes known, until the term became synonymous with evolutionary speculation), identified the stages of individual development with the successive creations of higher forms of life on earth (consistent with the fossil record) after a series of divinely invoked universal catastrophes. It was given the imprimatur of the eminent Swiss/American naturalist and special creationist Louis Agassiz and promoted by British geologists and naturalists into one of the leading manifestations of the Paleyan argument from design. It only became suspect and subject to interrogation after the runaway popular success of *Vestiges*.

Chambers structured the "development hypothesis" of *Vestiges* around the romantic gestation of nature as it was presented in French transcendentalist sources and the available embryological and physiological literature, and he defended it, to their mounting disquiet, by citing chapter and verse from some of the most respected progressionists, including Miller and Agassiz (see Chapter 2). This presented problems not only for orthodox special creationists whose progressionist views suddenly seemed too close for comfort to the heretical *Vestiges* but also for those, like Owen, who were currently drawing on the same complexity of sources in developing their own ideas on the naturalistic origin of new species.

More to my point, it also presented problems for the historian. For one thing, Owen became far more circumspect in presenting his heterodox views. He worked to distance himself publicly from the notorious *Vestiges*, only emerging openly as an evolutionist after the publication of the *Origin of Species*. His relation to *Vestiges* and his own early (deliberately ambiguous) evolutionary views became issues of dispute among historians, issues that Chapter 2 seeks to clarify.

If Owen's views were contentious, putting Charles Darwin into the picture was an even more challenging exercise for the historian. An intense controversy centred on Darwin's embryological views and their place in his evolutionary theorising. This dispute was bedevilled by the highly influential categories and arguments established in 1916 by the British biologist and vitalist, E. S. Russell, in his campaign against the recapitulationist interpretations of the German

Darwinian and materialist monist, Ernst Haeckel (Russell 1916; Gliboff 2008, 21–4). Haeckel's "fundamental biogenetic law" (first put forward in Haeckel's *Generelle Morphologie der Organismen* of 1866), popularised as the unqualified slogan "Ontogeny recapitulates phylogeny", dominated embryological research and the search for evolutionary genealogies until the end of the nineteenth century. It then went into decline, subject to a growing barrage of criticism, largely generated by inter-disciplinary and institutional conflict, until it was rejected (though not refuted) in the 1930s (Rasmussen 1991).

Russell, for his part, distinguished two sharply distinct, opposing earlier versions of the law of embryonic resemblances: one linear, progressive one, which he associated with a suspect German *Naturphilosophie* and French transcendentalism (dubbed the Meckel-Serres law, after its two supposed leading exponents) and the "correct" version of divergent development or differentiation put forward by von Baer, which precluded resemblances between embryos and adult ancestors. Russell connected Haeckel's theory of recapitulation with the earlier romantic, idealist, progressivist camp and Darwin's embryological understandings with (more acceptable) von Baerian branching and specialisation of the developing embryo. This set the scene for a nigh on obsessive trawling and cherry picking through the literature by assorted scientist critics and historians (notably De Beer 1962; Gould 1977; Bowler 1988, 51, 82–90), who worked hard to dissociate Darwin (and Darwinism) from a "pernicious" romantic-Haeckelian tendency to relate embryos to adults of lower present-day organisms or their adult ancestors and to identify him, rather, as a prescient von Baerian. Dov Ospovat offered the more refined interpretation that Darwin followed the linear recapitulation model in his earlier work, but by the mid 1840s had adopted the divergent von Baerian version of embryonic resemblance only to both present-day and ancestral embryos (Ospovat 1981).

In point of historical fact, the conventional view that von Baer's ideas were quite different from and contradictory of recapitulation theory did not become influential until late in the period of the biogenetic law's decline and is an artefact of that period (Rasmussen 1991). Furthermore, the received wisdom that von Baer's version of divergent development only became available to British biologists with the publication of Huxley's 1853 English translation of the relevant section of von Baer's treatise (a view much promoted by Huxley) is readily controverted. In 1836, the Edinburgh educated and German trained physician Martin Barry published a detailed explication of von Baer's embryological law in the pages of the *Edinburgh New Philosophical Journal* (1836–37b).[4] Barry's influential interpretation, complete with its diagrammatic illustration of divergent von Baerian development, was the obvious original source of the *Vestiges* version (see Chapter 2); it also was picked up by Owen (and Darwin quite probably derived

---

4 Another important source of von Baer's embryology was the English translation of Johannes Müller's physiology text, which appeared in two editions (1838, 1840).

his early understanding of divergent development from Barry as well) and given wide circulation by the physiologist William Carpenter (also Edinburgh educated) in his *Principles of General and Comparative Physiology* (1839). Carpenter's *Physiology*, which went through four editions by 1854, was quickly adopted as the standard physiological text in medical schools. It was a major source of von Baer's embryology in Britain until Huxley's translation. Even then, British anatomists and physiologists continued to gloss von Baer's law with a linear, progressivist overlay.

For many years, I too worked within Russell's established categories, but I found instead that the majority of early British popularisers, anatomists, and naturalists, including Chambers, Owen, and Darwin, did not. They routinely conflated the two versions of linear and branching development, seeing no great contradiction between them. And indeed, there is not. They are reconcilable, and Darwin (and Haeckel) resolved them. Historical opinion is now cohering around this revised interpretation (R. Richards 1992, 111–15, 169–80; Gliboff 2008; Nyhart 2009; E. Richards 1976, 358–93, 2017, Ch. 9).

This leads me to the next hotly contested historical issue: Darwin's relation to romanticism. Much ink has been expended on this dispute, which currently divides Darwin scholars into two major camps: those like Philip Sloan and Robert Richards who invoke direct German idealist or Humboldtian influences, while Michael Ruse, Jonathan Hodge, and Bowler would prioritise Darwin's Anglo-centric background, inspiration, and sources. My position was – and remains – that in explicating his version of embryogenesis and its evolutionary implications, there is no need to collapse Darwin into German idealism or historicism or French transcendentalism. There were more than enough diverse interpretations of embryogenesis abroad in mid-nineteenth-century Britain – notably in the evolutionary writings of Darwin's Edinburgh mentor, Grant and those of his grandfather Erasmus (see also Hodge 1985, 2005, 115), not to mention the assimilation of the romantic gestation of nature into Paleyan progressionism and the argument from design – to provide ample resources for Darwin's early notebook interest in generation and its relation to ancestry.

It is necessary to explain that in the late 1970s my personal and professional commitments in Australia forced a change of research direction. In 1976, I had managed a brief (and discouraging) visit to the then largely uncatalogued Darwin collection in the Cambridge University Library archives. I had been invited to revise my dissertation by a major publisher on the understanding that I would sharpen my analysis with archival research. This, long before the advent of the internet, was difficult enough to do from the antipodes; however, the birth of my triplet daughters at the beginning of 1978 made it utterly impracticable. The Cambridge archives were too hard to reach and would require prolonged immersion to achieve anything worthwhile. Meanwhile my then-department at the University of Wollongong (where I had been teaching a course on the "Darwinian Revolution" for some years) was moving away from traditional approaches and areas to prioritise the social analysis of contemporary science and technology. I came under some pressure to accommodate my teaching and research to this development. Encouraged by the burgeoning SSK literature, I determined to put my medical background (another story) to good use and to focus on the cancer clinical trial as a largely sociologically untouched test case of the objectivity and neutrality of the evaluation of contentious medical knowledge. For the next fifteen years or so, with the aid of a number of grants, I worked primarily in this field, though I did not abandon evolutionary history. Rather, the two fields of study reinforced one another. My attempts at symmetrical analysis of the vitamin C and cancer dispute (together with its comparative analysis with ongoing controversies over conventional cancer treatments) offered valuable pointers for the even-handed treatment of the competing claims of "losers" and "winners" in nineteenth-century battles over evolution. This problematised the historiographic resistance to sociological explanation or, where it was invoked, its asymmetric application to "losers".

Over the years, as family responsibilities and professional opportunity permitted, I was able to enrich my earlier work on embryology and evolution through brief but intensive bursts of research in some of the major British archival collections. By the time I returned to serious evolutionary history in the late 1980s, the Darwin industry was in full gear, and others besides Young were challenging traditional interpretations of the history of evolutionary biology. For one thing, it was becoming easier to study Darwin and other nineteenth-century evolutionists. The Darwin collection at Cambridge had undergone considerable reorganisation, and the publication of authoritative, meticulously edited transcriptions of his notebooks, manuscripts, marginalia, and some 14,000 pieces of correspondence was well underway. The fine-grained textual analyses of Darwin's intellectual development by historical specialists, such as Janet Browne, Howard Gruber, Sandra Herbert, Hodge, David Kohn, Ospovat, Sylvan Schweber, James Secord, Sloan, and others, had transformed the field.[5] Others, such as John C. Greene (1977), Greta Jones (1978), and James Moore (1991), were working to break down the barriers between Darwin and Social Darwinism. It was becoming harder to deny the connection of Darwin's theoretical constructions with Victorian politics and ideology. While Young's early work was critical to this development, full-blown contextual histories, enriched by close textual analysis and inflected by ideas and approaches from cultural studies, literary criticism, and other areas cognate to the field, such as gender studies and SSK, only began to appear in the late eighties. As Darwin studies proliferated, historians turned to the examination of other figures in evolutionary history, to Darwin's colleagues – Huxley, Wallace, Lyell, and Spencer and to the "other" or "alternative" evolutionists, notably Lamarck and Chambers.[6]

---

5 The occasion of the 100th anniversary of Darwin's death in 1982 served to generate a variety of commemorative events and a slew of publications. *The Darwinian Heritage* of 1985, edited by David Kohn, showcased a wide range of this work.

6 Secord had just begun the series of studies of Chambers' *Vestiges* that were to lead to his magnificent work *Victorian Sensation* (2000).

A small sub-industry, concerned with the reassessment of romantic contributions to early British biology – primarily via the Scottish connection – emerged. Edinburgh trained or native born comparative anatomists, naturalists, and physiologists featured in the accounts offered variously by Stephen Jacyna (1983), Philip Rehbock (1983), and Adrian Desmond (1982, 1989). Jacyna grasped the nettle and specified a "romantic programme" in British biology; Rehbock preferred the less problematic term of "idealism"; while Desmond rang the changes with "romanticism", "Platonism", "transcendentalism", and "idealism". Whatever the choice of terminology, this upsurge of scholarship gave substance and context to my earlier attempts to analyse the dispersal and deployment of romantic embryological concepts in British biology.

Desmond's richly detailed contextual studies of lesser-known evolutionary theorists and advocates, culminating in his radically revisionist *The Politics of Evolution* (1989), were of particular import. This was contextual history with a much finer grain and with more respect for textual exegesis, institutions, and professionalisation than Young's work had ever shown. It also shifted focus from the cultural elite, the middle class and aristocratic, male intelligentsia who dominated British scientific society and institutions in the first half of the nineteenth century. Desmond reconstructed early nineteenth-century evolutionary thought in relation to the reformist agendas and proto-evolutionary doctrines of self-development of those locked out of the corridors of power – the radical anatomists and medical dissenters of the 1830s and '40s – and to the highly politicised promotion of such doctrines in the pauper press by an underbelly of assorted atheists, revolutionaries, and republicans. In Desmond's analysis, Darwin was relegated to an afterword: fearing his identification with such rabble-rousers, he put off publishing his own evolutionary views for another twenty years until he could attain scientific and social credentialing and Malthusian respectability in mid-Victorian capitalism.

The revival of scholarly interest in romantic biology, Desmond's de-centring of Darwin, the greater accessibility of archival resources, and the proliferation of secondary sources all played their part in my re-worked contextual evaluations of the wildly divergent evolutionary, political, and institutional associations of the romantic gestation of nature in British biology. These are summarised in the first article in this collection (Chapter 1) and developed in greater detail with particular attention to the interpretations of Owen and Robert Knox in the remaining chapters in this section.

Owen's purported evolutionary views and his relation to *Vestiges* and the *Origin of Species* had come under fresh historical dispute. Chapter 2 is an attempt to resolve this dispute primarily through analysis and contextualisation of Owen's embryological understandings. Well before the publication of the *Origin*, this leading British anatomist and best-known exponent of romantic morphology was committed to a naturalistic theory of preordained saltatory organic descent akin to that promoted by *Vestiges*. Like *Vestiges*, Owen's evolutionary application of the embryological law of development was based on the analogy between abnormal or monstrous development and the production of new species. In asserting his property rights to this interpretation, Owen, then doyen of the Royal College of Surgeons, attempted to lay claim to its embryological underpinnings, the "true law" of embryological development, challenging variously Barry, Carpenter, and the French physiologist, Henri Milne-Edwards, for priority. But his cautious, tentative expression of his views caused professional problems for Owen by threatening his conservative ideological and institutional alliances. As a consequence, he muted his evolutionism in response to public criticism and thereby compromised his post-*Origin* attempts to lay claim to his earlier evolutionary views.

Knox was in many respects a more challenging and historically interesting evolutionary theorist than Owen. A highly controversial figure, Knox was tarnished forever by his association in the public mind with the notorious Burke and Hare, body snatchers turned murderers, who sold their victims to Knox's then-thriving anatomy school in Edinburgh. After his death, his idiosyncratic views on species and race were turned to the service of a vehemently racist and proslavery agenda that Knox had never endorsed in his own lifetime. He thus was well-known to medical historians because of his involvement in the Burke and Hare scandal (which led to the passage of the Anatomy Act of 1832) and to historians of anthropology and race through his supposedly critical role in the development of a discriminatory, politically reactionary, late-Victorian scientific racism. But, although Knox played a leading part in the teaching and propagation of romantic morphology among early British biologists and anatomists, historians of biology knew little of the man or his views. Rehbock, although acknowledging Knox's morphological significance, compounded earlier misunderstandings by classifying this outspoken materialist and "savage radical" as a philosophic idealist and anti-evolutionist (Rehbock 1983, 50–1, 56).

Chapter 3 is directed to revising Knox's enduring historical reputation as a reactionary racist and anti-evolutionary polygenist. When given his proper contextual significance as a player of note among assorted political malcontents and disaffected medical men who promoted evolutionary ideas as a means of challenging the established power of church and state, Knox is identified as a political radical and materialist who developed a highly distinctive "moral anatomy". In essence, Knox adapted romantic morphology and embryology to a theory of radical, non-directional species change based on the analogy of monstrous development. He was accordingly a monogenist who argued that the human races were the result of such monstrous change from a common "generic" origin but claimed that they were stable entities with fixed natural differences adapted to their different geographic locales. His racially determinist account of human history was essentially compatible with his radical anticolonial views, but after his death, Knox's idiosyncratic theories of racial origins and inevitable antagonism and conflict were pressed into the service of the proslavery and imperialist Anthropological Society of London and became identified with its anti-Darwinian polygenist platform. Their institutional appropriation of his views in the struggle with the Huxley-led Ethnological Society for hegemonic control of Victorian anthropology accounts for much of the historical confusion that has accreted around Knox. This interpretation also calls into question the supposed anti-evolutionist stance of the Anthropological Society and helps to explain the subsequent reconciliation of polygenism with Darwinian monogenism, a reconciliation first attempted by Alfred Russel Wallace and Huxley and endorsed by Darwin in the *Descent of Man*.

The paper that follows, "The Political Anatomy of Monsters" (Chapter 4), was directed to extending my sociologically informed analyses beyond the rejected knowledge of Owen and Knox to successful or accepted Darwinian evolutionary knowledge. It employed recently developed analytical tools and strategies in a "sociology of monsters" (Haraway 1992; Law 1991; Star 1991) in the exploration of the divergent roles of monsters in evolutionary theory, their origins in German and French transcendentalism, and their relation to the different political ideologies of their British exponents. It provided the opportunity better to comprehend the slippery concept of ideology and its complex relations to the making and breaking of scientific knowledge. This was the first (and only) time I ventured on a definition, which may bear repetition here:

> [I]deology is not to be understood simply in its classical Marxist sense as entailing the suppression of contradiction, or as false consciousness or "bad" science as opposed to truth or "good" science; I intend it, rather, more in the critical sense associated with the writings of Foucault on power, as relational and productive. Understood in this way, ideology is indissolubly linked with power. . . . Ideologies, shared or contradictory systems of belief, locate actors and actants in the continuum of power that makes or breaks scientific knowledge in the different communities of practice within the shifting social order of the larger culture or society.
>
> (Richards 1994, 404)

Put simply, ideology not only works to obscure and oppress, but it also may creatively inform and even empower. While it brings newer sociological and cultural work to bear on this interpretation and its applications, the fundamental message is similar to that argued by Young more than twenty years earlier: that ideology constitutes an "inescapable level of discourse" in the production, acceptance, and applications of evolutionary biology (Young 1971).

Chapter 4, then, details a political anatomy of certain historical monsters as deviant anatomical objects that embodied deviant evolutionary interpretations and deviant political ideologies. These include: the "tiger arm" anatomised by Knox; Owen's interpretation of the famous craniopagus skull of the "Bengali boy", still extant in the Hunterian Museum in the Royal College of Surgeons; and their divergent views on the so-called "Azteque" children, who were exhibited in London in the 1850s. My immersion in their respective bodies of work had identified these "monstrous" anatomical structures (or, in sociological terms, "actants") as of particular interest and theoretical import to Knox and Owen.

They accrued added meaning through my own familiarity with anatomical preparations and illustrations and the practice of dissection: I could, for instance, situate and understand the anatomical relations of the anomalous human supra-condyloid process in the diagrammatic illustration of Knox's "tiger arm" and so comprehend its significance for Knox's evolutionary views, and I stood transfixed before the craniopagus skull in its glass case in the Hunterian Museum, studying the articulations of the bony parts of the two heads, visualising them in Owen's hands (as described by the travelling German romantic morphologist, Carl Gustav Carus) and seeking to interpret their precise relations in terms of Owen's theory of the vertebrate archetype and of the laws governing its development.[7] Following this, the underlying ideological and political differences between the overtly simi-lar non-directional, saltatory theories of Knox and Owen could be made explicit through analysis of their conflicting observations of the same set of "monsters", the "Azteque" children.

The next stage in this analytical process was to bring Darwin and Darwinism into the frame, to extend the analysis to Darwin's interpretation of monstrosity and his early rejection of sudden, monstrous change as a viable means of spe-cies production. In Darwin's theorising, monsters lost the central explanatory role accorded them by transcendental inspired saltationists and were relegated to the artificial and the unnatural, to the domestic production of fancy breeds and other curiosities that could only be perpetuated through the artifice of breeders. What I dubbed Darwin's "domestication of the monster" was consistent with his insist-ence on the evolutionary necessity of gradual, continuous, minute change and his rejection of radical or revolutionary change in both the political and biological senses.

Finally, the implications of this political anatomy of monsters for the recent revival of the debate over the evolutionary legitimacy of monstrous change or macromutations in the theory of punctuated equilibria were addressed. In particu-lar, my account challenged Michael Ruse's asymmetric attribution of ideological content to Stephen Jay Gould's promotion of punctuated equilibria theory that left unexamined the political/ideological contingency of conventional theories of gradualist, continuous evolutionary change (Ruse 1993).

If the elaborate theoretical superstructure of Chapter 4 (originally published in *Isis* in 1994) seems a bit too clever and its tone more than a little polemical, its location in contemporaneous historiographical disputes over the legitimacy of importing social ideology into Darwin studies should be considered. This was a period when those historians insisting that there was more than mere biology going on in Darwin's work were still doing battle with mainstream internalists who eschewed sociological/ideological explanation and viewed the social con-struction of scientific knowledge with considerable distrust. It was not until some ten years later that Hodge, impatient of such prevarication, might declare the

---

7 The chapters in this first grouping emphasise the fundamental necessity of rigorous contextual his-tory fully to engage with the intellectual and observational content as well as the institutional and social contexts of the science under investigation. This requirement may seem all too obvious, but it is not always carried through.

"campaign" for contextual historiography "now won" and "it is good that it is" (Hodge 2005, 119).[8]

My "political anatomy" paper marked the culmination of my contextual re-workings of the roles of monsters in pre- and post-Darwinian evolutionary theory. However, I had not done with embryos. Nor had I quite done with Knox. Owen might be left to the resurgence of interest in his life and work. Nicolaas Rupke and others have "rehabilitated" Owen, and his significance in nineteenth-century biology has undergone major reappraisal (Rupke 1994, 2009). Knox has attracted less attention in this process of rehabilitation (but see Bates 2010, 2014; Davie 2011; Dawson 2016, Ch. 8). However, although I examined his relation to Darwin's views on race, I failed to make much of what others later brought to notice: the significance of Knox's aesthetically inflected views on gender and beauty (Callanan 2006, 44–6, 70–4; Neher 2011). I subsequently was able to make a good deal more of the intersection of beauty, sexuality, and race in Knox's writings and their import for Darwin's theorising on sexual selection. We now know that Darwin's earliest datable use of his neologism "sexual selection" occurred on the final page of his notes on Knox's *Races of Men* (E. Richards 2017, 314–30, 325).

This oversight is all the more surprising in that I had written on sexual selection, gender, and evolution some ten years earlier. The paper on "Darwin and the Descent of Woman" (which leads us now to the collection of chapters in Part II) was my earliest foray into contextual evolutionary historiography. This came about as a result of my involvement in setting up one of the earliest women's studies courses in Australia. It was written as a response to what I perceived as the limitations of the existing literature. Further to this, it might be remembered that of the thirty-two contributing authors (initially gathered together to mark the occasion of the centenary of Darwin's death), to the landmark collection, *The Darwinian Heritage* (1985), which showcased the "present rich state of historical work on Darwin and Darwinism" (Kohn 1985, 1), only three were women. The focus was almost exclusively on natural selection, and only one paper in this collection dealt, albeit tangentially, with issues of gender or sexuality (Beer in Kohn 1985, 543–88).

What little had been written on sexual selection and Darwin's views was to my newly awakened mind unsatisfactory, being either subsumed under the neo-Popperian mantle of the hypothetico-deductive method as Darwin's "brilliant" value-free hypothesis (the scientist historian Michael Ghiselin was the leading exponent of this interpretation) or condemned by outraged feminists as redolent of Victorian sexist ideology and motivated by anti-feminism. Most early feminist studies fell into the category of "feminist empiricism", i.e., they maintained the standard view of science as objective and value-free, while condemning Darwin's ideologically inflected "bad" science. Their solution to such androcentric science

---

8 Even so, his declaration was somewhat premature, as Hodge himself subsequently acknowledged (2008, viii–ix).

was to ensure adherence to the proper scientific methodology. On the other hand, I could find no recognition of feminist concerns among those critical of the standard view and committed to explicating the necessarily social contingency of scientific knowledge. Young's evolutionary history was regrettably gender-blind, as was constructivist SSK as practiced by its then leading exponents. My 1983 paper, then (Chapter 5), was an attempt to move beyond both scientistic defences of Darwin's methodology and simplistic charges of sexism and anti-feminism, by locating the man and his theorising in their intellectual, cultural, and social contexts.

Written in the spirit of the times, this chapter deals only sketchily with sexual selection, being more concerned with examining Darwin's socially derived ideas on women and their place in his evolutionary theorising, nor does it pay sufficient attention to racial issues. But a number of aspects of this early analysis were to persist in my subsequent writings on Darwin and Darwinism. Darwin's domestic relations had been of interest to historians and biographers only in so far as they added human colour to the great man or his deference to Emma Darwin's religious beliefs might be invoked as a ready-made explanation for the famous twenty-year delay between the inception of Darwin's theory of evolution and its publication in 1859. Rather, I examined his lived experience as Victorian husband and father in order to bring it to bear on his theorising, by arguing that his close domestic relations did not challenge Victorian stereotypes but conformed to them. His home life and family relations were not just sources of comfort but offered models of normative behaviour and gender roles that entered into Darwin's evolutionary theorising.

Further to this, I was able to bring out Darwin's differences with John Stuart Mill, both in the pages of the *Descent of Man* and in their contrasting experiences and expectations of women and their roles. When I finally viewed them in the Cambridge Library archives many years later, I was riveted by Darwin's carefully preserved extended notes on Mill's *Subjection of Women* (1868), to see his repeated attempts to make absolutely unambiguous his biologically based refutation of Mill's egalitarian thesis that gender differences were not innate but socially and culturally induced. These notes brought home the intensity of Darwin's conviction of the rightness of his insistence on feminine intellectual inferiority and of the bearing of his theoretical views on this (E. Richards 2017, 441–9). When I wrote this early paper (Chapter 5), my prior immersion in the history of embryology and evolutionary theory alerted me to a significant embryological component in Darwin's thinking on sexual dimorphism, both animal and human. I could link this with Darwin's refutation of Mill and trace it back to the period of his early notebooks. Thus, from a very early period of his theorising Darwin had argued that human females, like female birds, were more embryonic or primitive, both developmentally and evolutionarily, than males, i.e., woman's anatomy was more juvenile or childlike than man's and therefore representative of an earlier ancestral stage. Hence Darwin's notorious claim in the *Descent* that feminine intelligence was less evolved or more primitive, akin to that of the "lower races". In other words, this claim was not simply contingent upon his culturally conditioned sexism and racism but a necessary consequence of his theoretical views, specifically his views on embryology and inheritance. Darwin's embryological argument for sexual selection was to assume increasing importance in my later account of the making of sexual selection (E. Richards 2017).

At this early stage, I could, after Young, talk in a larger sense about Darwin's fundamental commitment to naturalistic explanation of human mental and moral characteristics, then go on to link this to the debate on the "woman question" in the contexts of socio-economic and class considerations of the mid- to late Victorian period, but I lacked the tools and material for finer analysis. It was not until I was able to draw on subsequent archival research and further contextual analyses that I could make the move to more detailed explorations of the intersection of Darwinian science with issues of gender and sexuality. These explorations also necessarily involved racial issues; indeed, it was in the course of researching Robert Knox's relation to the conflict between the anti-Darwinian Anthropological and the Darwinian dominated Ethnological Societies that I happened upon Eliza Lynn Linton's impassioned eight-page petition to Huxley, arguing against his stated intention of excluding women "visitors" from Ethnological Society meetings.

Linton was one of the most successful woman journalists in Victorian England. She founded her journalistic success on her sensational "Girl of the Period" essays in the popular *Saturday Review*. These essays fully endorsed the Victorian conventions of womanhood and vehemently castigated just about everything represented by nineteenth-century feminism: women's suffrage, birth control, higher education, and entry to the professions. Linton underpinned her anti-feminism with Darwinian argumentation, insisting on the biological basis of the continuing intellectual inferiority and domesticity of women. Nevertheless, Linton did support some aspects of women's rights, notably women's right to an education including some training in science, and it was Linton who offered the only documented resistance to Huxley's exclusion of women from the Ethnological Society. Linton's petition, symbolically neglected by historians of evolution and anthropology, provided a rare opportunity to locate this prominent woman intellectual in precise relation to the institutional and wider socio-political contexts of Darwinian science and its practitioners and to explore its gendered character.

This led to the paper here reproduced as Chapter 6, "Huxley and woman's place in science", in which Huxley's reputedly "enlightened" position on the woman question is reassessed in the context of his leading role in the anthropological/ethnological conflict. Here, this conflict, conventionally viewed by historians in terms of the racial and intellectual differences between the two societies, is reconfigured to include a significant gender dimension. The admission of "Ladies" to Ethnological Society meetings was a major precipitating factor in the establishment of the breakaway Anthropological Society, whose leading lights were as misogynistic as they were racist. The Anthropologicals subsequently played a prominent role in the "scientific" refutation of the claims by nineteenth-century liberal feminists for social and intellectual parity. Huxley, by contrast, is conventionally viewed as promoting female education and entry to the professions. His position on the woman question is problematised by close analysis of his writings on the issue and by his leading part in excluding "amateur" women from the Ethnological Society in order to further the professional status of ethnology and his goal of amalgamation with the anti-feminist Anthropologicals. This was a time when public lectures on science constituted almost the only form of scientific education available to women. Linton's confrontation of Huxley not only exposes Huxley's own manipulations of the woman question, but also brings to the fore all the contradictions of her position as a woman and a Darwinian in Victorian society.

Chapter 7 deepens and extends this analysis by contrasting Linton's responses to Darwinian science and its institutions with those of the feminist, theist, and leading anti-vivisectionist Frances Power Cobbe. The two case studies were intended to uncover something of the diversity and complexity of Victorian feminism and its general reliance on Victorian stereotypes of femininity upon which Victorian science was also contingent. They also offered a means of exploring the various strategies adopted by the dominant Darwinians in redrawing the boundaries against the incursions by women into this most masculine of professions. While Linton and Cobbe drew the qualities they attributed to women from the same model of femininity, Cobbe, in contradistinction to Linton, rejected evolutionary justification of women's subordination. Rather, she promoted female agency through feminine spiritual and moral superiority in her sustained campaign against the "cruel" experimental practices of physiologists and medical men, which she likened to the domestic abuse of women. Her antivivisection crusade illustrates the ways in which women like Cobbe tested and extended the limits of the sphere of femininity and constructed political identities for themselves on a terrain different from that of the scientists. However, Cobbe's assumption of the higher moral ground on behalf of women antivivisectionists was in direct conflict with Huxley's professionalisation strategy and his promotion of the scientist as the appropriate moral arbiter of important social questions, while the emotionality of her campaign against animal suffering conduced to the Victorian feminisation of feeling and the masculinisation of reason. For Huxley and Darwin, Cobbe and her "foolish" followers were the "enemy" of rational science, exploiting the conventional feminine tools of sentimentality and concern for animal suffering to threaten the progress and prestige of British science and medicine. They strongly defended the right of the scientist to animal experimentation and against accusations of "ungentlemanly" cruelty and spearheaded the organised scientific and legislative opposition to the antivivisectionist campaign.

Linton, as we saw, was debarred from the scientific society she craved by Huxley, who both excluded women from science in the name of science and redefined that science to ratify their exclusion. Cobbe's theistic ideology was more compatible with the leadership role she assumed for women in the antivivisection campaign but could not be sustained in a context of the growing authority and prestige of a science geared to the needs of a capitalist economy and the gendered nature of the public sphere. In this period it was the new secular "Priesthood", epitomised by Huxley, who articulated the dominant scientific constructions of femininity and sexuality and naturalised the barriers against feminine intellectual and social equality in the process of protecting Darwinian institutional and social interests against the threat posed by the burgeoning women's movement.

This Darwinian redrawing of traditional boundaries was made all the more effective by the problem that the views of many feminists were inflected by their enthusiasm for secular science and its methods. The appeal of a scientifically endorsed, naturalistic theory that emphasised difference, both individual and sexual, as the basis of evolutionary progress, was considerable, notably among "Gilded Age" American feminists, such as Antoinette Brown Blackwell (Hamlin 2014). Where Linton took this to the extremes of anti-feminism and the undermining of her own advocacy of the participation of women in scientific societies, Blackwell and others sought to enlist Darwin's theory of evolution in the service of women's rights by claiming a "complementary genius" for woman, one rooted in her innate maternal and womanly qualities. Social progress was dependent upon the full expression of this peculiarly female genius. While many women drew inspiration from these attempts to reshape evolutionary theory and utilise it as a feminist tool, their retreat from the egalitarian ideal had a dangerous tendency to reinforce traditional stereotypes and cater to the drawing of biological limits to feminine potentiality (Alaya 1977; E. Richards 1983).

The more radical feminist appropriations of Darwinism exploited the potential within Darwin's notion of female choice for female agency, for females as sexual selectors and the main agents of social progress. This form of Darwinism was promoted by some socialist feminists – notably by the American visionary Eliza Burt Gamble (1894) – and Alfred Russel Wallace, co-founder with Darwin of the theory of natural selection. Wallace's is the best-known of these attempts to subvert female choice to radical ends (1890). But he was not the first to do so and to draw the censure of orthodox Darwinians. Some years earlier, a version of human progress based on the Darwinian principle of female choice had been offered in a particularly threatening context of feminism, free love, birth control, and radical politics. Their source was that persistent and unwanted hanger-on to Darwinian coat-tails, Edward Bibbins Aveling, anatomy lecturer, fervent Darwinian, Secularist, and incipient Marxist. But their primary inspiration came from Annie Besant, well-known radical, feminist, freethinker, Neo-Malthusian, and Darwinian.

Chapter 8 brings the extraordinary Annie Besant into my comparative analysis of the relations of notable Victorian women with Darwinism. Besant is by far the most complex and paradoxical of the three women examined. She went from notorious, bible-bashing, birth control advocate who confronted Darwin's opposition to contraception at the sensational Knowlton Trial, via a stint as a socialist and trade union activist, to become a leading Theosophist who repudiated her former belief in the reforming powers of birth control, materialism, and socialism. Besant's role in opening up the public discussion of contraception is well-acknowledged, in spite of her subsequent renunciation. Less-recognised is the part she played (along with Aveling) in the dissemination and promotion of evolutionary ideas relevant to late-Victorian debates on the woman question and eugenics, particularly in relation to sexual selection. Besant was among those assorted novelists, feminists, social purists, neo-Malthusians, eugenicists, utopians, political radicals, socialists, and sex reformers who kept the notion of female choice in play during a period when it was without serious support among professional naturalists and biologists (E. Richards 2017, 491–516).

The 1877 Knowlton Trial was initiated by Besant and Charles Bradlaugh – leading Secularists, Neo-Malthusians, and birth-control crusaders – in order to challenge the law that defined contraceptive literature as obscene. Contraception, they argued, was the only rational remedy for the twin Malthusian spectres of poverty and overpopulation. Their attempt to subpoena Darwin as a witness in their defence backfired when the famous author of the *Origin of Species* informed them that his evolutionary views had long caused him to hold a "very decided" opinion in opposition to theirs. The trial went ahead without Darwin, but, as her trial testimony reveals, the redoubtable Besant devoted a large part of her eloquent defence to contesting Darwin's interpretation of the effects of birth control on natural and sexual selection and on female chastity.

The trial brought a great deal of public opprobrium down on Besant and the Secularists who, as Darwin indicated, were widely interpreted as advocating female promiscuity in the name of contraception and undermining the foundations of society. Shunned by the respectable Darwinians, Besant went on to become a leading re-interpreter of Darwinism for the secular and birth control causes. Her trial testimony became the basis of her best-selling birth control pamphlet, which was translated into several languages and sold hundreds of thousands of copies. Her relationship with Aveling provided Besant with the opportunity to deepen her evolutionary understandings by qualifying and enrolling for a science degree at London University, which had just opened its doors to women. In return, her views on women and marriage and the benefits of birth control were incorporated into Aveling's prolific writings on Darwinism and his radical reinterpretation of the "revolutionary truths" of natural and sexual selection. Their radical secular brand of Darwinism featured in the popular science classes established and taught for many years by Besant and Aveling at the Hall of Science, the London headquarters of the National Secular Society.

This brought them under parliamentary scrutiny and into conflict with a puritanical Huxley, intent on dissociating "decent" Darwinism and all it stood for from any taint of sensualism and immorality. For Huxley, Besant personified the dangerous "new woman" who threatened the old agreed lines of gender and sexuality. More than this, an increasingly conservative Huxley was as convinced as Darwin that social struggle engendered by the "Malthusian serpent" of overpopulation was necessary to social advance. It ensured the dominance of that scientific meritocracy based in the masculine virtues of reason, self-restraint, duty, and propriety, which included men like Huxley but which, by definition, excluded women. Huxley's much vaunted apolitical "ethical man", beloved of philosophers, conventionally counterpoised to a ruthless, amoral nature, has been politicised, professionalised, and institutionalised by a succession of historians (Helfand 1977; Paradis 1989; Desmond 1997; White 2003; Hale 2014) and is now gendered as well, as promoting an ethical code based firmly in traditional Victorian domestic ideology and conventional family pieties (E. Richards 1995, 2017, 512–15).

Looking back over these chapters, the combative, propagandising Huxley, rather than the retiring and increasingly sacrosanct Darwin, seems to have emerged as the villain of the piece. It was far from my intention to rescript or recast the old morality play. Darwin was no hero but a man enmeshed in the attitudes and prejudices of his time, able to see only so far beyond them. As I have argued, he failed to discern, let alone to challenge or reject, certain commonly held assumptions about sexuality, gender, class, race, and nationality, and he fed them into his theorising. By the same token, Huxley was no villain but rather, as Desmond's coruscating biography represents him, the right man in the right place at the right time. Desmond tracks the brilliant young mover and shaker as he clawed his way to the top of the British scientific heap, seizing on Darwinism to leverage power from an old ecclesiastical elite to a new technocratic elite, positioning himself as self-anointed "Agnostic Pope" of the new secular creed of evolutionary naturalism, above all, the ultimate professional who harnessed state power to create the modern laboratory, turning science from a gentlemanly pursuit into a salaried career for middle class men (Desmond 1994, 1997). Huxley's trajectory is emblematic of male intelligence, energy, power, and achievement at a time when these seemed unproblematically progressive.

By Desmond's account, Huxley's great achievement was to market an ultra-respectable agnosticism, aligned with increasingly conservative political values – an evolutionary creed appropriate to the industrial and social needs of late Victorian capitalism and imperial expansion. But he didn't do it alone, and it is arguable that Huxley ever quite achieved control of the Darwinian narrative or that scientific agnosticism was ever the undiluted success story historians were wont to credit (Lightman 2002). By the 1880s, as Ruth Barton's recent work demonstrates, the leading scientific naturalists, Huxley's like-minded colleagues and those proselytising fellow travellers who comprised the exclusive X Club, were in decline, physically and institutionally (Barton 2018). Nineteenth-century scientific naturalism itself, so closely identified with Huxley and Darwinism and long an unchallenged historiographic category, has come under renewed scrutiny and re-evaluation, and the competing perspectives that constituted its late Victorian representations are being unravelled (Dawson and Lightman 2014). The anti-Malthusian evolutionary tradition in English biology and political thought is undergoing retrieval and reassessment (Hale 2014). Recent work on the popularisation of science is uncovering the ways in which historically neglected works of Victorian popular science shaped public perceptions of evolutionary science (Lightman 2007). Evolutionary theism and spiritualism are being restored to their prominent place in Victorian culture; Wallace, one-time deluded spiritualist, suspect socialist, and feminist, is being reinstated as the "first Darwinian" (Fichman

2004; Hesketh 2019). New biographical studies of Darwin (Browne 1995, 2002), Hooker (Endersby 2008), Spencer (Francis 2007), and others are complicating and challenging earlier notions of nineteenth-century scientific practice and identity. Gender and race, once relegated to the historiographic margins, have been reconstituted as essential analytic categories (Desmond and Moore 2009; Hamlin 2014; Montgomery 2013; Evans 2017). Embryology and developmentalism are enjoying renewed attention, as the practitioners of the new discipline of "Evo-Devo" seek to come to terms with its history (Amundson 2007; Laubichler and Maienschein 2007). The application of the tools of cultural history, visual studies, literary criticism, and linguistic analysis has brought fresh insights to the evolutionary debates of the nineteenth century and opened up new fields of investigation (Secord 2000; Smith 2006; Dawson 2007; Levine 2011). Before all, we have come to see Darwinism as a protean concept, as more fluid and mutable than historians once acknowledged; to view struggles over its definition and control as of the essence of its history; and to understand them as never without ideological or political content and import.

We may spare a thought for Huxley, as Desmond sympathetically evokes him towards the end of his life, old, depressive, and marked by personal tragedy, the madness and death of his beloved artist daughter, the talented and highly strung Marian (Mady) – one of those troublesome new women who daringly broke new ground in the campaign for women's right to study and represent the nude of high art, imperilling her father's much-vaunted sexual respectability, before succumbing to depression and dementia (Desmond 1997, 129, 172–9; E. Richards 2017, 239, 597). Huxley's gendered and politicised "Evolution and Ethics" (1893) was his final great effort in defence of the evolutionary naturalism he had done so much to shape and codify – he died within two years. But it was too little and too late. By the last decade of the nineteenth century, Darwinism was in disarray, beset on all sides by competing interpretations and warring factions. In the longer term and within the terms of the studies presented in this collection and their historiographical contexts, the triumphal historiography Huxley did his best to found and defend has been dismantled: romanticism has been retrieved as an influential component of nineteenth-century biological thought; monsters and embryos have been brought back to centre stage in early evolutionary thinking; Owen and Knox have been rehabilitated as significant pre-Darwinian evolutionary theorists; and the individual struggles of those important but neglected Victorian women who, from within the confines of the Victorian sphere of femininity, sought to contest and reshape the Darwinian narrative, have been reclaimed.

In conclusion, I should like to pay tribute to all those colleagues, friends, and family who, over the years, read, discussed, encouraged, critiqued, challenged, extended, and finessed the work presented here. As Young once wrote, the history of science, like science itself, is a "social activity, born of society, and mediating its structures and values" (Young 1985, 186). It is my hope these studies reach out and speak to a younger generation of scholars, who may, in their turn, debate, dispute, tweak, refine, and revise. The history of science, as historians know well, is unfinished business.

Bibliography

Alaya, Flavia. 1977. "Victorian Science and the 'Genius' of Woman." *Journal of the History of Ideas* 38: 261–80.

Amundson, Ron. 2007. *The Changing Role of the Embryo in Evolutionary Thought: Roots of Evo-Devo*. Cambridge: Cambridge University Press.

Barry, Martin. 1836–37a. "On the Unity of Structure in the Animal Kingdom." *Edinburgh New Philosophical Journal* 22: 116–41.

———. 1836–37b. "Further Observations on the Unity of Structure in the Animal Kingdom, and on Congenital Abnormalities, Including 'hermaphrodites' with Some Remarks on Embryology, as Facilitating Animal Nomenclature, Classification, and the Study of Comparative Anatomy." *Edinburgh New Philosophical Journal* 22: 345–64.

Barton, Ruth. 2018. *The X Club: Power and Authority in Victorian Science*. Chicago: University of Chicago Press.

Bates, Alan W. 2010. *The Anatomy of Robert Knox: Murder, Mad Science and Medical Regulation in Nineteenth-Century Edinburgh*. Brighton: Sussex Academic Press.

———. 2014. "Retrogressive Development: Transcendental Anatomy and Teratology in Nineteenth-Century Britain." *Medicina nei Secoli arte e Scienza* 26 (1): 197–222.

Beer, Gillian. "Darwin's Reading and the Fictions of Development." In *The Darwinian Heritage*, edited by David Kohn, 543–88. Princeton, NJ: Princeton University Press.

Bowler, Peter J. 1976. *Fossils and Progress: Paleontology and the Idea of Progressive Evolution in the Nineteenth Century*. New York: Science History Publications.

———. 1988. *The Non-Darwinian Revolution: Reinterpreting a Historical Myth*. Baltimore: Johns Hopkins University Press.

Browne, Janet. 1995. *Charles Darwin: Voyaging*. London: Jonathan Cape.

———. 2002. *Charles Darwin: The Power of Place*. London: Jonathan Cape.

Callanan, Laura. 2006. *Deciphering Race: White Anxiety, Racial Conflict, and the Turn to Fiction in Mid-Victorian English Prose*. Columbus: Ohio State University Press.

Chambers, Robert. [1844] 1969. *Vestiges of the Natural History of Creation*. Reprint, Leicester: Leicester University Press.

Davie, Neil. 2011. "Dissecting the Races of Men: Robert Knox, Anatomy and Racial Theory in Britain, 1820–1870." In *Sexe, race et mixité dans l'aire Anglophone*, edited by Michel Prum, 19–42. Paris: L'Harmattan.

Dawson, Gowan. 2007. *Darwin, Literature, and Victorian Respectability*. Cambridge: Cambridge University Press.

———. 2016. *Show me the Bone: Reconstructing Prehistoric Monsters in Nineteenth-Century Britain and America*. Chicago: Chicago University Press.

Dawson, Gowan, and Bernard Lightman, eds. 2014. *Victorian Scientific Naturalism: Community, Identity, Continuity*. Chicago: University of Chicago Press.

De Beer, Sir Gavin. 1962. "Darwin and Embryology." In *A Century of Darwin*, edited by S. A. Barnett, 153–72. London: Mercury Books.

Desmond, Adrian. 1982. *Archetypes and Ancestors: Paleontology in Victorian London, 1850–1875*. London: Blond and Briggs.

———. 1989. *The Politics of Evolution; Morphology, Medicine and Reform in Radical London*. Chicago: University of Chicago Press.

———. 1994. *Huxley: The Devil's Disciple*. London: Michael Joseph.

———. 1997. *Huxley: Evolution's High Priest*. London: Michael Joseph.

Desmond, Adrian, and James Moore. 2009. *Darwin's Sacred Cause: Race, Slavery, and the Quest for Human Origins*. London: Allen Lane.

Endersby, Jim. 2008. *Imperial Nature: Joseph Hooker and the Practices of Victorian Science*. Chicago: Chicago University Press.

Evans, Samantha, ed. 2017. *Darwin and Women: A Selection of Letters*. Cambridge: Cambridge University Press.

Fichman, Martin. 2004. *An Elusive Victorian: The Evolution of Alfred Russel Wallace*. Chicago: University of Chicago Press.

Francis, Mark. 2007. *Herbert Spencer and the Invention of Modern Life*. Ithaca, NY: Cornell University Press.

Gliboff, Sandar. 2008. *H. G. Bronn, Ernst Haeckel, and the Origins of German Darwinism: A Study in Translation and Transformation*. Cambridge, MA: The MIT Press.

Gode-von Aesch, Alexander. [1941] 1966. *Natural Science in German Romanticism*. Reprint. New York: AMS Press.

Gould, Stephen Jay. 1977. *Ontogeny and Phylogeny*. Cambridge, MA: Harvard University Press.

Greene, John C. 1977. "Darwin as a Social Evolutionist." *Journal of the History of Biology* 10: 1–27.

Hale, Piers J. 2014. *Political Descent: Malthus, Mutualism, and the Politics of Evolution in Victorian England*. Chicago: University of Chicago Press.

Hamlin, Kimberly A. 2014. *From Eve to Evolution: Darwin, Science, and Women's Rights in Gilded Age America*. Chicago: University of Chicago Press.

Haraway, Donna. 1992. "The Promises of Monsters: A Regenerative Politics for Inappropriate/d Others." In *Cultural Studies*, edited by Lawrence Grossberg, Cary Nelson, and Paula Treichler, 295–337. New York and London: Routledge.

Helfand, Michael, S. 1977. "T. H. Huxley's *Evolution and Ethics*: The Politics of Evolution and the Evolution of Politics." *Victorian Studies* 20: 157–77.

Hesketh, Ian. 2019. "The First Darwinian: Alfred Russel Wallace and the Meaning of Darwinism." *Journal of Victorian Culture* 20: 1–15.

Hodge, M. J. S. 1972. "The Universal Gestation of Nature: Chambers' *Vestiges* and *Explanations*." *Journal of the History of Biology* 5: 127–51.

———. 1985. "Darwin as a Lifelong Generation Theorist." In *The Darwinian Heritage*, edited by David Kohn, 207–43. Princeton, NJ: Princeton University Press.

———. 2005. "Against 'Revolution' and 'Evolution'." *Journal of the History of Biology* 38: 101–21.

———. 2008. *Before and after Darwin: Origins, Species, Cosmogonies, and Ontologies*. Variorum Collected Studies Series. Aldershot: Ashgate.

Huxley, Thomas Henry. [1893] 1989. "Evolution and Ethics." In *Evolution and Ethics*, edited by James Paradis and George C. Willliams, 104–174. Princeton, NJ: Princeton University Press.

Jacyna, L. Stephen. 1983. "Immanence or Transcendence: Theories of Life and Organization in Britain, 1790–1835." *Isis* 74: 311–29.

Jones, Greta. 1978. "The Social History of Darwin's *Descent of Man*." *Economy and Society* 7: 1–23.

Kohn, David, ed. 1985. *The Darwinian Heritage*. Princeton, NJ: Princeton University Press.

Laubichler, Manfred D., and Jane Maienschein, eds. 2007. *From Embryology to Evo-Devo: A History of Developmental Evolution*. Cambridge, MA: The MIT Press.

Law, John. 1991. "Power, Discretion, and Strategy." In *Sociology of Monsters: Essays on Power, Technology, and Domination*, edited by John Law, 165–91. London and New York: Routledge.

Lenoir, Timothy. 1982. *The Strategy of Life: Teleology and Mechanics in Nineteenth-Century German Biology*. Chicago: Chicago University Press.

Levine, George. 2011. *Darwin the Writer*. Oxford: Oxford University Press.

Lightman, Bernard. 2002. "Huxley and Scientific Agnosticism: The Strange History of a Failed Historical Strategy." *British Journal for the History of Science* 35: 271–89.

———. 2007. *Victorian Popularizers of Science: Designing Nature for New Audiences*. Chicago: University of Chicago Press.

Lovejoy, Arthur O. 1959. "Recent Criticism of the Darwinian Theory of Recapitulation: Its Grounds and its Initiator." In *Forerunners of Darwin, 1745–1859*, edited by Bentley Glass et al., 438–58. Baltimore: Johns Hopkins Press.

Macleod, Roy M. 1965. "Evolutionism and Richard Owen: An Episode in Darwin's Century." *Isis* 56: 259–80.

Mark Francis. 2007. *Herbert Spencer and the Invention of Modern Life*. Ithaca, NY: Cornell University Press.

Mason, Stephen F. 1953. *A History of the Sciences*. London: Routledge and Kegan Paul.

Montgomery, Georgina. 2013. "Darwin and Gender." In *The Cambridge Encyclopaedia of Darwin and Evolutionary Thought*, edited by Michael Ruse, 443–50. Cambridge: Cambridge University Press.

Moore, James R. 1991. "Deconstructing Darwinism: The Politics of Evolution in the 1860s." *Journal of the History of Biology* 24: 353–408.

Neher, Allister. 2011. "Robert Knox and the Anatomy of Beauty." *Medical Humanities* 37: 46–50.

Nyhart, Lynn K. 2009. "Embryology and Morphology." In *The Cambridge Companion to the Origin of Species*, edited by Michael Ruse and Robert J. Richards, 194–215. Cambridge: Cambridge University Press.

Ospovat, Dov. 1981. *The Development of Darwin's Theory*. Cambridge, MA: Harvard University Press.

Paradis, James. 1989. "*Evolution and Ethics* in Its Victorian Context." In *Evolution and Ethics*, edited by James Paradis and George C. Williams, 3–55. Princeton, NJ: Princeton University Press.

Rasmussen, Nicolas. 1991. "The Decline of Recapitulationism in Early Twentieth-Century Biology: Disciplinary Conflict and Consensus on the Battleground of Theory." *Journal of the History of Biology* 24: 51–89.

Rehbock, Philip F. 1983. *The Philosophical Naturalists: Themes in Early Nineteenth-Century British Biology*. Madison, WI: University of Wisconsin Press.

Richards, Evelleen. 1976. *The German Romantic Concept of Embryonic Repetition and its Role in Evolutionary Theory in England up to 1859*. PhD diss. University of New South Wales.

———. 1983. "Darwin and the Descent of Woman." In *The Wider Domain of Evolutionary Thought*, edited by David Oldroyd and Ian Langham, 57–111. Dordrecht: Reidel.

———. 1994. "A Political Anatomy of Monsters, Hopeful and Otherwise: Teratogeny, Transcendentalism, and Evolutionary Theorizing." *Isis* 85: 377–411.

———. 1995. "Gendering the Romanes Lecture: The Sexual Politics of T. H. Huxley's *Evolution and Ethics*." Paper delivered at T. H. Huxley, Victorian Science and Culture Conference, Imperial College, London.

———. 2017. *Darwin and the Making of Sexual Selection*. Chicago: University of Chicago Press.

Richards, Robert J. 1992. *The Meaning of Evolution: The Morphological Construction and Ideological Reconstruction of Darwin's Theory*. Chicago: University of Chicago Press.

———. 2002. *The Romantic Conception of Life: Science and Philosophy in the Age of Goethe*. Chicago: University of Chicago Press.

———. 2008. *The Tragic Sense of Life: Ernst Haeckel and the Struggle over Evolutionary Thought*. Chicago: University of Chicago Press.

Rupke, Nicolaas. 1994. *Richard Owen, Victorian Naturalist*. New Haven, CT: Yale University Press.

———. 2009. *Richard Owen: Biology without Darwin*. Chicago: University of Chicago Press.

Ruse, Michael. 1993. "Is the Theory of Punctuated Equilibria a New Paradigm?" In *The Darwinian Paradigm: Essays on Its History, Philosophy, and Religious Implications*, 118–45. London and New York: Routledge.

Russell, E. S. 1916. *Form and Function: A Contribution to the Study of Animal Morphology*. London: John Murray.

Secord, James A. 2000. *Victorian Sensation: The Extraordinary Publication, Reception, and Secret Authorship of* Vestiges of the Natural History of Creation. Chicago: University of Chicago Press.

Smith, Jonathan. 2006. *Charles Darwin and Victorian Visual Culture*. Cambridge: Cambridge University Press.

Starr, Susan Leigh. 1991. "Power, Technology and the Phenomenology of Conventions: On Being Allergic to Onions." In *Sociology of Monsters: Essays on Power, Technology, and Domination*, edited by John Law, 26–56. London and New York: Routledge.

Temkin, Owsei. 1950. "German Concepts of Ontogeny and History around 1800." *Bulletin of the History of Medicine* 24: 227–46.

———. 1963. "Basic Science, Medicine and the Romantic Era." *Bulletin of the History of Medicine* 37: 97–129.

Wallace, Alfred Russel. 1890b. "Human Selection." *Fortnightly Review* 48: 325–37.

White, Paul. 2003. *Thomas Huxley: Making the "Man of Science"*. Cambridge: Cambridge University Press.

Young, Robert M. 1971. "Evolutionary Biology and Ideology: Then and Now." *Science Studies* 1: 177–206.

———. 1985. *Darwin's Metaphor: Nature's Place in Victorian Culture*. Cambridge: Cambridge University Press.

Part I

Romantic embryos, radical monsters, and racial others in evolutionary theorising

1

"METAPHORICAL MYSTIFICATIONS"

The romantic gestation of nature in British biology

We know what a masquerade all development is and what peculiar shapes may be disguised in helpless embryos. In fact, the world is full of hopeful analogies and handsome dubious eggs called possibilities.

– George Eliot, *Middlemarch*, 1871

One of the most significant and distinctive features of the positivist historiographic tradition has been its denial of the positive contribution of Romanticism to science, particularly the life sciences – notoriously susceptible to Romantic contagion. This reached its nineteenth-century apotheosis in the writings of Thomas Henry Huxley, who was probably the single-most influential and destructive English-speaking critic of the "metaphorical mystifications" of *Naturphilosophie*. Huxley's denigration of Romantic science was not only designed to ridicule and undermine his professional rival, Richard Owen (whose morphology was well known to be tarred with the brush of a suspect *Naturphilosophie*), but also to promote the much vaunted and methodologically guaranteed "objectivity" and purity of the Darwinian programme. In the Huxley-led drive for Darwinian dominance of nineteenth-century science, he and the Darwinians capitalized on their collective image as "plain, prosaic inquirer(s) into objective truth", by contrasting this ideologically neutral and sober representation of themselves with the wild-eyed speculations, "oracular utterances" and general verbal gymnastics of the unruly Romantics – the adversaries of everything that Huxley's reliable and socially efficacious "natural knowledge" stood for (Huxley 1894, 2: 315; Desmond 1982). It was a stratagem that proved very effective in furthering the interrelated social and professional interests of the "young guard" Darwinians. It also left a deep and enduring impression on evolutionary history, where several generations of historians continued to toe the partisan line established by the dominant Darwinians. It is only comparatively recently that this Huxleyan historiographic heritage has been challenged, and a rich and multilayered history of the deployment of Romantic concepts in nineteenth-century biology is now emerging.

This essay is focused on the Romantic conception of the history of nature, construed as one long gestation analogous to a normal human pregnancy, and its central role in nineteenth-century evolutionary theorizing. It confronts Huxley's assessment of Romanticism head-on by arguing that it was through such "metaphorical mystification" that Romanticism made its most powerful impact on evolutionary biology. Further, I shall try to show that it was its very "mystification", so much cultivated by the Romantics and so deplored by Huxley, that gave the metaphor of gestation its explanatory power and led to its wide deployment in nineteenth-century biology.

The Romantic preference for analogy and metaphor as the means of conveying, or rather suggesting, ultimate truths is well known. The most important production of *Naturphilosophie* in the life sciences, Lorenz Oken's *Lehrbuch der Naturphilosophie* of 1809–11 (which went through three German editions in his lifetime and was translated into English in 1847), is also an exemplar of this Romantic fetishization of symbolism. To the perplexity of its English readers (and the near apoplexy of some), it was found to consist of 3,652 consecutive aphorisms or *Fragmente*, each of which could be read in isolation as a whole in itself, or as part of an extended argument. Symbols, with all their attendant ambiguities and imprecisions, thus tended to become explanations in Romantic science. And this, together with the reluctance of the Romantics to analyse the meaning of the images they evoked, their deliberate cultivation of mystery in even the simplest things, made much of *Naturphilosophie* almost incomprehensible to the uninitiated. But at the same time this very ambiguity rendered its concepts malleable to the purposes of those who appropriated and employed them in nineteenth-century biology. Ambiguity is manipulable, and it was this aspect of *Naturphilosophie* which made it such a rich and fertile source of ideas and concepts, especially to its British followers who were not bound by the constraints of the formal and systematic philosophy of nature to which Oken aspired, and who were free to combine its elements eclectically with those from a very different cultural tradition.

As it came to fruition within German Romantic philosophy, the metaphor of the gestation of nature encapsulated the organicism, the uncompromising developmentalism, anthropocentrism and insistence on the fundamental unity of all nature, of the Romantics. The Romantic universe being metaphorically an immense animal, it was animal-like in its functions and constitution, and even in its method of procreation. Romantic literature abounds with references to this "universal gestation of nature" – to the "impregnation of the terrestrial womb", the "pregnancy" of the world, the "generation", "gestation", "growth" or "development" of nature. Above all, through the metaphor of gestation, biological and historical thought could be united so that the rules and concepts of the one could be applied to the other, and this is epitomized in Oken's very definition of *Naturphilosophie* as the "generative history of the world".

It cannot be overemphasized that the task of *Naturphilosophie* was primarily historical, and Oken's definition makes this patent. *Naturphilosophie* had to demonstrate how the universe originated, and to reconstruct its development or

Entwicklung from the original Idea thought by God to its highest manifestation as man. The task of reconstruction was to be aided by the essential parallel between man's individual history, or gestation, and the universal history. "It is certain", affirmed Schelling in 1811, "that whoever could write the history of his own life from its very ground, would have thereby grasped in a brief conspectus the history of the universe."[1] This certainty was based on the fundamental Romantic tenet that man is the prototype and model of all existence – the microcosm. There is only one developmental tendency, that of producing man, who is, as Oken put it, the "summit, the crown of nature's development, and must comprehend everything that has preceded him, even as the fruit includes within itself all the earlier developed parts of the plant".[2] Thus man's individual development or ontogeny necessarily replicates the development of life on earth, the universal development reflected in that abstraction the Romantics called *Entwicklung*. Man and nature share a common *Entwicklungsgeschichte* – a history of development.

Oken, who combined the functions of Romantic ideologue and political and professional activist (he campaigned vigorously for the unification of the German states and of German science) with those of a working (and respected) embryologist, offered a detailed anatomical account of the ideal "perfect parallelism" he advocated between the forms assumed by the developing human embryo and the ascending sequence of mature forms constituting the animal series. And, typically, he presented the results of his embryological investigations in aphoristic form, invoking the metaphor of gestation: "Animals are only the persistent foetal stages or conditions of man. . . . A human foetus is a whole animal kingdom" (Oken 1847, 491–2).

To the modern eye, the metaphor has obvious and almost inevitable evolutionary or, more precisely, transmutationist implications. If animals are merely embryonic or foetal men, then a lower animal could be transmuted into a higher one through a simple prolongation of development beyond the normal termination of its ontogeny. In this way, metaphor could be realized and the gestation of nature become a literal historical process. According to Karl Ernst von Baer, its leading critic among embryologists, this was the common practice among his contemporaries. "By degrees", he wrote in 1828, "it became the custom to look upon the different forms of animals as developed out of one another, and then many appeared to forget that this metamorphosis was after all only a mode of conceiving the facts" (von Baer [1828] 1853, 186–8).

Von Baer's attack on the Romantic "law of parallelism" and the transmutationism he associated with it was based on his rejection of the animal series or taxonomy integral to both concepts. Like the eminent French comparative anatomist, Cuvier, and in opposition to the Romantic morphologists, von Baer held that there

1 Schelling 1942, 94. For the classic analysis of the relation of Romantic embryology to the rise of historicism, see Temkin 1950.
2 Oken 1847, 2. Alexander Gode-von Aesch has argued that the embryological law of parallelism was "part of the intellectual equipment of every good Romantic thinker"; Gode-von-Aesch 1966, 120.

were four basic types of organization. According to von Baer, the type is manifested in the very early stages of ontogeny, with the result that the simple unilineal parallel insisted upon by Oken and the Romantics no longer applies, and the
embryo diverges more and more from other animal forms. It therefore repeats not
the adult forms of lower animals, but their embryonic ones. However, von Baer
failed to dislodge the deeply entrenched embryological law of parallelism. In spite
of the very real differences between it and von Baer's "law of divergence", they
had an underlying continuity and essential similarity (recognized by von Baer
himself) which facilitated their confusion and conflation. As well, parallelism, as
conceived by its *Naturphilosophen* exponents was an idealization, as Oken made
clear. It was not, in this sense, amenable to von Baer's empirical investigation
and criticism, and the two laws therefore coexisted or, more often, were conflated, until their ultimate conflation in the post-Darwinian theory of recapitulation (Richards 1976; Russell 1916; Gould 1977).

Similarly, for reasons implicit in Romantic philosophy, the animal series
was not usually assumed to constitute a genealogical sequence. *Entwicklung*
was primarily an archetypal development, not an evolutionary one, and for the
majority of the *Naturphilosophen* it remained an abstraction like the Romantic
"archetype", not to be materially realized in nature. Earlier interpretations to the
contrary, Oken himself, as more careful scholarship has shown, was no transmutationist. He offered an elaborate alternative to transmutation of species by positing
the spontaneous generation of new and higher organisms from a universal organic
"infusorial mucus" or *Urschleim*, through a process of polar conflict which necessitated the interaction of material nature with the spiritual or Absolute. For all
the suggestive ambiguity of his expression, Oken rejected the idea that nature
could be explained in terms of material causality and therefore rejected the idea of
organic evolution as a real physical possibility (Oken 1847, 185–93). The problem of man's origin could not be resolved by making him emerge from an ape
as the transmutationists required, and the *Naturphilosophen* generally seem to
have been as unwilling as Bishop Wilberforce (in his celebrated confrontation
with Huxley at the "Oxford Debate" of 1860) to countenance a "miserable ape"
as grandparent. Indeed, as Richard Owen was shortly to demonstrate, *Naturphilosophie* was not incompatible with British natural theology. And it was through
its thorough assimilation into natural theology via the ideology of progressionism
that the metaphor of gestation assumed its prominent role in pre-Darwinian British biology and paleontology, rather than through direct evolutionary speculation.

Nevertheless, as von Baer attests, some of the less idealistic contemporaries of
the *Naturphilosophen* did give the metaphor of gestation a literal interpretation,
and they were aided in this by the ambiguity of Romantic symbolism and the
failure of the leading *Naturphilosophen* to clarify the meaning of this potent metaphor. The French transcendentalist, Etienne Geoffroy St Hilaire, for one, found
its evolutionary and materialistic implications irresistible. Geoffroy founded his
morphology (which emphasized serial development, parallelism, transmutation
and unity of composition), on the sovereignty of material laws, and directed it

against Cuvier and towards the young medical reformers and republicans of Paris. For Geoffroy, to view the stages of embryonic development was to see in summary form the "spectacle of the evolution of the terrestrial globe", to "catch nature in the act". He was convinced that in provoking the birth of monstrosities from hens' eggs by artificially varying the conditions of incubation, he had experimentally illustrated the way new species arose in nature, and he offered a materialistic account of this process. According to Geoffroy, mechanical and chemical changes in the environment (especially in the respiratory milieu), induced changes in the organism during the embryonic stage which were akin to monstrous development. Through their propagation by inheritance, these embryonic changes brought about the transmutation of species (Appel 1987).

Geoffroy's evolutionary speculations were taken up and adapted in various ways for different social and institutional ends by the three most prominent and influential British transcendental anatomists, Robert Edmond Grant, Robert Knox and Richard Owen. Each of these patterned his version of organic descent on variations of embryological development, and in ways consistent with his politico-institutional position. Grant and Knox, who were both attracted by the contingent radical political ties of Geoffroy's morphology, tried to produce self-consistent materialistic theories of life: Grant, the radical democrat of the University of London, modelled his progressivist transmutation on Geoffroy's traditional unilineal version of embryogenesis and made it do service to reformist institutional and social interests (Desmond 1984), while Knox, institutional and social outcast, developed a theory of "generic descent", emphasizing the abrupt non-linear embryogenesis of new species as a constitutive part of his anomalous ideology of radical racism (Richards 1989a). Owen, on the other hand, doyen of the Royal College of Surgeons (at that stage under heavy attack from the medical reformers) and aspirant to the coveted title of the "British Cuvier", harnessed Geoffroy's transcendental anatomy to conservative institutional and social needs by reconciling it with Cuvierian functionalist teleology. At the same time, in the 1840s, he also attempted to adapt Geoffroy's teratological speculations to these same conservative concerns, by devising a non-materialist theory of gross embryonic evolutionary change, analogous to "anomalous monstrous births", and divinely pre-programmed to a divergent von Baerian embryological model (Richards 1987).

However, in 1849 Owen's evolutionism, conservative as it was, was brought up short against the larger social forces of the day. The cautious evolutionary hints of his *On the Nature of Limbs* of the same year (cast in the form of the traditional Romantic ambiguity of an archetypal development) were lumped together with the recent translation (well known to have been instigated by Owen) of Oken's *Lehrbuch der Naturphilosophie*, and the popular evolutionary work, *Vestiges of the Natural History of Creation*. All were publicly castigated in the *Manchester Spectator* for their promotion of a "desolating Pantheism" in the form of the "THEORY OF DEVELOPMENT" which was undermining religious belief and contributing to the contemporary political and social unrest. In the socially troubled forties evolution was virtually synonymous with revolution (as indeed the theories of Grant and Knox were constitutive of their radical politics). The ambitious Owen reacted to these "hard epithets" by muting his evolutionism until the more liberal climate of the prosperous and secular sixties and the appearance of *The Origin of Species* made possible the voicing of his (by then socially and intellectually outmoded) views.[3]

Owen's major difficulty lay in the closeness of his views to those of the heretical and much maligned, but widely read, *Vestiges*. It was the *Vestiges*, first published anonymously in 1844 and reissued in at least twelve subsequent English editions, which brought the transmutationist implications of the "universal gestation of nature" before the middle-class British public to an unprecedented degree. Its author, Robert Chambers, the Edinburgh publisher and essayist, drew explicit inspiration for his "development hypothesis" and its mechanism from the idea that the "ordinary phenomenon of reproduction was the key to the genesis of species". In effect, Chambers demystified and "domesticated" the Romantic metaphor of the gestation of nature. The production of new species, he reassured his readers,

> has never been anything more than a new stage of progress in gestation, an event as simply natural, and attended as little by any circumstances of a wonderful or startling kind, as the silent advance of an ordinary mother from one week to another of her pregnancy.[4]

This naturalistic idea, of course, had not come "unpromptedly" into Chambers's mind as he claimed, but through his exposure to transcendentalist conceptions. By the time of the writing of the *Vestiges* these were common currency, not only in British biology (and especially so in Edinburgh, the major centre for the dissemination of French transcendental anatomy into British biology and medicine), but also in natural theology. For these malleable concepts appealed not only to the nonconformist medical men like Knox and Grant, who naturalized and "preached" them to their students, but equally to the orthodox followers of William Paley, who absorbed them into a progressivist natural theology and employed them to demonstrate God's providential design and continuing stabilizing presence in nature and the social order. Even before Owen's formal synthesis of Geoffroyan morphology and Cuvierian teleology, British natural theology had managed to reconcile *Entwicklung* with Christianity in a progressionist philosophy of nature which was both discontinuous and historical.

"Progressionism", or "developmentalism" as its exponents sometimes termed it, may be described as evolution without physical continuity. The continuity exists only in the mind of God who has created a succession of new and higher life forms after the extinction of all their predecessors through a series of divinely invoked universal catastrophes. Progressionism, that peculiarly British phenomenon,

---

3 Richards 1987. The radical political associations of transmutation in this period are brought out in Desmond 1987.

4 Chambers [1844] 1969, 223. See also the excellent discussion by Secord 1989; Hodge 1972.

formed a conceptual framework surprisingly congenial to many of the ideas, if not the Idea, of Romantic *Naturphilosophie*. This is particularly evident in the writings of Hugh Miller, the Scottish stonemason, whose popular works outsold the *Vestiges* many times over. In the tradition of *Naturphilosophie*, Miller construed the whole of nature as a pre-ordained progress towards man, but he gave this familiar Romantic theme revelatory significance, and posited a succession of special creations consistent with the fossil record. For Miller, fossils were "mute prophesies" pointing towards man's coming, and embryos the "epitome of geologic history". He found the Romantic metaphor of gestation an especially beguiling "evidence" of God's handiwork and revelation to man, and he delighted in the "wonderful analogies" he constantly invoked between the fossil progress and the individual history of organisms: "analogies that point through the embryos of the present time to the womb of Nature, big with its multitudinous forms of being".[5] Such transcendental concepts and expressions even found their way into that compendium of natural theology, the Bridgewater Treatises, in one of which the physician Peter Mark Roget also distilled Christian comfort from heterodox German Romanticism. Roget even went so far as to ensure man's spiritual progress via *Entwicklung*, finding in man's "spiritual constitution" some "embryo faculties which raise us above this earthly habitation", "the traces of higher powers, to which those we now possess are but preparatory": a notion which Tennyson, in the true Romantic tradition, was shortly to celebrate in verse.[6]

This progressionist religion of embryology culminated in the paleo-biology of the famous Swiss-American naturalist and devout teleologist Louis Agassiz, who, while a student at Munich from 1827 to 1829, cut his teeth on the concepts of *Naturphilosophie* under Oken's direct tutelage. Agassiz made explicit the identification of ontogenetic stages with the fossil sequence, generalized it into a law of nature and familiarized it by reiterating it again and again in his writings, applying it with a thoroughness that bordered on the obsessive to all the phenomena uncovered by his extensive paleontological and embryological researches. Like Owen, Agassiz adopted von Baer's law of divergent development, but the none-too-muted undertones of the familiar metaphor of gestation impressed a unilinearity on his conception of nature which persistently neutralized von Baer's divergent schema of development. For Agassiz, the mastodon was an embryonic elephant, and fossils generally were the "embryonic types" of the living organisms of the "present creation", foetal structures which prophetically announced man's own creation.[7]

It was this dedicated life-long opponent of evolution in all its forms who, with all the force of his powerful reputation, made ontogenetic development the

---

5 Miller 1841, 244–6. For a discussion of "transcendental progressionism", see Bowler 1976, 47–62.
6 Roget 1840, 2: 573. On Roger's transcendentalism, see Rehbock 1983, 569. Tennyson's reference to the Romantic law of parallelism is to be found in the Prologue to "In Memoriam".
7 L. Agassiz 1849, 26–7, 1850. On the relation between Agassiz and Oken, see E. Agassiz 1887, 1: 34, 52–4, 150–4; Lurie 1960, 20–63.

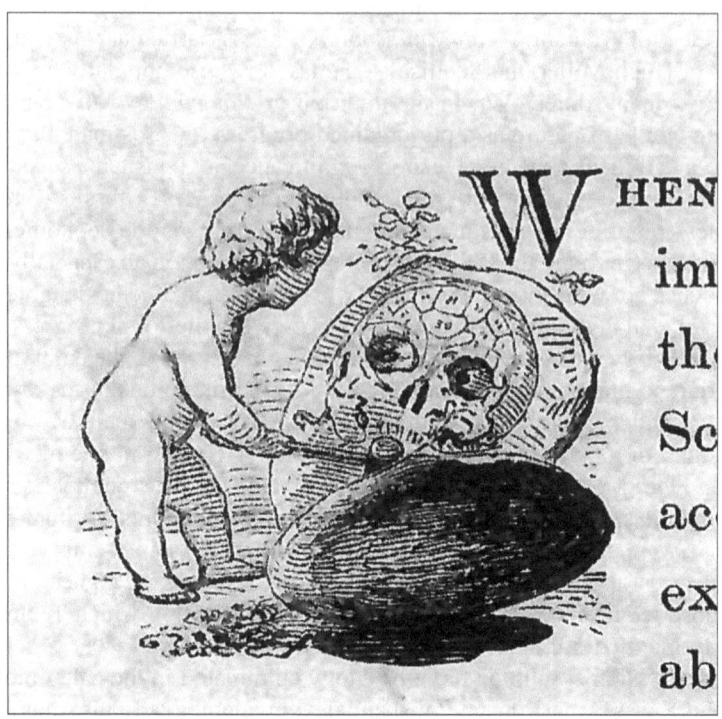

Figure 1.1 Nature's "gestation" revealed by geology. "Cutting into the stony womb of nature" the progressionist Hugh Miller found fossil testimony of man's own preordained creation. This romanticized image of nature's "gestation" as revealed by the geological hammer adorned the first page of Miller's *Footprints of the Creator* of 1849. Miller 1849, 1. University of Sydney Library, Rare Books and Special Collections.

paradigm of all of nature's history and the leading argument for premeditated design amongst his British devotees. But in the process Agassiz also brought progressionism too close for comfort to the "development hypothesis" of the *Vestiges*. Chambers could gather almost all the "vestiges" he required from this unimpeachable source, and he could, much to their consternation, cite chapter and verse from some of these most respected and influential natural theologians in support of his hypothesis. Adam Sedgwick, the Cambridge cleric and progressionist, who abhorred the *Vestiges*, this "foul book" which he brutally satirized as a "deformed progeny of unnatural conclusions" and "a filthy abortion", and held responsible for the imminent insurrection of the lower orders and the corruption of Victorian womanhood, belatedly was made aware of the pitfalls of ambiguity and metaphor. As he warned Agassiz in 1845, a great deal hinged on the troublesome term "development" and its expression: "Now I allow (as all geologists must do) a *kind progressive development*." But "Generation and creation are two distinct ideas, and must be described by two distinct words, unless we wish to introduce utter confusion of thought and language."[8]

For Sedgwick, the publication of the translation of Oken's *Lehrbuch der Naturphilosophie* by the Ray Society in 1847 was a most unwelcome revelation of the proximity of the progressionist faith to such ungodly (and almost unintelligible) notions. Although they had become so familiar with certain of its concepts in the guise of the argument from design, this was the first opportunity most English readers had had for any close scrutiny of the productions of German *Naturphilosophie*. A number, who deciphered "symptoms of unsound religious principle" in Oken's hieroglyphs, now identified the work as the "polluted source" of the *Vestiges*, and the Ray Society was censured publicly for its sponsorship of this "obscene and atheistical work". Sedgwick, who denounced its author as "a profane mystic, and a babbler", even going so far as to exonerate the author of the odious *Vestiges* from Oken's supposed excesses of atheism and materialism, expressed the pious hope that God might forever save the University of Cambridge from this "base, degrading, demoralizing, hopeless creed!" But as Sedgwick well knew, it was too late for divine intervention. The "Romantic tide" had already reached Trinity and spilled over into the progressionist faith. Canute-like, Sedgwick could only join forces with the *Manchester Spectator* in warning his colleagues and protégés like Owen of the pressing political need to clarify the ambiguity of their expressions, lest impressionable readers should misinterpret them and think (as Sedgwick himself had done in Owen's case!) that they were endorsing "some theoretical law of generative development from one type of animal to another along the whole ascending scale of Nature". Owen took the hint, while Sedgwick schooled himself in the latest von Baerian embryology and attempted to dissociate progressionism from the insidious and newly suspect embryological law of parallelism: "false at every step . . . an idle dream of the philosophy of resemblances" (Richards 1987, 163–5).

However, some, like Hugh Miller (who also did a good deal of hasty backtracking from the precipice of progressive development), refused to abandon such "wonderful" analogies to the infidel: "[T]his strange fact of the progress of the human brain is assuredly a fact none the less worth looking at from the circumstances that infidelity has looked at it first." For Miller such analogies gave "fair warning that, in tracing to its first principles the moral and intellectual nature of man, what is properly his 'natural history' should not *be* overlooked" (Miller 1849, 291–3). Charles Lyell, the eminent uniformitarian geologist, who had argued against both the progressionist and transmutationist implications of such transcendental embryological conceptions, now began to waver before the force of these

8 Sedgwick to Agassiz, 10 April 1845, in Clark and Hughes 1890, 2: 86.

same embryological arguments for the "transmutation hypothesis". As he thought-fully concluded while reviewing the species question in his private journal:

> There are none who so much aid the transmutationist as they who push the doctrine of progressive development farthest – since their assimila-tion of the successive appearance of species and genera to embryonic develop[ment] in an individual is the very opposite of that arbitrary fiat which at other times they invoke, it is creation working by law, accord-ing to a prescribed pattern & by a force analogous to that displayed in an individual from the embryo to the adult.[9]

One committed anti-progressionist who exploited the narrowing of the gap between transmutationists and progressionists to the full was the young and aggressively ambitious Huxley. During the fifties, he impartially lambasted what he called the "ideal quasi-progressionism" of Agassiz and Owen, the development hypothesis of the *Vestiges* and, above all, Owen's morphology and paleontology, opportunistically conflating them and capitalizing on the current distrust of the embryological law of parallelism and its associations with an ideologically sus-pect *Naturphilosophie* and a politically dangerous evolutionism. The anticlerical and iconoclastic Huxley found Oken, Owen and the *Vestiges* as obvious a target as had the *Manchester Spectator*, if for rather different reasons. Setting himself in professional opposition to Owen's Oken-inspired "osteological extravaganzas", Huxley aspired to become the British von Baer, and he determinedly applied von Baer's embryological criteria and methodology to the demolition of the perva-sive "popular notion" of the parallel between ontogeny and the fossil sequence in paleontology – a "fallacious doctrine" which he foisted onto his arch-rival Owen. However, the budding positivist who refused even to use the term "archetype", whose "connotation is so opposed to the spirit of modern science", did not escape baptism in the Romantic tide. In this crucial period, Huxley himself toyed with some very Romantic "mystifications", including this pertinent one: "The individ-ual animal is one beat of the pendulum of life, birth and death are the two points of rest. . . . The different forms which an animal may assume correspond with the different places of the pendulum."[10]

For Huxley before 1859 there was no evidence, embryological or paleonto-logical, to support any hypothesis of the "progressive development of animal life in time". After 1859 it was, of course, a different story, as Huxley became the chief expounder and champion of a Darwinian "world view" of "harmonious order governing eternally continuous progress". By the end of the sixties, Huxley and the Darwinians were virtually running British science from the epicentre of the influential X-Club and recruiting social support from an increasingly secular

---

9 Lyell, 27 May 1859, in Wilson 1970, 66–7. For Lyell's earlier views on transcendentalism, see Corsi 1978, 221–44.

10 Huxley 1851–4, 186, 1854–8. See Richards 1987, 167–71; Ospovat 1976.

and "progressive" middle class. By this time von Baer had disappointed Huxley's earlier expectation of his support for Darwinism, and Ernst Haeckel had emerged as Darwin's "German bulldog". Huxley relaxed his vigilance against the erstwhile "fallacious doctrine" once Haeckel and others (including Darwin and Herbert Spencer) had given it an acceptable Darwinian gloss and begun to put it to socio-political use. Translated into the theory of recapitulation in the form of Haeckel's triumphal slogan, "Ontogeny recapitulates Phylogeny", it went on to dominate embryological research until well into the twentieth century. And one of its leading British exponents was that selfsame Huxley, the prominent exorcist of Romantic influence from science.

By the seventies, Huxley was busily deducing ancestral structures from living embryos and using the theory of recapitulation to support the "natural" inferiority of women and blacks. While he relentlessly maintained his vendetta against Owen's "Okenism", ridiculing his embryogenetic theory of evolution and mocking the ambiguity of Owen's "continuous creation", Huxley was praising Haeckel's work for having "all the force, suggestiveness, and, what I may term the systematizing power, of Oken, without his extravagance".[11] A Darwinized Oken was permissible. Huxley was nothing if not selective in his Romantic targets. When the journal *Nature* was founded in 1869, it was Huxley who responded to the prestigious invitation to contribute the opening article with a few pet Romantic metaphors – a selection of "Aphorisms by Goethe", translated by himself – and who took a malicious delight in the ensuing mystification of the "British Philistines" who thought that he had "suddenly gone mad!"[12]

Historians increasingly see Darwinism as representing not so much a revolutionary break as a "subtle accommodation" with natural theology, and the history of the metaphor of gestation offers an exemplary instance of this historical contention. Its long assimilation into British natural theology via progressionism left its indelible impress on the post-Darwinian theory of recapitulation in the conflation of von Baer's law with the earlier and dominant Romantic law of parallelism, and in the progressionist trend of the whole of nature's gestatory history. Let us permit the older and mellower Huxley the last word on these complex interactions of German Romanticism and British evolution and natural theology, as in 1888 he complacently retools the leading article of the by then redundant progressionist faith for Darwinian use:

> It is quite certain that a normal fresh-laid egg contains neither cock nor hen; and it is also . . . as certain as any proposition in physics or morals, that if such an egg is kept under proper conditions for three weeks, a cock or hen chicken will be found in it. . . . Therefore Evolution, in the strictest sense, is actually going on in this and analogous millions and millions of

---

11 Huxley [1869] 1873, 301. For Huxley's use of recapitulatory arguments to endorse the inferiority of women and blacks, see Richards 1989b.
12 Huxley 1869. See Huxley to Anton Dohrn, 30 January 1870, in L. Huxley 1900, 1: 326–7.

instances, wherever living creatures exist. Therefore, to borrow an argument from Butler, as that which now happens must be consistent with the attributes of the Deity, if such a Being exists, Evolution must be consistent with those attributes. And, if so, the evolution of the universe, which is neither more nor less explicable than that of a chicken, must also be consistent with them.[13]

Bibliography

Agassiz, Elizabeth, ed. 1887. *Louis Agassiz, his Life and Correspondence*. 2 vols. Cambridge: Cambridge University Press.

Agassiz, Louis. 1849. *Twelve Lectures on Comparative Embryology*. Boston: H. Flanders.

———. 1850. "On the Differences Between Progressive, Embryonic, and Prophetic Types in the Succession of Organized Beings Through the Whole Range of Geological Times." *Edinburgh New Philosophical Journal* 49: 160–5.

Appel, Toby. 1987. *The Cuvier Geoffroy Debate: French Biology in the Decades Before Darwin*. Oxford: Oxford University Press.

Baer, Karl Ernst von. [1828] 1853. *Über Entwickelungsgeschichte der Thiere: Beobachtung und Reflexion* (Königsberg, 1828); English rendering by T. H. Huxley. "Fragments Relating to Philosophical Zoology, Selected from the Works of K. E. von Baer." In *Scientific Memoirs*, edited by A. Henfrey and T. H. Huxley, 176–238. London: Taylor & Francis.

Bowler, Peter J. 1976. *Fossils and Progress: Paleontology and the Idea of Progressive Evolution in the Nineteenth Century*. New York: Science History Publications.

Chambers, Robert. [1844] 1969. *Vestiges of the Natural History of Creation*. Reprint. Leicester: Leicester University Press.

Clark, John William, and Thomas M. Hughes, eds. 1890. *The Life and Letters of the Reverend Adam Sedgwick*. 2 vols. Cambridge: Cambridge University Press.

Corsi, Pietro. 1978. "The Importance of French Transformist Ideas for the Second Volume of Lyell's Principles of Geology." *British Journal for the History of Science* 11: 221–44.

Desmond, Adrian. 1982. *Archetypes and Ancestors: Palaeontology in Victorian London, 1850–1875*. London: Blond and Briggs.

———. 1984. "Robert E. Grant: The Social Predicament of a Pre-Darwinian Transmutationist." *Journal of the History of Biology* 17: 189–223.

———. 1987. "Artisan Resistance and Evolution in Britain, 1819–1848." *Osiris* 3: 77–110.

Gode-von Aesch, Alexander. [1941] 1966. *Natural Science in German Romanticism*. Reprint. New York: AMS Press.

Gould, Stephen Jay. 1977. *Ontogeny and Phylogeny*. Cambridge, MA: Harvard University Press.

Hodge, M. J. S. 1972. "The Universal Gestation of Nature: Chambers' *Vestiges* and *Explanations*." *Journal of the History of Biology* 5: 127–51.

Huxley, Leonard. 1900. *Life and Letters of Thomas Henry Huxley*. 2 vols. London: Macmillan.

Huxley, Thomas Henry. 1851–54. "Upon Animal Individuality." *Proceedings of the Royal Institution* 1: 184–9.

---

13 Huxley 1888, 2: 202. On the "subtle accommodation" of Darwinism with natural theology, see Young 1985.

―――. 1854–8. "On Certain Zoological Arguments Commonly Adduced in Favour of the Hypothesis of the Progressive Development of Animal Life in Time." *Proceedings of the Royal Institution* 2: 82–5.

―――. 1869. "Aphorisms, by Goethe." *Nature* 1: 9–11.

―――. [1869] 1873. "The Genealogy of Animals." In *Critiques and Addresses*, edited by T. H. Huxley, 303–19. London: Macmillan.

―――. 1888. "On the Reception of the 'Origin of Species'." In *Life and Letters of Charles Darwin*, edited by Francis Darwin, vol. 2, 179–204, 3 vols. London: John Murray.

―――. 1894. "Owen's Position in the History of Anatomical Science." In *The Life of Richard Owen*, edited by the Rev. Richard Owen, vol. 2, 273–332, 2 vols. London: John Murray.

Lurie, Edward. 1960. *Louis Agassiz, a Life in Science*. Chicago: University of Chicago Press.

Miller, Hugh. 1841. *The Old Red Sandstone*. Edinburgh: John Johnstone.

―――. 1849. *Footprints of the Creator: Or, The Asterolepis of Stromness*. London: Johnstone and Hunter.

Oken, Lorenz. 1847. *Elements of Physiophilosophy*. London: Ray Society.

Ospovat, Dov. 1976. "The Influence of K. E. von Baer's Embryology, 1828–1859: A Reappraisal in Light of Richard Owen's and William B. Carpenter's Paleontological Application of 'von Baer's Law'." *Journal of the History of Biology* 9: 1–28.

Rehbock, Philip F. 1983. *The Philosophical Naturalists: Themes in Early Nineteenth-Century British Biology*. Madison, WI: University of Wisconsin Press.

Richards, Evelleen. 1976. *The German Romantic Concept of Embryonic Repetition and its Role in Evolutionary Theory in England up to 1859*. PhD diss. University of New South Wales.

―――. 1987. "A Question of Property Rights: Richard Owen's Evolutionism Reassessed." *British Journal for the History of Science* 20: 129–71.

―――. 1989a. "The 'Moral Anatomy' of Robert Knox: The Interplay Between Biological and Social Thought in Victorian Scientific Naturalism." *Journal of the History of Biology* 22: 373–436.

―――. 1989b. "Huxley and Woman's Place in Science: The 'Woman Question' and the Control of Victorian Anthropology." In *History, Humanity and Evolution*, edited by James R. Moore, 253–84. Cambridge: Cambridge University Press.

Roget, Peter Mark. 1840. *Animal and Vegetable Physiology Considered with Reference to Natural Theology*. 2 vols., 3rd edn. London: William Pickering.

Russell, E. S. 1916. *Form and Function: A Contribution to the Study of Animal Morphology*. London: John Murray.

Schelling, Freidrich W. J. 1942. *The Ages of the World*. Translated by F. de Wolfe Bolman. New York: Columbia University Press.

Secord, James A. 1989. "Behind the Veil: Robert Chambers and *Vestiges*." In *History, Humanity and Evolution*, edited by James R. Moore, 165–94. Cambridge: Cambridge University Press.

Temkin, Owsei. 1950. "German Concepts of Ontogeny and History around 1800." *Bulletin of the History of Medicine* 24: 227–46.

Wilson, Leonard G., ed. 1970. *Sir Charles Lyell's Scientific Journals on the Species Question*. New Haven, CT and London: Yale University Press.

Young, Robert M. 1985. *Darwin's Metaphor: Nature's Place in Victorian Culture*. Cambridge: Cambridge University Press.

2

A QUESTION OF
PROPERTY RIGHTS

Richard Owen's evolutionism reassessed

"Questions of science," remarked Goethe, "are very frequently career questions. A single discovery may make a man famous and lay the foundations of his fortunes as a citizen. . . . Every newly observed phenomenon is a discovery, every discovery is property. Touch a man's property and his passions are immediately aroused."
— *Conversations with Eckerman*, 21 December 1823[1]

Upon my life I am sorry for Owen; he will be so d – d savage; for credit given to any other man, I strongly suspect, is in his eyes so much credit robbed from him. Science is so arrow a field, it is clear there ought to be only one cock of the walk!
— Darwin to Huxley, 28 December 1859[2]

When *Vestiges of the Natural History of Creation*, the anonymous evolutionary work which caused such a furore in mid-Victorian England, was published towards the close of 1844, Richard Owen, by then well-entrenched as the "British Cuvier", received a complimentary copy and addressed a letter to the author (Owen 1894, 1: 249–52).[3] This letter and how it should be interpreted have recently become the subject of historical debate, and this paper is directed at resolving the controversy. The question of Owen's attitude to the *Vestiges* argument is central to the larger historical problem of the views of this leading British morphologist and palaeontologist on the contentious issue of the "secondary causes" of species. Owen wrote so little directly on this subject prior to 1858, that the letter in

I should like to thank John Brooke, Adrian Desmond, James Secord and John Schuster for their criticisms of an earlier draft of this paper; and the following institutions and libraries for permission to study manuscript material: The British Library, British Museum (Natural History), Royal College of Surgeons of England and the Ray Society.

1 As cited in Hobsbawm 1973, 336.
2 *Correspondence of Charles Darwin*, 7: 459.
3 The location of the original letter is unknown.

question, together with his two letters of 1848 to the rationalist publisher John Chapman,[4] and the controversial conclusion to his *On the Nature of Limbs* (Owen 1849, 85–6), constitute the major evidence that Owen in this period subscribed to a naturalistic theory of organic change. On the basis of this evidence, historians of biology have generally concurred with Owen's biographer grandson that Owen had a "certain leaning towards the theories enunciated by Robert Chambers [the *Vestiges*' author]", but that his "official" anti-transmutationist stance of the 1840s did not permit full public expression of his own views.[5] As Ruse most recently summed up this historical consensus: Owen in the 1840s was "moving down a path not completely dissimilar from that followed by Chambers", and he "tried to have matters two ways, praising *Vestiges* to its author and condemning it to its critics".[6]

However, a recent analysis by John Hedley Brooke has fragmented this consensus and made Owen's equivocation over evolution problematic (Brooke 1977). Brooke offered a "more cynical" interpretation of Owen's *Vestiges* letter as the private attack on *Vestiges* that Owen declined to mount publicly, arguing on a number of grounds that Owen's letter should be construed as a "deft rebuke" to Chambers, rather than its "straightforward" conventional historical interpretation as one of "mild encouragement". Brooke has subsequently extended his analysis to argue that Owen's sympathetic references to a continuity of secondary causes should be taken in a theological rather than a scientific context, and should not be construed as evidence of Owen's positive commitment to evolution (Brooke 1979, 41).

Brooke's iconoclasm has created difficulties for those historians who continue to insist on Owen's pre-1858 commitment to a naturalistic theory of organic descent. For example, Ospovat in 1976 asserted that Owen believed early on in the production of new species by natural causes, probably in the form of "saltatory descent from pre-existing species" (Ospovat 1976, 22). But, by 1981, on the basis of Brooke's revision, Ospovat modified this to a more cautiously worded description of Owen's early views as "protoevolutionary" (Ospovat 1981, 138). It is significant, however, that in a footnote, Ospovat indicated some difference of opinion with Brooke:

> Owen himself took [his palaeontology and morphology] to be evidence of descent, though not of transmutation. . . . John H. Brooke, in discussing Owen's views, seems not to have considered the possibility of such a distinction.
>
> (Ospovat 1981, 285)

---

4 Owen 1894, 1: 309–11; Desmond has identified Owen's correspondent as Chapman (Desmond 1982, 29 n27).

5 Owen 1894, 1: 255; MacLeod 1965, 261; Hodge 1972, 133–4; Rudwick 1972, 207; Bowler 1976, 93; Ospovat 1976, 1978.

6 Ruse 1979, 125, 228. Ruse seems not to have known of Brooke's reassessment at the time of writing.

Again, Desmond, taking Brooke's revision into account in his assessment of Owen's reaction to *Vestiges*, interpreted Owen's position as anti-transmutationist (on the basis of Owen's prior attacks on the transmutationism of his social and professional rival, Robert Edmond Grant of University College). However, Desmond did dissent from Brooke's conclusion of Owen's "blanket condemnation" of the *Vestiges* argument:

> Owen might have lamented the mistakes, hated the transmutation, and even doubted aspects of Chambers' theodicy; yet he was in total agreement on the need for uniformity, and on its correct Providential interpretation.
>
> (Desmond 1982, 33)

Desmond went on to argue that Owen was quite sincere in his assertion to Chambers of the need for the "best naturalists" (meaning himself) to seek out the "secondary causes" of the production of new species. Desmond coupled this with Owen's 1848 claim to Chapman that he could, if pressed, come up with half a dozen such natural causes (including transmutation), and concluded that Owen himself at that stage favoured an explanation for the introduction of a new species via a process of descent akin to alternation of generations (a phenomenon that Owen had studied intensively in the 1840s, and which he termed "metagenesis").[7]

It is my intention in this paper to push the interpretations of Ospovat, Ruse and Desmond further in an attempt to clarify Owen's early (and later) views on the "secondary causes" of species. To this end, I shall argue, by reference to Owen's correspondence and papers, that his letter to Chambers cannot sustain the interpretation of a private attack on *Vestiges;* that far from being subtly ironic or destructive in intent, Owen's purpose in writing to Chambers was to indicate his own long-term interest in and superior understanding of certain major themes of *Vestiges*, and, in particular, to claim priority for the embryological basis of Chambers' central mechanism of species change – "the idea and diagram of page 212". I will argue further that Owen was neither theoretically nor historically opposed to a theory of organic descent based on the *Vestiges* diagram and that such an interpretation is consistent with those variously offered by Ruse, Ospovat and Desmond.

All this is essential to my major purpose of locating Owen's theoretical constructs and Owen himself in their larger institutional and social framework. It is only through such contextual analysis that we may arrive at a coherent historical explanation of Owen's response to the embryological argument of *Vestiges*, and his own views on the development of species. My analysis is indebted to Adrian Desmond who has stressed the "advantage of viewing Owen's science in political terms" (Desmond 1985a, 1982).[8] Desmond's studies of the institutional and social bases of Owen's early palaeontology and morphology offer a readymade

---

7 Desmond 1982, 33; see also E. Richards 1976, 313–15.
8 On the advantages of contextual historical analysis, see also Young 1973; Shapin 1982.

framework within which I have been able to locate my interpretation of Owen's embryology and its significance for his conception of organic descent. Desmond has been more concerned with explicating the extent to which social factors shaped Owen's early anti-transmutationism and how he structured his morphology and palaeontology in conformity with this socially derived perspective. My account emphasizes the significance of these developments for Owen's own version of organic descent.

My paper is organized in two parts. Part 1 analyses the relationship between Owen's and the *Vestiges'* versions of organic development; Part 2 deals with the institutional and social location of Owen's evolutionism.

Part 1. Owen's response to the embryological argument of *Vestiges*

According to Brooke, there were three points of substance in Owen's letter to the author of *Vestiges* which "ought to have made Chambers flinch" in spite of the veneer of courtesy (Brooke 1977, 135). These were Owen's negative conclusions on the issues of spontaneous generation and the ape-origin of humanity, and his references to the "idea and diagram" on page 212 of *Vestiges*. It is not my intention to dispute Owen's criticisms of two of the major links in Chambers' chain of reasoning, i.e., the ongoing emergence of animalcules from inorganic matter and the transmutation of apes into humans, but rather to argue that they do not constitute the destructive sneers Brooke argues them to be. To this end I will try to demonstrate that Owen's criticism of these two aspects of the *Vestiges* argument were quite consistent with his priority claim to the "idea and diagram" of *Vestiges*, and so shall proceed directly to the key issue of Owen's response to the embryological argument of *Vestiges*.

The critical passage in Owen's letter reads as follows:

> I take the liberty, in reference to the idea and diagram given in page 212, to request your attention to the concluding generalization in my twelfth lecture [of *Lectures on Invertebrates*], and to that on the "Metamorphoses of Insects," where there will be found, I believe, the first enunciation of the true law of the analogies manifested by the embryos of animals in their progress to their destined maturity.
>
> (Owen 1894, 1: 135)

The interpretation of the *Vestiges* letter hinges on this reference. As Brooke emphasized, the "idea and diagram" to which Owen referred constituted the central thesis of *Vestiges:*

> It delineated Chambers' mechanism – or the nearest he got to one – for the transmutation of species.
>
> (Brooke 1977, 137)

43

Thus, Owen's supposed dissent from such a crucial thesis underpins Brooke's overall analysis. It is his leading reason for finding Owen's supposed alignment with *Vestiges* "counterintuitive":

> It is difficult to believe that so accomplished a student of embryology [as Owen] could have swallowed Chambers's crudities.
>
> (Brooke 1977, 134)

But while Owen was certainly a well-educated embryologist, as historians have begun to emphasize in contradiction of the contemporaneous opinion of Thomas Henry Huxley, the embryology of *Vestiges* cannot be dismissed as mere "crudities" (Ospovat 1976, 1981, 129–35; Huxley 1894, 2: 320).

1.1 The embryological argument of *Vestiges*

It is true that in contrast to Owen, Chambers was no embryologist. His knowledge of embryology was entirely derivative, and, as he later acknowledged, he culled his material arbitrarily from "such treatises on physiology as had fallen in his way" (Chambers 1853, vi). At the time he wrote *Vestiges* there were two versions of embryological development current in British physiology. The first and older was embodied in the law of parallelism, sometimes known as the Meckel/Serres law, according to which a parallel was drawn between the two series: the ontogenetic sequence of forms and the taxonomic series of adult organisms. In this version, the taxonomic series was assumed to be uniserial, so development was also conceived as linear, and the embryo sequentially repeated the adult forms of animals lower in the scale.

The second and more complex version of development was integral to the law of divergence associated with the name of von Baer, who in 1828 conceived ontogeny as essentially a process of divergent differentiation by which the embryo becomes increasingly specialized or individualized, and therefore diverges more and more from other animal forms. Von Baer held that there were four basic types of organization and that the type is manifested in the very earliest stages. Hence the law of divergence is based not on a uniserial taxonomy, but on a multitypal or divergent one, with the result that the simple unilineal parallel no longer applies, and the embryo repeats not the adult forms of lower animals, but their embryonic ones. However, it is important to note that von Baer's law does allow for some similarity between embryos of "higher" or more differentiated animals and "lower" or less differentiated adults of the same type. Such similarities are explicable in terms of lack of differentiation in both cases, the lower animals having a more general, less differentiated organization, comparable in some ways to the embryonic condition of higher animals of the same type.[9]

---

9 The terminology and the distinctions drawn are indebted to E.S. Russell's classic analysis, *Form and Function: A Contribution to the Study of Animal Morphology*, 1916, reprinted, 1972. See also Richards 1976; Gould 1977.

Other factors aside, von Baer's law had the advantage of being more compatible with the conventional classificatory system of Cuvier. Yet, by the early forties, apart from an informed few, British comparative anatomists and physiologists either continued to promote the law of parallelism, or, more commonly, to conflate the two laws (Ospovat 1976; Richards 1976, 164–244). The embryology of *Vestiges* simply reflects the context in which it was constructed, for Chambers offered his readers both versions of development.[10] While some weighting must be given to his less than perfect comprehension of embryology, it must be remembered that his critics were, for the most part, initially as hazy on this point as Chambers. Nevertheless, his failure to discriminate properly between the two versions of development proved a godsend to committed opponents of his "development hypothesis", hot on the trail of fatal flaws and weaknesses in his mechanism. Adam Sedgwick for instance, after a quick schooling in embryology, was able to target those aspects of Chambers' embryology which could now be found wanting and ignore those parts of his hypothesis which were consistent with the "latest" embryology.[11]

These latter included the diagram in question, which Chambers employed to illustrate both the course of vertebrate embryology and "development". This was no "crudity", and in fact has a pedigree honourable enough to satisfy the most Whiggish of historians. We may trace its descent from the table utilized by von Baer in 1828 to illustrate his law of divergence[12] (reproduced with explanatory note in Figure 2.1), via Martin Barry's 1836/7 diagrammatic representation of von Baer's law[13] (Figure 2.2), to William Carpenter's simplified version of Barry's diagram in his 1841 exposition of von Baer's law[14] (Figure 2.3). Carpenter's diagram and description are the obvious sources of the *Vestiges* diagram and explanation of page 212[15] (Figure 2.4), the only changes Chambers introduced being to improve

10 Chambers [1844] 1969, 199–213; Ruse 1979, 102–5. Chambers drew on three major sources for the embryological information in the first edition of *Vestiges,* two of which he acknowledged. They were: the work of the transcendentalist John Fletcher (1837); a popular treatise on physiology by Perceval Barton Lord (1834) who seems to have been influenced by transcendental ideas while at Edinburgh; and the third and unacknowledged source was William Carpenter's *Physiology* of 1841. Both Fletcher and Lord propounded the law of parallelism, while Carpenter adopted von Baer's law.

11 Sedgwick (1845) devoted twelve pages to his review of *Vestiges* in the *Edinburgh Review,* to just such a selective critique of its embryology. Note that Sedgwick obtained his embryological information from the Cambridge professor of anatomy, William Clark, not from Owen (see text). Other early reviews which made much of the embryological confusion of *Vestiges* included those in *Blackwood's Magazine* (1845, *LVII:* 448–60), *Westminster Review* (1845, *XLIV:* 152–203), *British Quarterly Review*, (1845, I: 490–513). It is interesting to note that Carpenter wrote a very sympathetic review of *Vestiges* that did not dwell upon its embryological inadequacies (*British and Foreign Medical Review,* 1845, XIX: 155–81).

12 Von Baer 1828, 225, 229–30.

13 Barry 1836/7b, 346; see also Barry 1836/7a.

14 Carpenter 1841, 196–7.

15 This is also the opinion of Ospovat (1976, 13); Ruse (1979, 103); Gould (1977, 103).

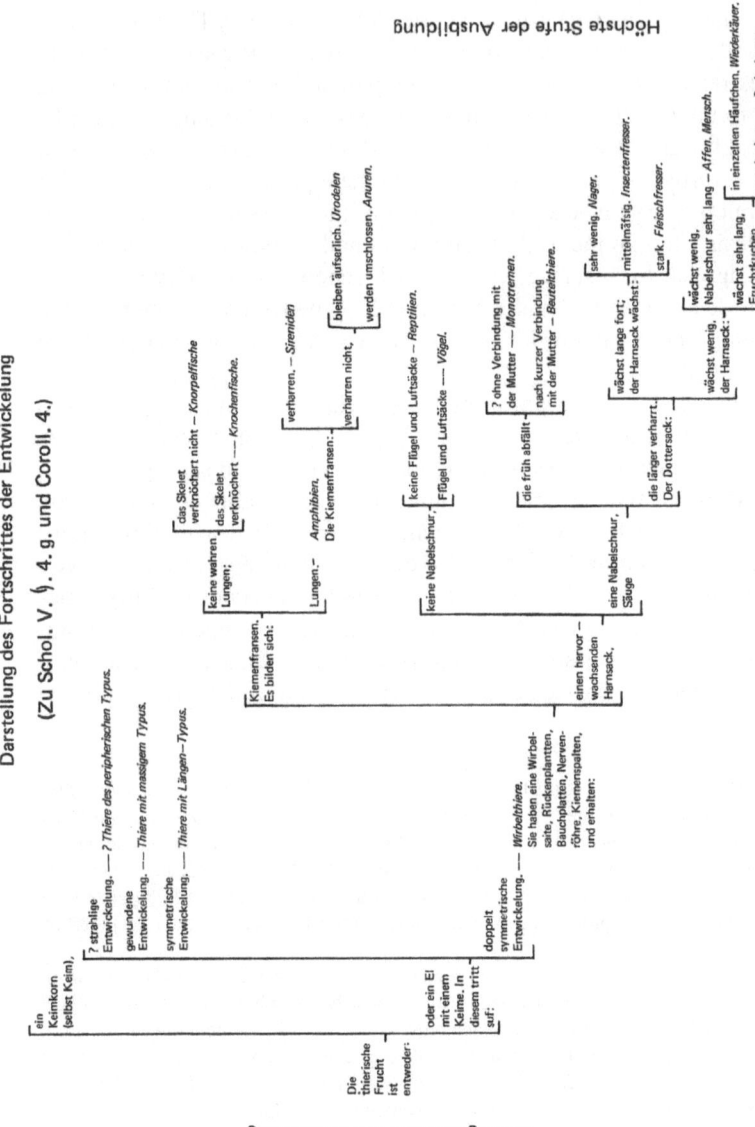

Figure 2.1 Von Baer's Table represents the progress of development as a dichotomous schema which each embryo must follow, branching off early or late according to the grade of development of its final form. At its earliest stage of development the animal is either a "germ granule (*ein Keimkorn*) or a ovum with a germ". At its next stage, it immediately takes on the distinguishing embryological characters of one of the four great types, so that the type is fixed from this stage on, and further development is a process of increasing divergence along the path of either radiate, spiral (Molluscous), symmetrical (Articulate), or double symmetrical (Vertebrate) development. Von Baer 1828, 225. University of Sydney Library, Rare Books and Special Collections.

The Tree of Animal Development ;
Shewing fundamental Unity in Structure, and the causes of variety; the latter consisting in *Direction* and *Degree* of development.

Fig. 12.

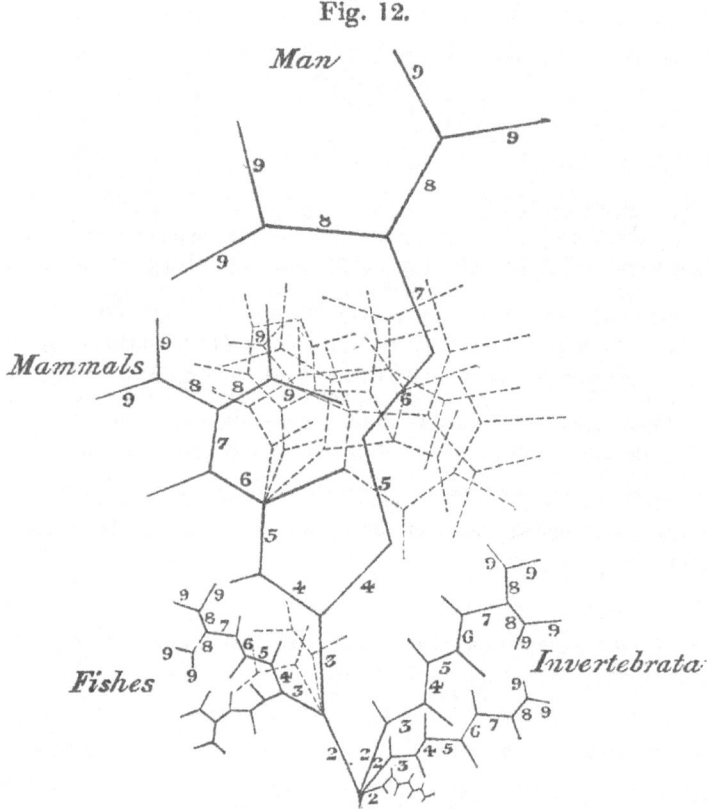

Figure 2.2 Barry's diagram to illustrate von Baer's law of development. "The whole figure represents the development of the entire animal kingdom. Any one of the primary divisions may rudely illustrate the development of a single organism, viz. – *Explanation* – to be read from below upwards. 9 The *Individual* character in its most special form. 8. The *Sexual* character obvious, but the *Individual* character obscure. 7. The *Variety* obvious, but *Sexual* difference scarcely apparent. 6. The *Species* manifest, but the *Variety* unpronounced. 5. The *Genus* obvious, but not the *Species*. 4. The *Family* manifest, but the *Genus* not known. 3. The *Order* obvious, but not the *Family*. 2. The *Class* manifest, but the *Order* not distinguishable. 1. No *appreciable* difference in the Germs of all *animals* (Fundamental Unity?). This illustration is but a coarse one, since it does not shew the *particular* direction, proper to the development of each individual germ" Barry 1836–37b, 346. University of Sydney Library.

47

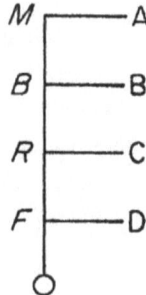

Figure 2.3 Carpenter's version of von Baer's law. "Let the vertical line represent the progressive change of type observed in the development of the foetus, commencing from below. The foetus of the Fish only advances to the stage F; but it then undergoes a certain change in its progress towards maturity, which is represented by the horizontal line FD. The foetus of the Reptile passes through the condition which is characteristic of the *foetal* Fish; and then stopping short at the grade R, it changes to the perfect Reptile. The same principle applies to Birds and Mammalia; so that A, B, and C, – the *adult* conditions of the higher groups – are seen to be very different from the *foetal*, and still more from the *adult*, forms of the lower; whilst beyond the embryonic forms of all the classes, there is, at certain periods, a very close correspondence arising from the law of gradual progress from a general to a special condition, already so much dwelt upon" Carpenter 1841, 196–7. University of New South Wales Library.

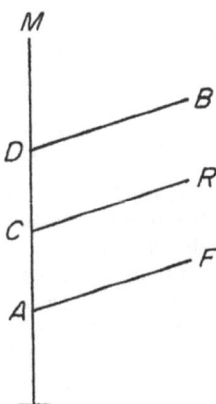

Figure 2.4 The *Vestiges* diagram. Chambers 1844, 212. University of New South Wales Library.

on Carpenter's lettering, and to give his horizontal lines an upward inflection to indicate the progressive nature of the process (more like Barry's original "Tree of Animal Development").

Chambers introduced his diagram with a decisive rejection of the law of parallelism (which he had just previously endorsed), and explained the real law of development as follows:

> It has been seen that, in the reproduction of the higher animals, the new being passes through stages in which it is fish-like and reptile-like. But the resemblance is not to the adult fish or the adult reptile, but to the fish and reptile at a certain point in their foetal progress. . . . It may be illustrated by a simple diagram. The foetus in all the four [vertebrate] classes may be supposed to advance in an identical condition to the point A. The fish there diverges and passes along a line apart, and peculiar to itself, to its mature state at F. The reptile, bird and mammal, go on together to C, where the reptile diverges in like manner, and advances by itself to R. The bird diverges at D, and goes on to B. The mammal then goes forward in a straight line to the highest point of organization at M.
>
> (Chambers 1844, 212)[16]

Chambers, of course, went well beyond Carpenter's description of von Baer's law by giving it, in Stephen Jay Gould's phrase, an "evolutionary translation" (Gould 1977, 109–22). It is obvious, he argued, that all that is necessary for the development of a reptile, say, from a fish (in the simplified terms of the diagram) is that the fish embryo should not diverge at A but continue onto C before diverging to reptilian maturity at R:

> To protract the *straight forward part of the gestation over a small space –* and from species to species the space would be small indeed – is all that is necessary.
>
> (Chambers 1844, 213)

Thus, for Chambers, embryogenesis was not merely a model for interpreting the history of life on earth, but provided him with the explanatory mechanism for species change. The "development" of a new species was the result of an abrupt deviation from the normal course of embryonic development, a deviation which he emphasized was not arbitrarily induced, but which was subordinate to a "higher law" of generation, referable in turn to "the original Divine conception of all the forms of being".[17] This was the essence of the "idea and diagram of page 212", which attracted Owen's attention.

16 The prior endorsement of parallelism occurs in Chambers [1844] 1969, 199–203.
17 Chambers argued by reference to Babbage's calculating machine, that the law that like produces like is subordinate to a higher law which on occasion permits the production of a higher form of

Chambers took as a crucial piece of evidence for his mechanism, the transcendental explanation for foetal abnormalities as arrests of development. According to this view, which was identified originally with the law of parallelism, abnormalities in higher animals represented states of organization which were permanent in lower ones. In the late 1820s, the French morphologist, Geoffroy St Hilaire, had made this thesis, which he substantiated with a series of teratological experiments, the backbone of his speculations on transmutation.[18] Chambers followed Geoffroy in arguing by analogy from the production of monstrosities to the origin of species. This was an analogy which, I shall argue, also especially interested Owen. In Chambers' interpretation, the length of the gestation period was the critical factor, with premature birth resulting in arrest of development and monstrosity, while prolonged gestation led to an advance of development and transmutation of species (Chambers 1844, 219).[19] This close association of his mechanism with the law of parallelism via the theory of arrest of development, and his invocation of prolonged gestation, impressed a unilinearity on his "development hypothesis" which predisposed Chambers to the conflation of von Baer's law with the law of parallelism and made his embryology and mechanism so vulnerable to critical attack. As well, following Agassiz (Ospovat 1976), Chambers related ontogeny to the palaeontological sequence which, in the earlier editions of *Vestiges*, he tended to depict as a unilineal progress, and this too predisposed him towards the law of parallelism.

In theory, as Chambers had proposed both the Meckel-Serres law and von Baer's law, he was faced with a choice of two possible mechanisms for transmutation, the distinction being based on the stage at which the mutation is assumed to take place, and depending on which version of embryological development is adopted. If the law of parallelism is given an evolutionary interpretation, then change is assumed to be linear or additive – as an additional step after the completion of normal development – which means that the mutation should logically occur when the organism is mature and recur in the offspring at a correspondingly late stage of development. Chambers was quite specific in his rejection of this version of species change, presenting his mechanism explicitly in terms of

---

life ([1844] 1969, 210). Thus new species are formed by law in accordance with a Divine plan (230–33). At the same time, as he insisted on a parallel between geological and biological uniformitarianism, he also admitted the direct action of the environment on the reproductive system (229). This latter was Geoffroy's thesis, to which Owen was opposed (see Note 42). However, in discussing the *Vestiges* thesis, Owen focused on the idea of a pre-ordained higher generative law (see text). In later editions of *Vestiges*, Chambers adopted the more Lamarckian explanation of an "inherent impulse" to advance, although this was, of course, still Divinely induced. For a discussion of Chambers' changing evolutionary ideas and their relation to the romantic conception of nature and, in particular, the ideas of Geoffroy, see Hooykaas 1957, 3–18; Hodge 1972.

18 Geoffroy St Hilaire 1828, 1833; see also Russell [1916] 1972, Ch. 5; Canguilhem et al. 1960; Gould 1977, 49–52.

19 It was this aspect of Chambers' mechanism from which Owen dissented, see Note 38 and Brooke 1977, 138. See text for Owen's conception of "transmutation", and the basis of his anti-transmutationism.

divergence, complete with diagram, and emphasizing that the vertebrate foetus repeats not adult but lower embryonic forms. Yet, although he was quite definite as to when the mutation occurs – *during* not after ontogeny – apart from the one passage on pages 212 and 213 where he detailed his mechanism, Chambers persistently blurred the distinction between the two versions of development.[20] It was not until the tenth edition of *Vestiges* in 1853, which was revised by William Carpenter, that they were disentangled and the confusion resolved entirely in favour of von Baer's law. In this edition also, following Carpenter and Owen, the palaeontology of *Vestiges* was made consistent with divergent embryological development.[21] In the meantime, Chambers was able to exploit the prevailing tension between the two theories of development and their regular conflation in the textbooks in rebutting the attacks of suddenly more discriminating critics (Chambers 1846, 103–9).[22]

Nevertheless, it was the "idea and diagram of page 212" that Owen singled out for special reference, and it must be acknowledged that its sources were impeccable and must have been readily recognized by Owen. Barry and Carpenter were acknowledged embryological authorities in the 1840s and Owen was very familiar with their work. He drew on Barry's 1836/7a,b papers for his own *Lectures on Invertebrates*,[23] and, so far from being critical of Carpenter's version of development, Owen fought a protracted priority dispute with Carpenter over its application to the fossil record.[24]

20 For instance, Chambers 1844, 226–7.
21 Chambers 1853, Preface, 147–62; Ospovat 1976, 13, n35, 1981, 216.
22 Richard Yeo has documented how Chambers was able to exploit a similar controversial situation with respect to the nebular hypothesis in dealing with his critics: Yeo 1984, 18–19.
23 Owen 1843, 24. Owen states only that Barry provided him with some notes for his Lectures, but he was clearly familiar with Barry's 1836/37a,b papers and these were the obvious source of Owen's embryological references in his Hunterian Lectures of 1837 when he first began advocating the embryological law of divergence in opposition to the transmutationism of Grant and Geoffroy (see text). See also Ospovat 1976, 9–10, 1981, 130–2; Desmond 1985a, 25–50, 46–9. Owen's relationship with Barry warrants closer investigation in view of their embryological and morphological affinities, and Barry's failure to protest Owen's apparent appropriation of his work (see text). They corresponded frequently and Barry regarded Owen as his mentor, dedicating his final work to him [British Museum (Natural History), hereafter referred to as BM(NH), Owen Correspondence, vol. 2, ff. 253–309; vol. 9, f. 214]. It was Owen who confirmed Barry's observation of the presence of spermatozoa "on, and apparently in, the fallopian ovum of a rabbit" in 1842 (Royal College of Surgeons, Stone Watson Papers, Mss and Autographs).
24 BM(NH), Owen Correspondence, vol. 6, ff. 333–8. Carpenter's earliest letter to Owen on this problem of priority is dated 20 August 1851. Although Carpenter put up a good fight, Owen wore him down over the following years, until in a final exasperated retort, dated 11 February 1854, Carpenter conceded defeat: "You will see that I have *differentiated* the principle which is *properly yours* from that which I *thought to be mine,* having been in the habit of stating it in Lectures for the last *12* or *14* years, and not being aware that it had been enunciated by anyone else. But I do not intend to make any claim to it whatever. . . . In fact, I do not intend to make *any* claims to originality, being quite satisfied that my labours are doing good, and caring little about anything else". See also Desmond 1982, 92–3; Ospovat 1976, 18–19.

In fact, Owen was prepared to go to considerable lengths in protecting what he regarded as his property rights to the law of divergence, even to the extent of claiming precedence over von Baer himself (Owen 1851a, 430n). But his most serious rival for embryological honours was undoubtedly the French zoologist, Henri Milne-Edwards. At the very time he wrote his *Vestiges* letter, Owen was contesting ownership of von Baer's law with Milne-Edwards. As it is my contention that Owen's *Vestiges* letter must be interpreted in the context of this dispute, I will here briefly reconstruct it.

1.2 Owen's priority dispute with Milne-Edwards

It began when Owen noticed a paper by Milne-Edwards in the February 1844 issue of the *Annales des Sciences Naturelles* in which Milne-Edwards offered his own views on comparative embryology, views essentially the same as those of von Baer.[25] Owen immediately wrote to Milne-Edwards, drawing his attention to the similar generalizations in the final lecture of his own prior-published *Lectures on Invertebrates*, and a translation of the relevant section of this work subsequently appeared in the *Annales* later in the same year (Owen 1844). Milne-Edwards, however, proved a formidable opponent, refusing to relinquish his own claims to such a significant embryological discovery. With editorial licence and Gallic *élan* he cut the ground from under Owen's feet with a bland footnote on his pleasure in finding himself in agreement with so eminent an authority, and proceeded to assert his own precedence over Owen by a good 10 years (Owen 1844, 162, n)!

Unfortunately, Owen's immediate response to this audacious counterclaim is not available, but we may judge of its tenor and the importance Owen attached to the issue, by the detailed case he presented for his own precedence over Milne-Edwards (and Barry and Carpenter to boot), some seven years later, i.e., in 1851. This is to be found among Owen's papers in the British Museum (Natural History), in the unpublished portion of the proof of a review of Owen's collected works. This review was written for the *Quarterly Review*, ostensibly by Owen's close associate William Broderip, but extensively revised by Owen himself.[26]

---

25 Milne-Edwards 1844. For a discussion of Milne-Edwards' views on development and their relation to those of von Baer and Owen, see Ospovat 1976, 10–12, 1981, 117–40.

26 R. Owen, "Proofs of Reviews, 1847–1882," BM(NH), OC, vol. 39. There are two relevant proofs in this collection, one dated "Quarterly Review" 1851, and the other "Review of *Archetype* in 'Quarterly' for 1852?" Both have been extensively revised and amended in Owen's own hand, and much-pruned versions were published as Broderip 1851–1852, 1853. Their production was a collaborative process between Owen and William Broderip, naturalist and jurist, extending over almost 4 years, and documented in Broderip's correspondence with Owen. Lockhart, the editor of the conservative *Quarterly*, first approached Broderip in August 1849, requesting an article on Owen [BM(NH), OC, vol. 5, f. 176, Broderip to Owen], and the final revised proof of the second article was not received until May, 1853 (ibid., ff. 236–7). Close study of the various proofs and their final published versions reveals a lengthy process of arbitration between Owen and Lockhart (who argued that Owen did not require "minute vindication" of his priority claims; Broderip to

Owen clearly intended that Broderip's review should do full justice to his morphological researches and, in particular, legitimize his embryological claims.

In brief, Owen's case for priority over Milne-Edwards hinged on his dismissal of Milne-Edwards' claims to have previously propounded the law of divergence in his unpublished lectures at the Parisian *Faculté des Sciences*, by insisting on publication date as the only acceptable criterion:

> The date of publication is that alone which science accepts as the grounds for her awards of priority. . . .[27]

On this basis Owen could establish his own precedence by a narrow margin. Two extracts from his *Lectures on Invertebrates* were cited in support of this. The first occurring "at the conclusion of the 18th. lecture . . . delivered May 6th, 1843, and published in the same month in Part VII of the entire work"; the second and more comprehensive statement of Owen's version of von Baer's law comprises the extract which Milne-Edwards had published at Owen's insistence in 1844. This extract, as Owen now precisely dated it, occurs in the "concluding lecture of the course, delivered May 20th, 1843, and published in *Part* XII, in July, 1843."[28] (My emphasis).

The point, as Owen was making it, was that his *Lectures on Invertebrates* was issued by the publisher in twelve parts, and by exploiting this he could increase his publication margin over Milne-Edwards by a couple of months. *My* point is, that from about February of 1844 when he first became aware of Milne-Edwards' similar views, Owen was intent on establishing his own priority, and his letter to Chambers of late 1844 must be interpreted within the context of this dispute.

One further step is necessary before we are in a position to evaluate Owen's response to the "idea and diagram" of *Vestiges* which so interested him, and that is to examine Owen's version of the law of divergence and determine whether it differs essentially from von Baer's law as exemplified in the *Vestiges* diagram.

1.3 Owen's "True Law" of embryological development

In the concluding summary to the final lecture of his *Lectures on Invertebrates*, on which Owen laid so much emphasis in his dispute with Milne-Edwards, Owen specifically rejected the law of parallelism and enunciated his own version of the

---

Owen, 23 May, 1852, ibid., ff. 213–14), with Broderip acting as go-between. Rupke has discussed the contents of the final published versions of these papers in some detail, but attributes their production entirely to Broderip and overlooks Owen's guiding hand: Rupke 1985.

27 [R. Owen and W. Broderip], "Review of Archetype". proof (see note 26), 11, 7–12. This statement was edited out by Lockhart in the published version, and the whole section on Owen's embryological researches was "pruned" on Lockhart's insistence: Broderip 1853; Broderip to Owen, 26 May 1853, BM(NH), OC, vol. 5, ff. 236–7.

28 "Review of Archetype" (see note 26), 9. See also Broderip 1853, 62–3; Owen 1851a, 430, fn., where Owen gave the same references.

law governing the analogies between "lower" animals and the embryonic forms of "higher" ones as follows:

> The extent to which the resemblance, expressed by the term "Unity of Organization", may be traced between the higher and lower organized animals, bears an inverse ratio to their approximation to maturity.
>
> (Owen 1843, 368)

This means, he explained, that all animals resemble one another at the earliest period of their development, but development is a progress from the general to the special, and each animal, while it approximates the forms of lower animals in its early stages, must complete its development by diverging to its own special form, and this precludes resemblance to lower forms in the later stages of development. Further, the only permanent animal form represented by all embryos is that of the "infusorial monad" which is to be regarded as the "fundamental or primary form". The vertebrate embryo, for instance, begins with the monad form, followed, through the processes of fission and assimilation, by a "vermiform apodal" form, but, "this is distinguished from the corresponding form of the Insect by the Vertebrate characteristics of the nervous centres, – viz. the spinal chord and its dorsal position; whereby it is more justly comparable to the apodal fish than to the worm". Hence its unity of organization with the invertebrate is restricted to this initial fleeting stage (Owen 1843, 369–70).

As his discussion makes clear, Owen's rejection of a uniserial taxonomy and adoption of Cuvierian typology underlay his conception of development, which is remarkably similar to that of von Baer. Like von Baer, he held that there were four types of organization, and that the type is manifested in the very earliest stages and becomes increasingly specialized throughout the course of development. An examination of Figures 2.1, 2.2 and 2.3 will demonstrate the compatibility of Owen's conception of development with the various published formulations of von Baer, Barry and Carpenter, all of which predated Owen's publication. Owen's emphasis on the very early divergence of all four types from an initial fundamental form is explicit in von Baer's and Barry's representations of the course of animal development; it is not indicated in Carpenter's diagram which represents only the course of vertebrate development (as does the *Vestiges* diagram), but is implicit in his divergent conception of all development.[29]

In the face of all this, it is difficult to perceive just where Owen's insistent claim to originality lay. Perhaps he saw it in his explicit formulation of the restriction of Geoffroy's unity of organization to the monad stage,[30] which was of considerable

---

29 This interpretation is consistent with that of Ospovat 1976, 10–12, 1981, 117–40.

30 This seems to be the basis of his claim for precedence over von Baer (Owen 1851a) and Barry (Broderip 1853, 61–2). Both claims, however, are specious. It is significant that Owen refers neither to von Baer's 1828 publication (which contains the relevant 5th Scholion) nor to Barry's second paper of 1837 (which contains Barry's diagram of divergent development) in this connection

significance in his early anti-transmutationism (see below). Perhaps, it lay in his neat mathematical formulation of the relationship. Whatever the exact basis of his claim, he quietly dropped his crusade for priority about the same time the Fifth Scholion of von Baer's *Entwickelungsgeschichte* was made available in English translation by Huxley in 1853, although he held to his independent formulation of the law of divergence.[31]

Nevertheless, throughout the forties and early fifties, Owen single-mindedly promoted himself as the originator of von Baer's law on the basis of his 1843 formulation, minimizing von Baer's contribution, ignoring Barry's, riding roughshod over Carpenters' claims, and contesting ownership with Milne-Edwards. Given his determination to appropriate this discovery and its professional rewards to himself, it was inevitable that Owen in late 1844 should have viewed the *Vestiges* "idea and diagram" from this proprietary perspective; that he should have directed Chambers to his own prior-published *Lectures on Invertebrates*, where Chambers would find "the *first* enunciation" (my emphasis) of the law of divergence.

In the light of the foregoing analysis, it is clear that Owen's own views on development were compatible with the *Vestiges'* "idea and diagram"; that his priority dispute with Milne-Edwards had alerted him to the need for vigilance in this matter; and that in drawing Chambers' attention to the relevant sections of his own prior-published lectures he was indicating his precedence over the *Vestiges* author; not, as Brooke interprets the reference, "undermining the one semblance of a mechanism that Chambers had to offer" (Brooke 1977, 137). At the same time, it is possible that Owen also may have had the intention of setting Chambers straight as to which version of development he should adhere, for his own extended discussion makes explicit the distinction between the two versions, as I have indicated above. This would be consistent with his earlier statement to Chambers that "[t]here are a few mistakes where you treat of my own department of science, easily rectified in your second edition" (Owen 1894, 1: 251).[32] Owen most certainly regarded embryology, and particularly the law of divergence, as peculiarly his "own department of science", and if Chambers were to tidy up

---

(Barry 1836–37b). Nor did he ever cite Carpenter's 1841 publication throughout his dispute with Carpenter. The only conclusions possible are either that Owen was guilty of duplicity and deliberately suppressed the relevant publications, or, that he was not aware of their existence (which seems most unlikely in the case of Barry and Carpenter). With respect to von Baer, I have formed the opinion that Owen's claim to precedence was based on the simple fact that he did not read the relevant Scholion until its translation into English by Huxley in 1853 (see Note 31). Owen's German was not as good as he liked to make out, and he often required assistance with the translation of his German correspondence.

31 The second edition of Owen's *Lectures on Invertebrates* of 1855, reproduced substantially the same summary on development as that of 1843, but Owen conceded in a footnote that his propositions were "well supported and illustrated by von Baer", and cited von Baer's *Entwickelungsgeschichte* of 1828 in this and other notes appended to this edition (Owen 1855, 645, fn). However, on the basis of the similarity between Owen's quotations from von Baer and Huxley's translation, his source was most likely the latter: Huxley 1853.

32 Brooke concedes that this sentence "sticks in the gullet" (Brooke 1977, 138).

his discussion of embryology and make the whole consistent with his "idea and diagram", his overall argument could be strengthened. It was good advice, but Chambers was not to take it for another nine years.

One of the reasons for his long delay may be that Chambers failed to locate the all-important summary in Owen's *Lectures on Invertebrates*;[33] because, for some reason, which I shall here attempt to unravel, Owen misdirected Chambers. He referred him, not as we would expect to the final lecture (i.e. the twenty-fourth) on which Owen laid so much emphasis in his dispute with Milne-Edwards and which had just been published in the *Annales*, but to the "concluding generalization" of the "twelfth lecture". This, on inspection, turns out to be a statement on the ova of Annelata to the effect that they replicate structures and phenomena "characteristic of mature animals widely separated in the natural scale" (Owen 1843, 147).[34]

However it is analysed, this vague and ambiguous statement cannot bear the interpretation of Owen's "true law". Owen clearly intended to direct Chambers to the relevant twenty-fourth lecture, and his misdirection not only may have misled Chambers, but most definitely misled Brooke. For Brooke, on consulting the conclusion to the twelfth lecture and assuming it to be Owen's "true law", incorporated its limitations into his interpretation of Owen's references as implied criticisms of the *Vestiges* diagram:

> Owen's "true law" was considerably more restricted and subtle than anything implicit in Chambers's diagram. . . . This was far from the general identity of foetal development implicit in the lower stretches of Chambers's vertical line.
>
> (Brooke 1977, 137)

While this is a reasonable enough interpretation of Owen's conclusion to lecture twelve, this was *not* what Owen meant by his "true law". Moreover, while the conclusion to lecture twelve does not incorporate the principles of von Baer's law, it is not inconsistent with it (nor with the *Vestiges'* diagram). As I indicated

---

33 In his *Explanations,* in rebutting the embryological criticisms of *Vestiges,* Chambers did refer to Owen's *Lectures on Invertebrates,* but not to the summary (Chambers 1846, 106–7). The passage cited by Chambers is the second reference discussed in Note 34 – and indicates that Chambers did follow up Owen's references. Significantly, Chambers interpreted Owen's reference to the early "vermiform" stage of the human embryo as meaning that the human embryo at this stage was comparable to an invertebrate, and cited this passage in opposition to Sedgwick's criticism (see Note 11). Nevertheless, although he misconstrued Owen's embryology, he evidently did not construe Owen's reference to "the idea and diagram" of *Vestiges* as criticism.

34 Owen's second reference to the conclusion of the lecture "On the Metamorphosis of Insects" (see text) is the same as the first reference cited in the 1851 account of his priority dispute with Milne-Edwards (see text). It is to the effect that the human embryo passes through the earlier forms of the vertebrates, and although it is superficially similar to an articulate in its very early stages, it does not actually represent one (Owen 1843, 249). This is compatible with the *Vestiges* diagram, but hardly warrants the description of Owen's "true law".

above, von Baer's law allows for a limited resemblance between embryos and adult structures of the same type.

Hence, Brooke's interpretation, although not unreasonable in the circumstances, is incorrect, and hinges on Owen's apparent misdirection; "apparent", because it seems to me possible that Owen, with his priority dispute with Milne-Edwards very much in mind, was referring to the fact that his *Lectures* had been published in twelve *parts*, and that the section he regarded as the major statement of his "true law" was contained in the twelfth part, as he emphasized in his 1851 account. If this is so, it strengthens my interpretation.[35]

In any case, irrespective of the reasons for his [mis]direction, Owen intended no subtle criticism of the *Vestiges* diagram, but, rather, was indicating his concurrence with it in the most "straightforward" way possible, i.e., by claiming priority for it. This being the case, was Owen also receptive to Chambers' "evolutionary translation" of von Baer's law? Or, to put it another way, did Owen's property claim extend to the evolutionary application of von Baer's law?

1.4 Owen's mature theory of saltatory descent

Owen made explicit his preference for the *Vestiges* explanation to that of natural selection in his well-known 1860 review of *The Origin of Species*. There are a number of statements to this effect, and I here reproduce one in order to demonstrate Owen's uncritical endorsement of the *Vestiges'* embryology, his good understanding of Chambers' mechanism of transmutation and the evidence Owen adduced in support of it.

[The author of *Vestiges*] came to the task of attempting to unravel the "mystery of mysteries", when a grand series of embryological researches had brought to light the extreme phases of form that the higher animals passed through in the course of foetal development, and the striking analogies which transitory embryonal phases of a higher species presented to series of lower species in their permanent or completely developed state. He also instances the abrupt departure from the specific type manifested by a malformed or monstrous offspring, and called to mind the cases in which such malformation had lived and propagated the deviating structure. The author of "Vestiges", therefore, speculates – and we think not more rashly or unlawfully than his critic has done – on other possibilities, other conditions of change, than the Lamarckian ones; as, for example, on the influence of premature birth and of prolonged foetation in establishing the beginning of a specific form different from that of the parent. And does not the known history of certain varieties, such

---

35 Other interpretations are, of course, possible. Owen may merely have been careless, or the letter may have been mistranscribed for publication.

as that of M. Graux's cachemir-wooled sheep, which began suddenly by malformation, show the feasibility of this view?

(Owen 1860, 504–5, see also 497, 503, 508)

Elsewhere in the same review, Owen emphasized that the *Vestiges* mechanism depended on "preordained exceptions in the long series of natural operations, giving rise to the introduction of new species", a teleological dependency which made it more acceptable to Owen than natural selection.[36]

But what is more striking than this expressed preference for the *Vestiges* mechanism to that of the *Origin*, is that when Owen finally publicly articulated his own evolutionary views, his "derivative hypothesis" of the sixties incorporated the preordained embryonic deviation thesis of *Vestiges*, and, like *Vestiges*, was based on the analogy between abnormal or monstrous development and the production of new species:

> I deem an innate tendency to deviate from parental type, operating through periods of adequate duration, to be the most probable nature, by way of operation, of the secondary law, whereby species have been derived one from the other . . . no explanation presents itself for such transitional changes, save the fact of anomalous, monstrous births.
>
> (Owen 1868, 3: 807–8)

Owen went on to draw an explicit parallel between ontogeny and the fossil sequence, and he illustrated his hypothesis by reference to the fossil series he had established in 1851 (Owen 1851a, 448–50) from the three-toed Paleotherium of the Eocene, via the Miocene Hipparion with its greatly reduced inner and outer toes, to the single-toed modern horse:

> What, then, are the facts on which any reasonable or intelligible conception may be formed of the mode of operation of the derivative law exemplified in the series linking on *Palaeotherium* to *Equus?* A very significant one is the following:- A modern horse occasionally comes

---

36 Owen 1860, 508. Note Owen's discussion of the distinctions and similarities between his conception of the forces controlling the "development of organized beings" and those of the *Vestiges* author, and his related reference to his "true law" of embryological development of 1843 (Owen 1860, 506–7). As Owen explained it, the teleological principle ("the specific organizing principle" which shapes the living thing to its functions) is opposed by a "general polarizing force" which brings about repetition of parts (Owen's "irrelative repetition") and similarity of forms – all the signs of unity of organization and of the archetype. The extent to which the teleological principle overcomes the general polarizing force is an index of the grade of the species. These antagonistic forces not only control individual development, but the development of life on earth. Owen first enunciated his conception of the principles of polarity in development in his major theoretical work *On the Archetype and Homologies of the Vertebrate Skeleton* (1848, 171–2). See also Russell [1916] 1972, 111–12.

into the world with the supplementary ancestral hoofs. . . . In relation to actual horses such specimens figure as "monstra per excessum;" but, in relation to miocene horses, they would be normal, and those of the present day, would exemplify "monstra per defectum".

(Owen 1868, 3: 794–5)

According to Owen then, the modern horse had arisen as a preordained "monstrous" deviation from the parental Hipparion, which had been similarly derived from the ancestral Paleotherium.[37] Like Chambers (and in opposition to Darwin), Owen emphasized that such deviations would be "sudden and considerable" (Owen 1868, 3: 795), but, unlike Chambers, he laid greater stress on the role of premature birth, i.e., the birth of a "monstra per defectum", rather than prolonged foetation in the origin of a new species.[38] This, I would suggest, was consistent with the parallel he drew between such saltatory mutation in vertebrates and "metagenesis" or alternation of generation in invertebrates. In the latter process, as Owen finally articulated it in 1860, the parasitic fluke ordinarily goes through its cycle of changes – egg, ciliated "monad", gregarina, cercaria, etc. – without any of the intermediate forms becoming permanent:

> But circumstances are conceivable, – changes of surrounding influences, the operation of some intermittent law at long intervals, like that of the calculating-machine quoted by the author of "Vestiges", – under which the monad might go on splitting up into monads, the gregarina might go on breeding gregarinae . . . etc., and thus four or five not merely different specific, but different generic, and ordinal forms, zoologically viewed, might all diverge from an antecedent quite distinct form.
>
> (Owen 1860, 502–3; see also Owen 1849, 3, 1858, lxxv)

In other words, the emphasis here is on a kind of premature birth or arrest of development at each stage, by means of which the cycle is broken and four or five new genera or even orders might be established. Around the same time, Owen specifically referred the "whole assemblage of alternate-generative phenomena" in invertebrates to the "fact of anomalous, monstrous births" in vertebrates

---

37 In Owen's own words, the series evidences "(preordained) departures from parental type, probably sudden and seemingly monstrous, but adapting the progeny inheriting such modifications to higher purposes"; Owen 1868, 3: 797.

38 This would conform to Owen's conception of the antagonistic forces controlling development. The production of a new species is dependent upon the agency of the teleological principle, the "specific organizing force", which "subdues and moulds" the "general polarizing force" in "subserviency to the exigencies of the resulting specific form". See Note 36. The three toes of the Palaeotherium are the signs of irrelative repetition and are due to the action of the general polarizing force. The development of the extra toes is suppressed or arrested through the agency of the teleological principle which thus brings about the birth of the Hipparion, and ultimately, through further arrest of development, the single-toed modern horse.

(Owen 1868, 3: 807).[39] It was not so much that Owen in the sixties was "moving insensibly from metagenesis to saltation" as Desmond states (1982, 78), but rather that he regarded them as variants of the one generative process, i.e., of preordained premature or "monstrous" births.

Clearly, Chambers' invocation of "a higher [generative] law" in order to "explain" embryonic deviation held considerable appeal for Owen, and his *post-Origin* "derivative hypothesis" was remarkably similar to the *Vestiges* mechanism of 1844.

Lest it be argued that Owen, forced by the Darwinian disputes of the sixties to declare his own views on the origin of species, in his desperation appropriated the mechanism he had scorned in the forties, there exists one crucial piece of evidence to show that Owen was not dependent on *Vestiges* for his "derivative hypothesis" and which at the same time forecasts his subsequently favourable attitude to the *Vestiges* "idea and diagram". This we have on the authority of Charles Darwin himself. In his transmutation notebook for 1837, Darwin, ever on the alert for relevant information, noted:

> Mr. Owen suggested to me, that the production of monsters (which Hunter says owe their origin to very early stage and which follow certain laws according to species), present an analogy to production of species.[40]

There is more than a touch of irony in the picture this brief entry conjures up of the two future adversaries speculating together on the origin of species. Owen, engaged in cataloguing the Hunterian collection and manuscripts in the Royal College of Surgeons, had obviously been giving some thought to the implications of Hunter's ideas on abnormal development, and clearly he was also influenced by Geoffroy St Hilaire's more recent teratological experiments and associated speculations on transmutation. It was in this same year that Owen edited and published a collection of John Hunter's papers on animal physiology, and an examination of this collection makes Darwin's references to Hunter's (and Owen's) views more explicit.

1.5 Owen's early views on saltation

In his lengthy introduction, Owen discussed Hunter's attempt to classify congenital defects and to explain them by reference to the "transitory structures or metamorphoses of foetal life", which Hunter referred in turn to "permanent" structures

---

39 In all his writings on the topic, Owen seems, in the light of hindsight, to have been fumbling towards some conception of paedomorphosis: (literally, "shaped like a child"), the retention of youthful ancestral characters in later ontogenetic stages of descendants. On paedomorphosis, see Gould 1977, 221–8. Owen's 1849 Hunterian Lectures on Generation [BM(NH), OC, vol. 38], warrant further study.

40 Darwin's *Notebooks*, B 161; Barrett et al. 1987, 210–11. Note also Darwin's reference to Owen's contemporaneous criticism of the Meckel-Serres law (Notebooks, B 163). Darwin also had recently been reading some of Geoffroy's work (Notebooks, B 133–5; Barrett et al. 1987, 202–3).

in lower animals. Owen compared Hunter's teratological work with that of Geoffroy St Hilaire, and twice made the point that Hunter attributed the production of monsters to the "condition of the original germ", in contrast to Geoffroy, who, on the basis of his experiments, attributed them to exogenetic or mechanical causes (Hunter 1837, xxiv–xxvii, 44–5). Hunter's reason for assigning an endogenous cause to foetal malformations was described by Owen as "one of the most remarkable laws of aberrant formations" and reproduced by him as follows:

> I should imagine that monsters were formed monsters from their very first formation, for this reason, that all supernumerary parts are joined to their similar parts, as a head to a head, etc. etc.
>
> (Hunter 1837, xxvi)

I suggest that the major significance of Hunter's teratological views for Owen, lay in their emphasis on endogenous causation and the evidence Hunter adduced in support of this. This gave Owen the empirical basis[41] he needed to confront the opposing view of Geoffroy that foetal malformations were the result of chance variations in the environment. This, as indicated above, was crucial to Geoffroy's speculations on transmutation.

Geoffroy was convinced that in provoking the birth of monstrosities from hen's eggs by artificially varying the conditions of incubation, he had experimentally illustrated the way new species arose in nature. By extending this experimental model of exogenetic causation of foetal abnormalities to species change, Geoffroy could attribute hereditary change in the organism to the direct action of the environment. Thus, in his view, mechanical and chemical changes in the environment (especially in the respiratory milieu) induced changes in the organism during the embryonic stage which were akin to monstrous development, and which, through their propagation by inheritance, brought about the transmutation of species.[42]

Owen, like Geoffroy, saw in teratology the potential explanation of how new species might have arisen, but, by focusing on Hunters' endogenous explanation of teratological change, he was able to oppose Geoffroy's materialistic emphasis on external causation by invoking teleological causation. As all foetal malformations are inherent in the "germ", so all saltatory change was Divinely pre-programmed, and not referable to any chance external causation. This was the thrust of Owen's subsequent "derivative hypothesis". The tendency to deviate from parental type

---

41 There are many instances of such "double monsters" in the extant Hunterian Collection. See also "Ms. Catalogue of Physiological Series of Hunterian Museum", original fascicules partly in John Hunter's hand corrected by William Clift, 1816, Royal College of Surgeons, 49.d.5.

42 Geoffroy St. Hilaire 1828, 1833. In his 1833 Memoir, Geoffroy referred to monstrosities as "êtres ébauchés" – preparatory or precursory beings (85). Geoffroy also expressed organic change as the resolution of polar conflict. He distinguished between two conflicting influences on the developing organism: a conservative factor inherent in the germ, which tends to produce an offspring exactly like the parent; and a factor for change – the external influence of the environment. Any alteration in the environment resolves the conflict in favour of change (ibid., 214–16).

was "innate" and in 1868 Owen argued this by specific reference to "anomalous, monstrous births" which, according to Owen, demonstrated that:

> a species might originate independently of the operation of any external influence; that change of structure would precede that of use and habit; that appentency, impulse, ambient medium, fortuitous fitness of surrounding circumstances, or a personified "selecting Nature," would have had no share in the transmutative act.
>
> (Owen 1868, 3: 795)

Hunter's teratological views held another and related significance for Owen in their emphasis, as Darwin reported, on the "very early stage" of the origin of foetal malformations. This allowed Owen to dissociate them from the embryological law of parallelism and bring them into line with divergent development. For it was in 1837 also, that Owen began to incorporate the law of divergence into his Hunterian lectures at the Royal College of Surgeons, the lectures that formed the basis of his *Lectures on the Comparative Anatomy of the Invertebrates*.[43] Owen's suggestion to Darwin that a new species might result from a "monstrous" deviation from the normal course of development in the early stages of embryonic formation accords with von Baer's law of divergence and the *Vestiges* plan which subsequently so interested Owen. It does not, as explained above, fit in with the evolutionary application of the embryological law of parallelism, where the progress to a new species would take place as an additional step once the normal course of development had been completed. Owen was well aware of this implication, and explicitly rejected it in his *Report on British Fossil Reptiles* of 1841.[44]

All this goes to show that Owen consistently looked to teratology for the aetiology of the production of new species; and that from around 1837 on, he was feeling his way towards a theory of saltatory descent, as Ospovat argued (1976, 22). This was related to his concurrent promotion of divergent development, which provided embryological evidence of fundamental discontinuities between the great classes of living things, but at the same time indicated how new species might have arisen by divergence from the normal course of ontogeny. Viewed in this light, his published work of the forties elicits an underlying pattern which Owen himself identified in 1860 when he drew together those aspects of his work which had "tended to impress upon the minds of the most exact reasoners

---

43 Hunterian Lectures, MS 42.d.4, ff. 95–8; see Desmond 1985a, 48.
44 To wit: "If the progressive development of animal organization ever extended beyond the acquisition of the mature characters of the individual, so as to abrogate fixity of species by a transmutation of a lower into a higher organization, some evidence of it ought surely to be obtained" (Owen 1841, 197). Note the similarity between Owen's speculations "with all due diffidence" on the relationship between the changing oxygen content of the atmosphere and the succession of the vertebrate classes in time (202–4), and Geoffroy's idea that as the respiratory medium changes it brings about a corresponding change in the species (see Note 42).

in biology the conviction of a constantly operating secondary creational law" (Owen 1860, 500).

1.6 Owen in the forties: testing the waters

The 1840s represent Owen's period of greatest intellectual activity. It was during this period that he published his major morphological works and forged the synthesis between the transcendental anatomy of Geoffroy and the typology and teleology of Cuvier, which, with some introjection of German *Naturphilosophie*, formed the theoretical framework or system within which he structured his palaeobiology for most of his working life. As Macleod recognized, Owen's system, eclectic as it was, imparted a satisfying unity and consistency to his morphology, and during the forties Owen was intent on drawing all morphological knowledge within the bounds of his system (MacLeod 1965, 264–70; see also Ruse 1979, 116–25). This included not only his major theoretical preoccupation with the construction of the vertebrate archetype, but his work on metagenesis and teratology and his palaeontological application of the embryological law of divergence. Towards the close of the decade, a number of these themes were drawn together by Owen in the work in which he came closest to implying publicly the idea of organic descent.

It is necessary to emphasize at this point that I do not mean to suggest that Owen's morphology of this period was dominated by the idea of descent. Rather, it was dominated by the idea of the archetype. As Russell stressed:

> Pure morphology is essentially a science of comparison which seeks to disentangle the unity hidden beneath the diversity of organic form. It is not immediately concerned with the causes of organic diversity.
>
> (Russell 1916, 312)

However, Owen was not a pure morphologist, and, while his primary concern of this period was to establish the archetypal phenomena, he was also concerned with explaining the diversity of nature, and not only in the teleological sense. Ospovat has described Owen's *On the Nature of Limbs* of 1849 as "an extended argument against teleological explanation in biology" (Ospovat 1978, 36). While this is perhaps an overstatement, it does focus attention on the extent to which the "British Cuvier" had adopted the morphological priority of structure to function by the end of the forties. Moreover, Owen's commitment to the morphologists' *Bildung* encompassed its emphasis on historic process.[45] His vertebrate archetype, as is well known, was not a mere static abstraction, but was historically realized in nature by a process of continuous divergent development, culminating in man.

---

45 Morphology is not the science of fixed form or *Gestalt,* but of the formation and transformation of organic forms or *Bildung.* For a discussion of the historic process implicit in romantic morphology see Gode-von Aesch 1966; Temkin 1950.

This archetypal development, which Owen conceptualized in the symbolism of *Naturphilosophie* as a process of polar conflict,[46] was, of course, patterned upon the embryological law of divergence, which Owen, in competition with Carpenter, was making the basis of a general theory of palaeontological succession, which he elaborated during the fifties (Ospovat 1981, 133; Desmond 1982, 42–8, 50, 65–72).

Even allowing for a certain carefully cultivated (and typical) ambiguity in Owen's expression, his conclusion to *On the Nature of Limbs*, makes it clear that Owen conceived the archetypal development as a real process in nature, one governed by "natural laws or secondary causes" in accordance with a Divine plan; and that the divergent nature of the process resolved his major objection to the hypothesis of "transmutation of specific characters", which, as he wrote to Chapman in the previous year, was "always coupled with the idea of a specific direction – viz. *upwards*" (Owen 1849, 85–6; Ospovat 1978, 47–8). When we link this conception of vertebrate development with Owen's other major theoretical work of 1849, his *On Parthenogenesis*, which contained his theory of metagenesis, the implications for a theory of saltatory descent based on embryological divergence for both vertebrates and invertebrates are strong, and, moreover, they are consistent with both his earlier conversation with Darwin and his later *post-Origin* claims.

Owen's constant auditor Sedgwick, for one, certainly got a whiff of where all this might be tending. In 1850, while he was tooling up for another "rub at the Vestiges" and associated dangerous works such as Oken's *Naturphilosophie* (see Part 2), Sedgwick required some reassurance from Owen that his recent morphological researches did not lend support to the pernicious doctrine of "progressive development".[47] I shall return to this exchange between Owen and Sedgwick later in my analysis. My immediate purpose is to demonstrate the evolutionary implications of Owen's morphology of the forties and to show that these were evident (and embarrassing) to one of his conservative allies.

Owen's proprietary attitude towards von Baer's law during the forties and early fifties was not limited to his perception of its morphological and palaeontological significance, but most clearly extended to his recognition of its implications for a theory of saltatory descent. On the basis of this interpretation, we might speculate that the "idea and diagram" of *Vestiges* may have reinforced Owen's prior recognition of the relation between divergent development and saltatory descent and, in view of his later advocacy of it, catalysed his perception of how such occasional "monstrous" deviations could be at once subjected to some intermittent natural law and given an acceptable teleological explanation as manifesting "Intelligent Will" and according with a divine plan.

---

46 See Notes 36, 38 and 42.
47 Sedgwick to Owen [early 1850], BM(NH), Owen Correspondence, vol. 23, ff. 283–4. This letter is undated, but I have assigned a date of early 1850 to it on the basis of Sedgwick's references to his forthcoming book (Sedgwick 1850), and to his previous letter of February 1850, ibid., ff. 249–50; see also ff. 306–7.

Ospovat has stressed that for those biologists who in the forties and fifties were willing to entertain the idea of descent "the problem confronting them was principally one of redefining for themselves and those of their colleagues who remained teleologists the relationship of the new biological doctrines to natural theology" (Ospovat 1978, 46). Owen's particular brand of natural theology has been the subject of considerable debate. Much of the interpretative difficulty has arisen from the fact that Owen was as adept as most of his contemporaries, as Brooke has demonstrated, in exploiting the ambiguity of design arguments for socially and professionally diplomatic purposes (Brooke 1979, 47). Nevertheless, I agree with Desmond and Ospovat that we may discern in his writings from the early 1840s on, a cautious advocacy of a position close to that openly adopted by Chambers, i.e., that God's design was manifest in nature's uniformity and that creation had to be explained naturally (Desmond 1982, 29–48, 62–4; Ospovat 1978). This culminated in Owen's public suggestion of 1849 that animal species, including man, had originated by "natural laws or secondary causes" rather than by divine intervention. Although the ensuing "Puseyite" (see text below) and other reactions (such as Sedgwick's) to his conclusion to *On the Nature of Limbs* caused Owen, as I shall argue, to retreat almost immediately from its more far-reaching implications, by way of compensation he went way beyond Darwin's "truckling" statement in the *Origin* that "life was first breathed" into "some one primordial form" by openly confronting the problem of the naturalistic origin of life on earth (Darwin 1859, 488–90; Gruber 1974, 209). In 1860, stung by Darwin's attribution of naive creationism to the opponents of natural selection, Owen advocated the thorough-going uniformitarian solution of the spontaneous generation of the primordial and all other "monads" through the "operation of secondary and continuously operative creative laws". This brought him even closer to Chambers' position of 1844.[48]

Part 2. Oxbridge connections and "Puseyite reptiles"

Ospovat has argued a cogent case for grouping naturalists in the pre-Darwinian period into teleologists and those who sought to explain organic structure and succession by natural biological laws, rather than into the Huxleyan categories of creationists and evolutionists (Ospovat 1978). Ospovat's useful dichotomy permits the location of Owen, Chambers, Carpenter, Darwin and others on the same side of the teleological divide through their common commitment to naturalistic scientific explanations, in contrast to those like Sedgwick, Buckland and Lyell, who insisted on functional teleological explanations of organic structure and succession. But while this classification clarifies the relation of Owen's biology to that of Chambers, Darwin, etc., and avoids the historical distortions of Huxley's

---

48 Owen 1860, 514; see also 1866, 92, 1868, 3: 809, 814–25. By this stage, Owen seems to have been, like Chambers in 1844, advocating a form of continuous parallel evolution; see Hodge 1972.

categories, it obscures the professional and social alliances which reached across the teleological divide – which, for instance, absorbed Lyell into the Darwinian camp and made uneasy bedfellows of Sedgwick and Owen. In other words, it falls short of explaining why Owen did not, in the crucial pre-Darwinian period, consolidate his obvious biological affinities with Chambers, Carpenter, Darwin *et al.*, why he, in fact, continued to be identified, and largely identified himself, with those in the more conservative category. In order to answer these questions, it is necessary to locate Owen's biology in the wider institutional and social context as developed by Desmond.[49] This will make more explicit the social and intellectual constraints within which Owen operated and how these shaped his ideas and the articulation of his views on the organic origins question. The institutional and social location of Owen's biology will also help us make sense of Owen's simultaneous concurrence with the central thesis of *Vestiges* and his rejection of Chambers' associated speculations on the ape-origin of humanity and spontaneous generation.

2.1 Ideological constraints and professional alliances

Owen had, as he stated to Chambers, most carefully investigated the possibility of an ape ancestry for mankind, but whether he had done so "without . . . the slightest prejudice against such a relationship", as he claimed, is debatable (Owen 1894, 1: 249–52). As Desmond has demonstrated, Owen's investigations were inspired by his confrontation with Robert Edmond Grant, who was concurrently promoting transmutationist views in an overtly materialist form and in a radical political context. According to Desmond's analysis, Owen in the 1830s, from his institutional base of the Royal College of Surgeons (at that stage under heavy attack from the medical reformers), aligned himself with Peelite conservatism and structured his anatomy and palaeontology in opposition to the transformist assumptions of Grant and his mentor Geoffroy St Hilaire, thus securing the patronage of the powerful Oxbridge scientific network, of which Whewell and Sedgwick were prominent members (Desmond 1982, 115–21, 1985a).

Grant and Geoffroy both assumed a serial development of life from monad to man. In confronting this, Owen specifically opposed the sequence from ape to man, which Geoffroy had established in terms of decreasing facial angle. On the basis of the craniology of the adult chimpanzee, Owen had argued in 1835 for "impassable generic distinctions between Man and the *Ape*", but had failed to convince the eminent French morphologist. As Desmond points out, it was all a matter of interpretation. Owen's evidence was equivocal and could have engendered pro-transformist deductions as well as the anti-transformist interpretation

---

49 Desmond 1982, 1984, 1985a, 1985b. Desmond offers by far the most perceptive and detailed contextual reconstruction of Owen's early intellectual and institutional milieu. But see also Rupke 1985.

Owen chose to give it. Nevertheless, Owen's interpretation was an important plank in his anti-transmutationist platform of the thirties, which brought him rich professional and social rewards.[50]

In the same period, Owen undermined Grantian and Geoffroyan transmutation at a more fundamental level by attacking its interrelated embryological and morphological underpinnings – the embryological law of parallelism and Geoffroy's morphological precept of unity of organization. Owen's ideological objections to both principles were resolved by his adoption of a von Baerian embryology. As we saw, his "true law" of embryonic divergence restricted Geoffroy's unity of organization to the initial "infusorial monad", after which there was fundamental divergence, which precluded the possibility of serial transmutation.

It is essential to emphasize at this point that what Owen understood by "transmutation" was the process he had committed himself to oppose, i.e., as advocated by Grant and Geoffroy. "Transmutation" for Owen thus connoted materialism, gradualism and unilineal progressive species change, and, with its necessary consequence that man was a transformed ape, threatened bestialism, which was anathema to Owen and his Anglican contemporaries.[51] If, at the very time he was publicly confronting transmutation and wresting the coveted title of "British Cuvier" from Grant, Owen was privately committed to a naturalistic[52] explanation of the production of species by saltatory descent, as his 1837 conversation with Darwin suggests, then his advocacy of a von Baerian embryology could resolve this problem for him as well. A preordained divergence from some lower vertebrate form, consistent with the *Vestiges* mechanism, could account for man's animal origin without having him emerge directly from an ape. And in fact, a few years later, Owen was to hint at such a possibility in the conclusion to his *On the Nature of Limbs*, as indicated above (see also Ospovat 1978, 47–8). It was therefore theoretically consistent for Owen to endorse the *Vestiges* mechanism but dissent from Chambers' speculations on the ape-origin of humanity; and in this case, ideologically essential as well.

Similarly, spontaneous generation, by its presumed association with the doctrines of materialism and transformism, was ideologically and scientifically suspect, and Owen was professionally committed to its opposition (Desmond 1982, 37; Farley and Geison 1980). At the same time, his 1840 Hunterian lectures on

---

50 Desmond 1985a, 41–5, 49–50. Note Gould's remarks on Geoffroy's relation of the young ape to man, with its implication of a paedomorphic theory of human origins (Gould 1977, 353–55), and refer to Note 39.

51 Desmond stresses these aspects of Owen's anti-transmutationism over and over in his writings. This is also the thrust of Ospovat's criticism of Brooke's interpretation of Owen's *Vestiges* letter (Ospovat 1981, 285). Owen himself drew these distinctions between his own position and that of the "transmutationists" in his 1860 review of *The Origin* (Owen 1860, 500–6).

52 "Naturalistic" in the sense that Owen subscribed to the belief that natural laws were the basis of Omnipotent design and that all causes were hence "secondary". See text, Part 1.

generation indicate a certain ambivalence on Owen's part, and he explored the issue in considerable detail. While, on the whole, Owen thought the evidence to be against spontaneous generation, he asked rhetorically:

> Is then the Generatio equivoca to be exploded from Physiology without further enquiry? To this conclusion the sincere lover of truth will hardly accede while such Philosophers as Carus, Burdach and von Baer warmly defend it. And independently of the scientific standing of its supporters, the physiological principles which it involves are too important, and the natural phenomena which immediately relate to it, are too interesting for the Theory of the Generatio equivoca to be lightly set aside.[53]

Chambers' deistic belief in the spontaneous generation of low forms of life and his continued support for the questionable experiments of Messrs Crosse and Weekes (who claimed to have generated mites from an inorganic solution by means of electricity), have often been represented as the outstanding instance of Chambers' naivety and credulity, and the most scientifically vulnerable aspect of *Vestiges* (Millhauser 1959, 93, 94, 99; Gillispie 1959, 104). But Chambers, as Owen was well aware, was in the best of company on this issue. Spontaneous generation was not the exclusive property of dangerous materialists and transformists, but was one of the doctrines fundamental to the romantic conception of nature and deployed in various ways within the teleological framework of early nineteenth century German transcendental physiology. The viability of this research programme has been attested by the recent studies of Timothy Lenoir.[54] While Owen in the forties formally opposed the suspect doctrine of spontaneous generation, he was not inclined to dismiss the views of the best continental physiologists lightly, and he was sufficiently ambivalent about the issue to have investigated the experiments of Crosse and Weekes closely. We may take at face value his assertion to Chambers that he had seriously explored the possibility of spontaneous generation "in every department of animated nature . . . but hitherto in vain". He was not intent on "putting [Chambers] in his place" and knocking out the key pin of an objectionable deistic theodicy, as Brooke would have it (Brooke 1977, 135–6). Rather, as a committed uniformitarian, he was not ideologically opposed to spontaneous generation (provided it was located within an acceptable teleological framework),

---

53 R. Owen, "Manuscripts, notes and synopses of lectures, 1828–1841," *Hunterian Lectures,* 1840, Lecture 2, BM(NH), OC, vol. 38, ff. 28, 23.

54 As late as 1876, von Baer was still arguing for the spontaneous generation of complex organisms in opposition to natural selection (Lenoir 1982, 263). Owen discussed (and rejected) such speculations within the context of German physiology in his 1840 lectures (ff.27–9. Note Owen's references to Oken). This was almost certainly one of the six possible "secondary causes" of the production of new species Owen claimed to know in his 1848 letter to Chapman (Owen 1894, 1: 249–52; see Note 4). Note also William Carpenter's review of *Vestiges* where he stated his own preference for such an explanation and defended the experiments of Crosse and Weekes (Carpenter 1845, 169–81).

but his acceptance of it was contingent upon its demonstrable ongoing operation. (Which is precisely why Chambers had so much riding on the experiments of Crosse and Weekes.)

If Owen in 1844, for a combination of professional and scientific reasons, was unable to endorse the experiments of Crosse and Weekes, those of Pouchet in 1859 were another matter. From the first, Owen publicly championed Pouchet's theory of heterogenesis, which supposed that a "plastic force" organized molecules into germ-cells or eggs which subsequently developed into mature organisms. While Pouchet denied the possibility of evolutionary transformation, and insisted on the compatibility of his views with the doctrine of successive creations, Owen, as indicated in Part 1, incorporated heterogenesis into his "derivative hypothesis" and defended it on "inductive" grounds and against theological objections and the charge of materialism. For Owen in 1860, Pouchet's experiments comprised the "best observations" and the "most carefully conducted and ingeniously devised arrangements for insuring success" (Owen 1860, 514; Farley and Geison 1980). After Pouchet's defeat at the hands of Pasteur, Owen, from his standpoint of 1868, showed a perceptive understanding of the political dimensions of their celebrated debate, observing that Pasteur, like Cuvier in his 1830 debate with Geoffroy, had the "advantage of subserving the prepossessions of the 'party of order' and the needs of theology" (Owen 1868, 3: 814).[55]

Owen should have known, and his own bitterness and sense of betrayal were acute. For more than 30 years, as the British Cuvier, he had moulded his science to the same ends, and his achievement was all the greater for having reconciled an ideologically suspect transcendental morphology with Cuvierian teleology and made the whole a mainstay of natural theology. But by the sixties, the "mediating functions of natural theology were becoming . . . redundant" in a climate of secular naturalism, and Darwinism, in the hands of Spencer and others, was in the process of assimilation to the capitalistic requirements of a new "party of order" (Brooke 1979, 53; Young 1980). What seems to have particularly rankled with Owen, in those years of reassessment as he faced his professional eclipse by the triumphant Darwinians, was his perception that he had paid a heavy price for his earlier success, and that was the muting of his evolutionism. I consider that much, if not all, of Owen's legendary equivocation over evolution in the sixties (MacLeod 1965) is explicable in terms of his attempts to demarcate and lay claim to a socially and intellectually outmoded conception of organic descent which he had never publicly articulated nor theoretically developed to any extent; which had been geared to earlier social needs, was repressed by them, and finally rendered redundant by changed social needs.

---

55 Owen's expression "party of order" is interesting. The "party of order" was the name the French gave to the union of conservative and formerly moderate forces with the old regimes against the revolutionary forces of 1848 (Hobsbawm 1975, 17). See Farley and Geison (1980) for a discussion of the political dimensions of the Pasteur-Pouchet Debate.

In other words, I want to suggest that Owen's early speculations on saltatory descent were shaped by the same conservative institutional and social forces to which he adapted his anti-transmutationist biology and palaeontology; that, as Desmond argues, his adoption of a von Baerian embryology was not directly based on biological phenomena but was to some degree ideologically motivated by his anti-transmutationism; that, at the same time, von Baerian embryology offered a model for the naturalistic origin of species which was ideologically acceptable to Owen in that he could dissociate it from the radical, materialist, and bestial transmutationist speculations of Grant and Geoffroy and, through Hunter's endogenous explanation of teratological change, assimilate it to a conservative, non-materialist, teleological framework. Thus, all external causation was excluded and organic development could only occur according to a preconceived Divine plan. All organic change was inherently lawful and conservative – nothing new or radical could arise. The ideological potential of such a theory is evident. Not that I am suggesting for a moment that Owen explicitly and opportunistically set about constructing a theory of evolution to serve conservative social and political interests. He would have done better by far to have endorsed the miraculous origin of species! Rather, I am trying to show that Owen selectively deployed available biological concepts and incorporated them into a conception of organic development that satisfied his commitment to both natural law and natural theology; and further, that this conception of development might have served conservative needs, just as his reconciliation of Geoffroy's transcendental anatomy with teleology was welcomed by conservatives such as Whewell in the post-*Vestiges* period. Owen's archetypal theory resolved the teleological problem of the explanation of structural similarities between organs and organisms which do not serve similar functions, and his "conserving reform" came to the aid of natural theology and the social order in a period when they were both under considerable stress (Desmond 1985a, 50; Ruse 1979, Ch. 5). I think we gain a better understanding of Owen's early evolutionism if we view it in similar terms, as an attempt by Owen to reconcile Geoffroy's teratological speculations with conservative teleological concerns. In other words, Owen adapted not only Geoffroy's morphology but also his theory of organic descent, and his appropriation and promotion of a von Baerian embryology was crucial to both adaptations.

But while his revised Geoffroyan morphology was welcomed by his conservative colleagues and readily subsumed within natural theology, his cautious attempt of 1849 to give voice to his related views on saltatory descent was not. Had it not been for the dangerous proximity of those views to the central mechanism of *Vestiges*, Owen might have been less cautious and his audience more receptive. For, if my interpretation is correct, the publication of *Vestiges* in 1844 and its continuing hostile rejection by his colleagues and erstwhile patrons presented Owen with a serious problem – a problem which he initially attempted to resolve by adopting a strategy of dissociating himself from such a source of anathema without himself being forced to join in the conservative chorus of denunciation and repudiation.

70

By the mid 1840s, Owen's professional reputation was secure, to the extent that he was able to extend his own patronage to aspiring younger zoologists. Even so, he clearly had no intention of jeopardizing his establishment backing and his prestigious Oxbridge connections. If Darwin, in the process of establishing his own scientific credentials and creating a receptive liberal audience for his own version of descent theory, perceived the need to dissociate himself from such a notorious and patently amateurish work as the *Vestiges* (Herbert 1977; Hodge 1972, 134), how much greater was Owen's need to do the same in order to maintain his established, but apparently irrevocably anti-*Vestiges*, audience?

2.2 Owen's strategy

This was made easy for him by the varied, and, in some cases, contradictory embryological menu that Chambers proffered his readers. Owen had only to select from its least palatable offerings and focus his requisite criticisms on these, thus satisfying the Oxbridge clique's expectation of his necessarily critical response to Vestiges (Owen 1894, 1: 252–6). Owen's letter to Whewell of 3 February 1845, on which Brooke places so much emphasis in his interpretation of Owen's response to *Vestiges*, may therefore be read as an extended criticism of Chambers' delineation and palaeontological applications of the law of parallelism, but *not* of the "idea and diagram of page 212", which is left untouched by this critique. As Brooke points out, Owen did, however, criticize, on factual grounds, Chambers' invocation of prolonged gestation as a factor in species change (Brooke 1977, 141, 138). But, as I have emphasized, such criticism was consistent with Owen's opposition to the transmutationary application of the embryological law of parallelism, in which change is assumed to be additive, rather than divergent, as Owen believed it to be, and as the *Vestiges* "idea and diagram" depicted it. Owen himself, as I have argued, placed the emphasis on premature birth in both his "metagenesis" mechanism for invertebrates and his "derivative hypothesis" for vertebrates. To clinch the matter, Owen concluded by directing Whewell, in terms with which we have by now become familiar, to his "*concluding* [my emphasis] Lecture on the Invertebrata" of 1843, where Whewell would find Owen's "precise" rendering of the embryological law in question (Brooke 1977, 142). Owen could thus manipulate the ambiguity of his morphology by hinting at its implications for a theory of descent, then, when necessary, as in this case, fall back to a Baconian concern for the "facts" and eschew all speculation.

So by exploiting Chambers' failure to discriminate properly between the two versions of development, Owen, for a while, could continue to run with the evolutionary hare and hunt with the anti-transmutationist hounds, as Ruse suggested, without compromising either position (Ruse 1979, 124). This is the light in which I would interpret both his letters to Whewell and his blanket refusal to either review *Vestiges* with his own "*master hand*", as Murchison urged him to do, or to "appear as having directly or personally aided in any thing that may be regarded as a refutation or antidote to 'Vestiges'".[56] If Owen hoped in the long term to convert his colleagues to his idealized version of organic descent – a version which was uncomfortably close to the *Vestiges* "idea and diagram" – then it stands to reason that he would not want to be cited in connection with any "formal and direct refutation" (note his choice of words) of the work, and that he would do his best to dissuade those same colleagues from this course.[57] At the same time, he had no objection to the citation of his criticisms of the law of parallelism and its applications in any informal and indirect refutation that Whewell cared to publish, for this was quite consistent with a strategy of dissociating himself from such a notorious work without being forced into direct confrontation with it on the strength of his professional anti-transmutationist reputation – where such confrontation was potentially embarrassing.

Nor was this strategy inconsistent with his argument to Whewell that a direct refutation would lend undue importance to a "superficial" work (Brooke 1977, 142). It was quite in keeping with Owen's purposes that *Vestiges* should die a quick and natural death; and if I were required to sum up Owen's attitude to *Vestiges* in one word, that word would be *patronizing*. Throughout his *Vestiges* letter Owen is the expert, condescending, although tolerantly, towards the amateur, whose collation of the evidence for a theory of descent from such a wide variety of sources is impressive, but who is unable to distinguish the wheat from the chaff he has so diligently gathered together; who, above all, does not fully comprehend the implications of the diagram he has appropriated and who persistently compromises it with an outmoded law of embryological development, the backbone of an ideologically objectionable transmutationism which had been the major target of Owen's professional expertise for the past ten years. How else could Owen be expected to perceive the *Vestiges*, but as a "superficial" and amateurish work? But not one "so superficial as to be beneath contempt," as Brooke suggests (1977, 134).

For all his transmutationism and lamentable lack of discrimination, the unknown author (unlike the godless and dangerously radical Grant), shared Owen's commitment to explanation through Providential law, and, if my interpretation is correct, Owen himself had, as he told the author, "profit[ed]" from his reading of *Vestiges*. Owen was, to this extent, addressing the author as one "true searcher after truth" to another (Owen 1894, 1: 249–52), and, secure in his own superior understanding, could indulge in a little constructive criticism and hint at his own long-term interest in the idea of descent, and, above all, with his reference to the "idea and diagram", establish his priority.

This same air of patronage and superiority towards the author of *Vestiges* is evinced in Owen's remarkable letter to Chapman of 1848 (which letter is a further illustration of how Owen tailored his discussion of the introduction of species to

56 Owen's second letter to Whewell of 14 February (1845); Owen 1894, 1: 254; Brooke 1977, 142
57 Owen to Whewell, 22 February 1845, quoted in Brooke 1977, 140.

his audience – in this case a rationalist who did not share Whewell's abiding teleological concerns):

> Transmutation of species in the ascending course is one of six possible
> secondary causes of species apprehended by me, and the least probable
> of the six. When I remarked to the [reputed] author of "Vestiges," the
> last time he visited the museum, how servilely the old idea had been fol-
> lowed by De Maillet, . . . Lamarck, and the author of "Vestiges" – viz. of
> "progressive development" – and that there were five more likely ways
> of introducing a new species, he asked suddenly and eagerly, "What are
> they?" I declined to give him the information, but shortly after brought
> prominently under his notice the facts that might have suggested one,
> at least, of the more likely ways. He saw nothing of their bearing, and
> I shall refrain from publishing my ideas on this matter till I get more
> evidence.
>
> (Owen 1894, 1: 309–10)[58]

Desmond has suggested that these significant specimens in the Hunterian Museum
bore on the phenomena of metagenesis, which Owen must then have been studying in preparation for his publication of the following year (Desmond 1982, 35;
Owen 1849). But what is more germane to my immediate purpose is Owen's
consistent attitude to the author of *Vestiges*. Again Owen was intent on informing
the author (and Chapman!) of his own superior understanding of the matter and
throwing out a few helpful hints. He even condescendingly puts certain of the
relevant "facts" under the presumed author's very nose, "facts" which, like the
conclusion to his *Lectures on Invertebrates* and perhaps connected with it, should
have alerted the author of *Vestiges* to the error of his "transmutation of species
in the ascending course" and reoriented him to the correct path – of divergent
descent? It is also possible that Owen may have been warily testing the embryological understanding of the supposed author with his "facts". But as his visitor
was most likely the phrenologist George Combe, whom Owen regarded at this
stage as the *Vestiges* author (Desmond 1982, 210–11), it is not too surprising that
the putative author "saw nothing of their bearing", and Owen's consciousness of
his own superiority was reinforced to the almost overbearing extent of his letter
to Chapman.

In spite of Whewell's warning of 1846 (cited in Brooke 1977, 139) that "men
of real science" did not venture to speculate on the natural origin of species (an
argument that Owen himself deployed on occasion), Owen was cautiously willing to indulge in this with an appropriately responsive audience (as, for example,
with Darwin). However, if such views were ever to achieve scientific credibility,
and, more importantly for Owen, recognition, then it was essential to extend this

---

58 See also Owen's second letter to Chapman (Owen 1894, 1: 310–11).

audience beyond amateurs to those like Whewell and Sedgwick who comprised the influential Oxbridge network. Perhaps Owen was encouraged by their favourable response to his archetypal theory; perhaps he, like Ruse, detected a "slight softening" in Whewell's position on natural law and organic origins (Ruse 1979, 127). Whatever his reasoning, his decision to test the waters in *On Limbs* did not elicit the response he must have hoped for. Whewell and Sedgwick were in the process of divorcing the organic origins question from science (Ruse 1979, 270), and gave him no encouragement to pursue this line of enquiry (Sedgwick positively discouraged him).

His failure to palliate the teleological scruples of Whewell and Sedgwick was bad enough, but much more devastating in its immediate impact was the "Puseyite" attack on his views in the pages of the *Manchester Spectator*. This, for Owen, marked the end of his non-confrontationist strategy with respect to *Vestiges*, as he was forced into publicly dissociating himself from the work in refuting the charge of Pantheism which had been levelled against him. More seriously than this, however, Owen's burgeoning evolutionism was brought up short against the larger social forces of the day. As the "Puseyite reptile"[59] of the *Manchester Spectator* made clear to him, in a blanket condemnation of the theories of Oken, Owen and the *Vestiges*, such dangerous doctrines were out of place in a context of Chartist agitation and the interests of the Manchester captains of industry in maintaining low wages, long working hours and the preferential employment of women and children. These were the "hungry forties", when Britain came close to political revolution. If in England, at least, revolution did not eventuate, it broke out all over Europe in 1848, and its reverberations were felt, even by Cambridge fellows such as Whewell and Sedgwick, well into the fifties (Ruse 1979, 129; Hobsbawm 1975, Ch. 1, 1979, 77).

It was not only that Owen was treated to a "thinly-veiled charge of atheism" (Brooke 1979, 47) by the *Manchester Spectator*, but to a not so thinly-veiled charge of fomenting political revolution.

2.3 The "Puseyite assault"

The enduring impact on Owen of this public censuring of his early views on organic descent, and just how keenly he felt it, is evidenced by his rebutting of

---

59 Owen's expression, Owen to Powell, 26 January 1850, cited in Desmond 1982, 46; cf. Brooke's commentary on this letter (1977, 40–1). The "Puseyites" were the Anglo-Catholic followers of E.B. Pusey who denounced science for its incompatibility with revealed religion. Although Owen associated his critic with the Puseyites, the doctrinal affiliations of the author of the articles are not so clear cut. He seems, if anything, to have been an admirer of Francis W. Newman and, when challenged by Owen, claimed that it was not the policy of the *Spectator* to "set a mere theologic dogma in opposition to an established scientific fact" ("Christianity and Civilization," *The Manchester Spectator*, 1849: 24 November, p. 4; 1 December, p. 4; 8 December, pp. 4–5; 22 December, p. 4).

this "Puseyite" assault of 1849 in all his major writings on the species question during the 1860s. Here he is on the subject in 1868, still smarting in retrospect:

> Even in his partial quotation from my work of 1849. . . [Darwin] might have seen ground for apologising for his preposterous assertion in 1859: – that "Professor Owen maintained, often vehemently, the immutability of species". . . . The significance of the concluding paragraphs of [*On the Nature of Limbs*] was plain enough to BADEN POWELL . . . and drew down on me the hard epithets with which Theology usually assails the inbringer of unwelcome light [*Manchester Spectator*, 1849].
> (Owen 1860, 3: 796, n, see also, 1866, 90, 1868, 511)

More telling evidence of the deep and immediate effect of this criticism on Owen is to be found in his personal copy of *On the Nature of Limbs*, the final pages of which are interleaved with lengthy excerpts from the offensive *Spectator* articles, Owen's letter of rebuttal to the editor, the *Spectator's* response, and some notes in Owen's hand on the teleological implications of his morphology (Owen Correspondence, 18, 85–92). An examination of these newspaper cuttings in their original context of the relevant issues of the *Spectator* is illuminating, and makes Owen's long-term reaction explicable.

The *Manchester Spectator* was a short-lived weekly, whose policies reflected the commercial and political interests of the cotton manufacturers.[60] Like other middleclass reform papers of the period, it was devoted to the principles of free trade and championed the parliamentary and financial reforms that would enhance the political and economic power of the provincial manufacturers. Such moderate liberalism was, in the wake of 1848, fully alive to the dangers of revolution and the threat it posed to middle-class aspirations and values. As the *Spectator* saw it, the grievances and discontents of the new industrial working classes were best remedied through education and religion which would teach them to understand their place in the new society and ensure their self-improvement and personal salvation, not through dangerous and misconceived mass movements aimed at overturning the social order and levelling society. Education, however, for all its remedial powers was not without its own dangers. Its "primary element" was the inculcation of spiritual and moral values, not the mere provision of "useful knowledge". And above all, this knowledge must not undermine the religious faith on which social stability depended and so promote the very outcome it was designed to circumvent. These were the underlying concerns of the series of articles published under the rubric of "Christianity and Civilization", in the course of which Owen was attacked.

---

60 The *Manchester Spectator* was founded in 1849 and ceased publication in 1851. Judging by its leaders, it was competing for the same middle-class provincial readers as its more successful rivals, the *Manchester Guardian* and the *Manchester Examiner and Times* (Read 1961; Koss 1981, 1: 60–2).

The series was primarily occupied with exploring the implications of the more recent scientific publications for the moral progress of civilization (i.e., for Christianity, without which there could be no civilization). Among those singled out for particular attention were the recently translated *Elements of Physiophilosophy* by Lorenz Oken, and Owen's *On the Nature of Limbs*, which were lumped together with *Vestiges* and castigated for their promotion of a "desolating Pantheism". The thrust of the article of 8 December 1849, that so alarmed and outraged Owen, was that civilization was endangered by the diffusion of this "scientific Pantheism" which could be detected in his and Oken's works; and that this "Pantheism" was insidiously undermining religious belief among intelligent, rational working men, not merely among "half-witted atheists, who have got a smattering of chemistry, and a smattering of astronomy, and a smattering of phrenology, conjoined with some incoherent notions about SOCRATES, priestcraft, and the rights of man, with a glimmering of communism, and sundry other *isms* . . ." (*Manchester Spectator*, 8 December 1849, 4). The danger the author had in mind was clearly political revolution, and the spectre of 1848 loomed large over all the rhetoric on civilization and moral progress.

Civilization ("that protection from violence which secures to industry the fruits of its labours . . .") was contingent upon religious faith, especially among the potentially insurrectionary working classes ("those vast industrious classes who are daily growing in strength and intelligence") (*Manchester Spectator*, 24 November, 4; 8 December, 4). Religion for the masses was the antidote to revolution, and the speculations of Oken and Owen were sapping the foundations of religion and contributing to moral degeneracy and political and social unrest. Oken had gone too far in speculating on the naturalistic origins of man, but Owen had gone further still and brought "all his profound scientific knowledge and demonstrative skill, in support of what is called the THEORY OF DEVELOPMENT, and which has become popularly known by its introduction into the book called the *Vestiges of Creation*". According to the *Spectator*, Owen's morphology had led him to conclude that "God had not peopled the globe by successive creations, but by the operation of general laws", and this was an instance of that "modern Pantheism", which "assumes that it has detected certain general laws, as governing and evolving all things", and from there it was but a short step to atheism and revolution:

> Now, scientific speculations like these may do no harm to self-balanced minds; but the notions filter downwards, until the ignorant get a crude idea of them, and their minds are at once demoralized.
> (*Manchester Spectator*, 8 December, 4)

In short, it was not only that Owen was accused of adopting the development hypothesis of *Vestiges*, but his views were linked with those of Oken, and the whole was capped with an apposite quotation from Coleridge. Transcendental morphology as a whole was the target, identified with transformism and incipient atheism, and construed as morally and socially threatening.

The *Spectator's* linking of Owen with Oken was charged with significance. It was well known that Owen's archetypal theory was strongly influenced by Oken's *Naturphilosophie*,[61] and Owen had already received strong intimations of the problems attendant upon the morphological affinity of their respective theories.

2.4 The Oken connection and the Ray Society affair

Two years earlier he had weathered a storm within the Ray Society, when, at his instigation, the Society undertook the translation and publication of Oken's *Lehrbuch der Naturphilosophie*.[62] This was one of the few occasions where Owen's enthusiasm, or, more correctly, his opportunism (for this translation of Oken's major theoretical work was undoubtedly very useful in the preparation of Owen's own major morphological productions of 1848 and 1849), overrode his characteristic caution. Oken safely obscure in German was one thing; made accessible in English quite another.

After two members of the Ray Society Council had tendered their resignations and Oken's *Physiophilosophy* and the Society had been denounced in the pages of the *Zoologist* (which carried the Society's advertisements), the Council moved to counter the "injury which that [extremely objectionable] publication is calculated to do to the Society". An anonymous reader was hurriedly commissioned to scrutinize the *Physiophilosophy* for "symptoms of unsound religious principle". The reader, somewhat bemused by Oken's aphoristic style, hedged his bets, deciding that there was "no clear evidence of [Oken's] being a believer in revealed religion; nor yet of the reverse, tho' his details resemble a little the 'Vestiges of Creation'". He concluded that the Council had been unwise in authorizing this publication.[63]

The Council (of which Owen was a member, although he discreetly absented himself while these deliberations were in progress) tacitly agreed, and, belatedly decreeing that in future all works proposed for publication were to be vetted by some competent person, hastily closed the books on the matter.[64] Through all this Owen lay low, and although he emerged without personal censure from the affair, he did not emerge unscathed. While there is nothing in the minutes of the Society to identify him as such, he was widely (and correctly) assumed to have been responsible (Huxley 1894, 2: 313).

---

61 See, for instance, Owen 1848, 8, 73–5. Owen later wrote a laudatory *Britannica* article on Oken (8th edn., 1858–1859); see Rupke 1985, 249, 252.

62 English translation of 3rd German edn., Zurich, 1843, by A. Tulk for the Ray Society (Oken 1847). See Oken to Owen, 12 January 1847, BM(NH), Owen Correspondence, vol. 20, ff. 362–3.

63 *The Zoologist,* (1847), 5, Preface, v–x; "Minutes of the Ray Society, from its commencement in 1844 to 31 December, 1847," 16 October and 5 November 1847, Ray Society Archives, BM(NH).

64 Ray Society Minutes, 19 November and 17 December 1847. See also copy of letter of 20 December from Thomas Bell to William [Thompson].

It has been asserted that there was nothing in Owen's archetypal theory to offend the "ontological inhibitions" of his contemporaries (Hodge 1974). However, Owen trod a very fine line in his reconciliation of transcendentalism with teleology,[65] and at times, with his obvious admiration for Oken, seemed to some at least of his contemporaries (Sedgwick for one) to be in some danger of stepping off into the dark and dangerous waters of German *Naturphilosophie*. A major difficulty lay in the suspicious degree of correlation many readers now detected between the contents of Oken's *Physiophilosophy* and the *Vestiges*.[66] Sedgwick, while impressed by the similarity between the two, was inclined to exonerate the author of *Vestiges* from Oken's supposed excesses of atheism and materialism:

> The materialism of the Author of The Vestiges is not so horrible as this, because he assumes the being of a God, and accepts many of the great conclusions drawn from our moral and immaterial nature.
>
> (Sedgwick 1850, 230, 283)

If even the odious *Vestiges* began to look less objectionable to Sedgwick by comparison with Oken's philosophy, how then was he to approach Owen's archetypal theory with its acknowledged morphological debt to Oken? The result was a series of contortions in which he conferred praise on those aspects of Oken's morphology endorsed by Owen's appropriation of them, and castigated those not so "inductively" guaranteed. At the same time, he indicated his disquiet at Owen's association with such an ideologically unsound transcendentalism and its allied "Vestiginarianism", and sounded his warning to the "British Cuvier". He quoted the conclusion to *On the Nature of Limbs* and commented:

> Had I not known the opinions of this great comparative anatomist, as they are expressed in many of his recent works, I should, perhaps, have thought that in this passage he meant to indicate some theoretical law of generative development from one animal type to another along the whole ascending scale of Nature.
>
> (Sedgwick 1850, Preface, ccxiv)

This served the double purpose of letting Owen off the hook while warning him that Sedgwick would have none of this insidious verging on the offensive doctrine of continuous organic development, and that this carelessness of expression on the part of his colleague was becoming an embarrassment to one of his more prestigious allies.

---

65 Desmond makes this point with respect to Owen's leanings towards Geoffroyan transcendental anatomy (1984, 218–19).

66 Hugh Miller referred to the author of *Vestiges* as "the most popular contemporary expounder of Oken's hypothesis" (Miller 1851, 274). See also the *Athenaeum* review of Oken's *Physiophilosophy*, 2 October 1847.

2.5 Owen's letter to the *Manchester Spectator*: the repudiation of *Vestiges*

Sedgwick's alarms and excursions lay in the immediate future. In the meantime, Owen had to deal with the "hard epithets" of the *Manchester Spectator*. It was not only that they were upsetting his Lancashire relatives, as he complained to Baden Powell (Desmond 1982, 46), but his connection with Oken (and, by implication, with *Vestiges)*, had made him vulnerable to such charges, and he had had fair warning in the Ray Society affair that such attitudes were not confined to "Pusey-ite reptile[s]" in Manchester. However outrageous the imputation of atheism and consequent social disruption might be, he was not in a position to treat it lightly.

It was a valuable political lesson for Owen, and he moved quickly to establish his Christian credentials. Within the fortnight the *Spectator* had his angry retort in hand, and published it in the context of a lengthy commentary by the unrepentant "Puseyite". Owen claimed that the intent of the contentious conclusion to *On the Nature of Limbs* was not to support the development hypothesis, but rather to argue that his morphology demonstrated the pre-existence of the archetypal ideal in the mind of the Creator, and thus refuted the argument of the old Pantheists and atheists that mind did not precede matter. He demanded the insertion of the relevant paragraph in the *Spectator* and turned the tables on his accuser by charging him with deliberate and un-Christian, if not downright atheistic, misrepresentation (*Manchester Spectator*, 22 December 1849, 4).[67]

If Owen intended to emerge from the encounter as the better practising Christian, this intention was thwarted by his wily accuser, who seized on the opportunity to reiterate his claim:

> Now, we admit that we fell into a mistake in saying that Professor Owen teaches the theory of development. He does not teach it; he repudiates it; and yet a man of very general knowledge . . . to whom we lent the "Nature of Limbs," . . . returned it to us, with the significant remark, "After all, the theory of development *is* here."

Even though Owen disclaimed the theory of development, his work was open to this interpretation, and to half-informed minds "unaccustomed to rigorous demonstration" might reinforce belief in the very theory that he repudiated. In this way "even eminent scientific men might *unwittingly* strengthen Pantheistic notions" (*Manchester Spectator*, 22 December 1849, 4).[68]

The moral was clear, and was shortly to be reinforced by Sedgwick. If Owen was to avoid such "misinterpretations" in the future, he must be more careful in his phraseology. In fact, as a close reading of his *Spectator* letter makes clear,

---

67 Brooke has republished Owen's letter entire (Brooke 1977, 143).
68 Owen preserved the entire "Puseyite" response along with his own letter to the editor in his personal copy of *On the Nature of Limbs* (see text).

Owen had been very careful of his phraseology indeed in dissociating himself from *Vestiges:*

> Of the nature of the creative acts by which such successive races of animals were called into being, I have never presumed to offer an opinion, save in refutation of some inadequate hypotheses, more especially of that stale, but lately revived, one of Development and Transmutation.
>   (See "Report on British Fossil Reptiles," p. 196.)

Once more he retreated to the position of the eminent Cuvierian who did not indulge in the scientific sin of speculation; he flourished his credentials as an established slayer of transmutationist dragons; and (without mentioning it by name), he relegated *Vestiges* to the same "stale" category of transmutationism as the "inadequate" speculations of Grant and Geoffroy, which he had demolished three years before its publication. In this roundabout fashion, he publicly established his opposition to *Vestiges* without ever openly confronting the work. It was being borne in upon Owen that in the socially troubled forties, evolution was synonymous with revolution (see Desmond 1987), and the *Manchester Spectator* was not concerned with fine discriminations such as those he was elsewhere intent upon making.

Neither, however, was Sedgwick. Only a few months after he had fended off the *Spectator* attack, Owen was compelled to replay the same scenario with Sedgwick. In the face of Sedgwick's pointed questions about his *Parthenogenesis*,[69] Owen had to reassure Sedgwick of his continuing antitransmutationist stance and suppress the evolutionary implications of his metagenesis theory, and then endure Sedgwick's public criticism of his supposed looseness of expression and his associated attack on Oken and transcendentalism. Sedgwick was also writing in the shadow of 1848, and intent on pointing the moral in characteristic heavy-handed and long-winded fashion:

> Men are the fools of fashion, and schemes of development are the fashion of the hour. Constitutions that once came to life only after long gestation and many a mortal pang, are now to be developed into full stature, while the sun is making one of its daily rounds in the sky. Law is to develope its true supremacy by the dissolution of the elements of order. Nations are

---

69 Sedgwick wanted information on the incidence of larval sexual precocity (paedogenesis), and asked: ". . . with low animals which have more than one larva type, and exhibit the parthenogenesis of certain forms, the ultimate form, or perfect animal is always of one specific type. Is it not so?" (Sedgwick to Owen, 1850, Owen Correspondence, 23: 283–4). To which Owen, as the strict inductivist, responded, keeping to the facts and giving no hint of his heretical speculations: "not any of these facts of larval fecundity . . . support . . . 'progressive development'. The most complex cases of Parthenogenesis only show a more roundabout way of arriving at the specific structure from which the generative process started" (Owen to Sedgwick, ibid., 285; cf. Owen 1860, 502–3).

to be developed into riches, power, and strength, and happiness, by the abolition of the rights of property. On a revolving mechanism, all things, once thought great and glorious, are to descend into the kennel; and out of it are to rise the elements of another system which are to be twirled and developed into something newer and more glorious. [Etc., etc.]

(Sedgwick 1850, Preface, ccxxvii)

It was all too much,[70] and Owen made no more public forays into such politically troubled waters until the denouement of natural selection brought him out into the open once more. Then, in an 1863 address to the YMCA, he whipped up a theological storm with a denunciation of denominational opposition to "continuous creation" that quite astounded the Darwinian party (Desmond 1982, 41, 79–80). Was Owen's public emergence as a "free thinker", which amazed friend and foe alike, connected with a pent-up resentment at his selfcensorship of his evolutionism in response to the "hard epithets" of all those years ago? I think it was. All the available evidence points to this. The *Vestiges* letter, the letters to Chapman, the conclusion to *On Limbs*, all pre-date 1850. After this we hear nothing more that might be construed as evidence of Owen's commitment to a theory of descent until another decade had passed and Owen was confronting *The Origin of Species* and (what must have been the unkindest cut of all), passionately defending himself against the charge of subscribing to a naive creationism. Even then he cloaked his expression in the ambiguity which had served him so well in weathering the theological-cum-political criticism of 1849–1850, but which now merely made him the butt of Huxley's ridicule and further compromised his position (Desmond 1982, Ch. 2).

Conclusion: whose property?

Owen's *post-Origin* claims have been in the past too readily dismissed as the anguished cries of a loser, who projected an evolutionism distilled from hindsight onto his earlier work in a desperate attempt to claim some of Darwin's glory. The revisionist historiography of the Darwinian debates now under construction takes a more sympathetic and discerning view of Owen's earlier morphology and its intellectual and ontological strengths. No longer can historians toe the partisan line established by the triumphant Darwinians that there was only one viable path for nineteenth-century naturalists and that was along the route that led to natural selection. Rather there were a number of paths that ran along side and sometimes intersected with Darwinism. Owen's was by no means the only nineteenth-century attempt to found a theory of evolution on embryogenesis. Grant, Geoffroy and Chambers all saw embryogenesis as an obvious and appropriate model for

---

70 Political considerations aside, Desmond makes the point that Owen's "timing was inopportune". His transcendentalism was out of phase with the growing positivism and secular naturalism of contemporary science (1982, 48).

evolutionary change, and we may add Kölliker and von Baer himself to the list (Desmond 1982, 78; Lenoir 1982, 263–70).

A close analysis of Owen's pre- and post-Darwinian views on the "secondary causes" of new species demonstrates a surprising degree of consistency. If we are to deduce coherent historical explanations for the failure of Owen's views, rather than adopt the simplistic solution that Owen got it wrong and Darwin got it right, we might do better to look at Owen's inability either to tailor a theory of descent to fit the social and institutional concerns of his conservative audience, or, with their professional and social eclipse, to relocate his views in a more liberal context. There is still a good deal of debate about the extent to which contextual factors shape theory content, but there can be little remaining doubt about the need to establish a professional audience for the acceptance of one's views, and this Owen, in contrast to Darwin, failed to do.[71]

In other words, I have suggested that Owen, having earlier aligned himself with conservative forces and established himself professionally on the basis of his anti-transmutationism, was unable openly to articulate his ideas on organic descent without alienating himself from his conservative institutional and social base; that his one tentative attempt to do so in 1849 exposed him to theologico-political criticism which had a deep and enduring impact on him, and that throughout the fifties he muted his evolutionism in the face of this criticism.

By this I do not mean that he abandoned it. He revised his palaeontology, making it more consistent with von Baer's laws (and with a divergent theory of evolution), and he pursued his priority disputes with Milne-Edwards and Carpenter with undiminished vigour during the early part of the fifties. At the same time, he continued his exploration of the phenomena of heterogeneous generation, writing on metamorphosis and metagenesis (Owen 1851b). But we hear no more of "continuous creation" until the end of 1858, i.e., after Darwin's and Wallace's papers had been presented to the Linnean Society (Owen 1858). As Desmond has put it, throughout the fifties, "Owen edged forward hesitantly, desperate lest he offend" (Desmond 1982, 62). One of his major problems, as I hope I have established, was that his views lay too close to those of the heretical *Vestiges*, and he could not lay claim to them without provoking odious comparisons with this continuing source of anathema. Meanwhile, under the impact of the avalanche of criticism, Chambers unwittingly exacerbated Owen's demarcation problem by continually refining his arguments, until, with the tenth edition of *Vestiges* in 1853, his embryology and palaeontology were finally revised along von Baerian lines. This brought Huxley out in full cry, and he used his review of this edition of *Vestiges* not only to lambast *Vestiges*, but also to attack Owen's palaeontology (Huxley 1854; Ruse 1979, 143). Oken, Owen and the *Vestiges* were as distasteful a blend to Huxley as they were to Sedgwick, if for very different reasons. Huxley

---

71 See Desmond's remarks on the London "idealist community" of the 1860s (those like Argyll and Mivart who championed Owen's views) and their lack of an effective power base by comparison with the proDarwinian scientific naturalists (Desmond 1982, 176–7).

had no time for creation by law, and nothing but contempt for Owen's "Okenism" (Huxley 1894, 312–20).

Any chance Owen may have had of finding a more liberal audience for his views on descent was continually thwarted by Huxley, who throughout the fifties devoted himself singlemindedly to undermining Owen's morphology and palaeontology, making his theory of metagenesis and his vertebrate archetype his special targets.[72] Owen's energies were dissipated in safeguarding his morphology from Huxley's increasingly aggressive attacks, and in the face of this mounting professional threat he could afford even less to alienate his thoroughly conservative, but increasingly redundant, "party of order". In short, Owen could find no receptive audience for his idealized version of descent theory, and was reduced to debating its teleological implications with himself.[73]

Had he received any encouragement from Whewell and Sedgwick, Owen might have withstood the *Manchester Spectator* attack more resolutely, but by 1850 it was clear that they were in full flight from the organic origins question and, in his response to the contentious conclusion to *On the Nature of Limbs*, Sedgwick had more or less allied himself with the *Manchester Spectator* (for which treachery he finally received a serve from Owen in his 1860 review of *The Origin*; Owen 1860, 511). Only Baden Powell gave him positive encouragement, picking up the ball Owen had by then publicly dropped by unambiguously embedding the archetype in a developmental scheme of divergent descent.[74] In return he received little support from Owen, who around that time was even extending his self-censorship to his friend Martin Barry, by striking out Barry's manuscript reference to "the possible organic origin of all things".[75]

It was only in the sixties that Owen was prepared to acknowledge that Powell had interpreted him correctly, and belatedly to set about articulating his long-held views on descent. But what had been too revolutionary to express openly in 1844 was deemed reactionary in 1868. The more liberal climate of the prosperous and secular sixties, which made the voicing of Owen's views possible,

---

72 Huxley 1894, 320–30; Desmond 1982, Chs. 1, 2; Ruse 1979, Ch. 6. Both Desmond and Ruse offer a revisionist interpretation of the Owen-Huxley clashes of the fifties, and Ruse suggests that Huxley's "baiting of Owen" in the fifties pushed Owen into opposition to Darwinism (ibid., 144–5).

73 See Owen's comments facing pp. 84, 85 in his personal copy of *On the Nature of Limbs* (Owen Correspondence, 18: 85–92). Note also the quotation inside the front cover.

74 Baden Powell 1855, 333, 365–411, 456–7; see also Desmond 1982, 46–7; Bowler 1976, 103.

75 "I am much obliged to thy striking out the passage on the possible organic origin of all things" (Barry to Owen, 11 September, [1854], BM(NH), OC, vol. 2, ff. 303–4). I am aware of the ambiguity of this passage, and I have been unable to trace the publication to which it refers. Nevertheless, it seems likely that Barry, who died not long after this was written, was another of those von Baerians who saw in embryogenesis a model for the derivation of species. Darwin's famous diagram of divergence in *The Origin* could be superimposed over Barry's 1837 "Tree of Animal Development". Barry, like Owen, was also influenced by Geoffroy, and certain of his 1837 statements could be given an evolutionary interpretation (Barry 1836–1837b, 362–3). In any case, Owen's censorship of this (admittedly ambiguous) passage, illustrates my point of the degree of caution he had adopted in the post-1849 period.

was more attuned to the emerging Social Darwinism, which made unobstructed competition the guarantee of continuous social progress without revolutionary or radical change (Richards 1983, 87–97). Owen's invocation of "continuous creation" could not possibly match the powerful ideological appeal of such a scientific endorsement of mid-Victorian capitalist enterprise. More to the point, by the time Owen finally staked his full property claim, his claim had been jumped. Historians are now generally agreed that it was onto Owen's von Baerian palaeontological model of divergence that Darwinism was to be mapped, and that in the process Owen's concept of the archetype was transmuted into that of common ancestor.[76] But Owen, who suffered a particularly bad press at the hands of the victorious Darwinians, especially Huxley, received little recognition of his palaeontological contribution and still less for his morphology. In any case, as Ospovat demonstrated, Darwin drew his inspiration for his divergent model of evolution, not from Owen, but from Owen's old rival for embryological honours, Milne-Edwards, and, moreover, from the very paper in the *Annales* that had so aroused Owen's proprietary instincts (Ospovat 1981, 159–65, 174–76).

Finally, Huxley delivered the *coup de grâce* to Owen's embryological pretensions, when he, of all people, was commissioned to write the authorized assessment of his old enemy's place in the history of anatomical science. Here, Huxley, who was intent on staking a property claim of his own, casually dismissed Owen's embryology in a footnote to the effect that up to the time of the publication of his own translation of exerpts from von Baer's works in 1853, von Baer's ideas "were hardly known outside Germany" (Huxley 1894, 299n). Huxley followed this up by attributing Owen's later eclipse largely to his failure to understand the significance of von Baer's embryology for morphology (Huxley 1894, 320).[77] As Owen, in an earlier period, had capitalized upon his title of the British Cuvier, throughout the fifties, Huxley aspired to become the British von Baer, and had as determinedly wrenched scientific capital from Owen by opposing his von Baerian methodology to Owen's "osteological extravaganzas".[78]

We might think, in view of Owen's earlier cavalier treatment of Barry and Carpenter, not to mention Chambers and von Baer, that he got his desserts when the expropriator was expropriated by the ruthless Huxley. But, perhaps in the final analysis, Owen was the better von Baerian than Huxley. For, in the last year of his life (1876), von Baer subjected Darwin's theory to a detailed critique, and in the process elaborated what Lenoir describes as a "limited theory of evolution", which he had first proposed in 1834, and which, in its final version, was remarkably like

---

76 Bowler 1976, Ch. 5; Ospovat 1976, 1981, Ch. 5; Desmond 1982, Chs. 1, 2, 1985a, 49.

77 In a period when embryology was assuming increasing importance as an arbiter in settling questions of comparative anatomy, Owen fought a rear-guard action against this trend. Thus Owen, in contrast to Huxley, refused to regard embryology as the criterion of homology (see Russell 1972, Ch. 10). However, although they received short shrift from Huxley, Owen had cogent reasons for his stance, and they were consistent with his conception of development: Owen 1866, 1: xxi–xxv.

78 See Huxley to Leuckart, 1859 (Huxley 1900, 1: 162–3); see also Huxley to Darwin 1859 (ibid., 1: 175).

Owen's, incorporating the phenomena of heterogeneous generation which had all along so interested Owen: spontaneous generation, the alternation of generations, paedogenesis and metamorphosis (Lenoir 1982, 263–70).

Perhaps, after all, the property rights in this complex history should revert to von Baer himself.

Bibliography

Baer, Karl E. von. 1828. *Über die Entwickelungsgeschichte der Thiere: Beobachtung und Reflexion*. Konigsberg: Bornträger.

Barrett, Paul H. et al., eds. 1987. *Charles Darwin's Notebooks, 1836–1844*. Cambridge: Cambridge University Press.

Barry, Martin. 1836–37a. "On the Unity of Structure in the Animal Kingdom." *Edinburgh New Philosophical Journal* 22: 116–41.

———. 1836–37b. "Further Observations on the Unity of Structure in the Animal Kingdom, and on Congenital Abnormalities, Including 'hermaphrodites' with Some Remarks on Embryology, as Facilitating Animal Nomenclature, Classification, and the Study of Comparative Anatomy." *Edinburgh New Philosophical Journal* 22: 345–64.

Bowler, Peter J. 1976. *Fossils and Progress: Palaeontology and the Idea of Progressive Evolution in the Nineteenth Century*. New York, NY: Science History Publications.

Broderip, William. 1851–52. "Professor Owen – Progress of Comparative Anatomy." *Quarterly Review* 90: 363–413.

———. 1853. "Generalizations of Comparative Anatomy." *Quarterly Review* 93: 46–83.

Brooke, John Hedley. 1977. "Richard Owen, William Whewell and the *Vestiges*." *British Journal for the History of Science* 10: 132–45.

———. 1979. "The Natural Theology of the Geologists: Some Theological Strata." In *Images of the Earth: Essays in the Environmental Sciences*, edited by L. J. Jordanova and R. S. Porter, 39–64. Chalfont St Giles: British Society for the History of Science.

Canguilhem, Georges et al. 1960. "Du développment à l'évolution au XIXe siècle." *Thales* 11: 3–63.

Carpenter, William B. 1841. *Principles of General and Comparative Physiology*. 2nd edn. London: John Churchill.

———. 1845. "Vestiges of the Natural History of Creation." *British and Foreign Medical Review* 19: 155–81.

Chambers, Robert. [1844] 1969. *Vestiges of the Natural History of Creation*. 1st edn. Reprint. Leicester: Leicester University Press.

———. 1846. *Explanations*. 2nd edn. London: John Churchill.

———. 1853. *Vestiges of the Natural History of Creation*. 10th edn. London: John Churchill.

Darwin, Charles. 1859. *The Origin of Species by Means of Natural Selection*. London: John Murray.

———. 1985. *The Correspondence of Charles Darwin*. Edited by F. Burkhardt et al. Cambridge: Cambridge University Press.

———. *Notebooks*; see Barrett et al.

Desmond, Adrian. 1982. *Archetypes and Ancestors: Palaeontology in Victorian London, 1850–1875*. London: Blond & Briggs.

———. 1984. "Robert E. Grant: The Social Predicament of a Pre-Darwinian Transmutationist." *Journal of the History of Biology* 17: 189–223.

———. 1985a. "Richard Owen's Reaction to Transmutation in the 1830's." *British Journal for the History of Science* 18: 25–50.

———. 1985b. "The Making of Institutional Zoology in London 1822–1836: Parts 1 and 2." *History of Science* 23: 153–85, 223–50.

———. 1987. "Artisan Resistance and Evolution in Britain, 1819–1848." *Osiris* 3: 77–110.

Farley, John and Gerald L. Geison. 1980. "Science, Politics and Spontaneous Generation in Nineteenth-Century France: The Pasteur-Poucher Debate." In *Darwin to Einstein: Historical Studies on Science and Belief,* edited by C. Chant and J. Fauvel, 107–33. Harlow and New York: Longman.

Fletcher, John. 1837. *Rudiments of Physiology.* Edinburgh: Carfrae.

Geoffroy St Hilaire, Étienne. 1828. "Mémoire où l'on se propose de rechercher dan quels rapports de structure organique et de parenté sont entre eux !es animaux des ages historiques, et vivant actuellement, et les espèces antediluviennes et perdues." *Mémoires du Muséum d'Histoire Naturelle* XVII: 209–29.

———. 1833. "Le degré d'influence du monde ambiant pour modifier Jes formes animales; question interessant l'origine des espèces téléosauriennes et successivement celle des animaux de l'époque actuelle." *Mémoires de l'Académie des Sciences* XII: 63–92.

Gillispie, Charles C. 1959. *Genesis and Geology.* New York: Harper Torchbooks.

Gode-von Aesch, Alexander. [1941] 1966. *Natural Science in German Romanticism.* Reprint. New York: AMS Press.

Gould, Stephen Jay. 1977. *Ontogeny and Phylogeny.* Cambridge, MA: Harvard University Press.

Gruber, Howard E. 1974. *Darwin on Man: A Psychological Study of Scientific Creativity.* London: Wildwood House.

Herbert, Sandra. 1977. "The Place of Man in the Development of Darwin's Theory of Transmutation. Part 2." *Journal of the History of Biology* 10: 155–227.

Hobsbawm, Eric J. 1973. *The Age of Revolution: Europe 1789–1848.* London: Cardinal.

———. 1975. *The Age of Capital 1848–1875.* London: Weidenfeld and Nicolson.

———. 1979. *Industry and Empire.* Harmondsworth: Penguin Books.

Hodge, M. J. S. 1972. "The Universal Gestation of Nature: Chambers' *Vestiges* and *Explanations.*" *Journal of the History of Biology* 5: 127–51.

———. 1974. "England." In *The Comparative Reception of Darwinism,* edited by Thomas F. Glick, 3–31. Austin: University of Texas Press.

Hooykaas, R. 1957. "The Parallel Between the History of the Earth and the History of the Animal World." *Archives International History of Science* 10: 3–18.

Hunter, John. 1837. *Observations on Certain Parts of the Animal Economy,* with notes by Richard Owen. London: Longman, Rees, Green, and Longman.

Huxley, Leonard, ed. 1900. *Life and Letters of Thomas Henry Huxley.* 2 vols. London: Macmillan.

Huxley, Thomas Henry. 1853. "Fragments Relating to Philosophical Zoology, Selected from the Works of K. E. von Baer." In *Scientific Memoirs,* edited by A. Henfrey and T. H. Huxley, 176–238. London: Taylor & Francis.

———. 1854. "The Vestiges of Creation." *British and Foreign Medico-Chirurgical Review* 13: 425–39.

———. 1894. "Owen's Position in the History of Anatomical Science." In *The Life of Richard Owen,* edited by the Rev. Richard Owen, vol. 2, 273–332, 2 vols. London: John Murray.

Koss, Stephen E. 1981. *The Rise and Fall of the Political Press in Britain: The Nineteenth Century*. Chapel Hill: University of North Carolina Press.

Lenoir, Timothy. 1982. *The Strategy of Life: Teleology and Mechanics in Nineteenth-Century German Biology*. Dordrecht: Riedel.

Lord, Perceval B. 1834. *Popular Physiology*. London: John W. Parker.

Macleod, Roy M. 1965. "Evolutionism and Richard Owen, 1830–1868: An Episode in Darwin's Century." *Isis* 56: 259–80.

Miller, Hugh. 1851. *Footprints of the Creator*. 4th edn. Edinburgh: Johnstone and Hunter.

Millhauser, Milton. 1959. *Just before Darwin: Robert Chambers and Vestiges*. Middletown, CT: Wesleyan University Press.

Milne-Edwards, Henri. 1844. "Considérations sur quelques principes relatifs à la classification naturelle des animaux." *Annales des Sciences Naturelles* 1 (3): 65–99.

Oken, Lorenz. 1847. *Elements of Physiophilosophy*. London: Ray Society.

Ospovat, Dov. 1976. "The Influence of Karl Ernst von Baer's Embryology, 1828–1859: A Reappraisal in Light of Richard Owen's and William B. Carpenter's Palaeontological Applications of 'von Baer's Law'." *Journal of the History of Biology* 9: 1–28.

———. 1978. "Perfect Adaptation and Teleological Explanation: Approaches to the Problem of the History of Life in the Mid-Nineteenth Century." *Studies in the History of Biology* 2: 33–65.

———. 1981. *The Development of Darwin's Theory: Natural History, Natural Theology and Natural Selection, 1838–1859*. Cambridge: Cambridge University Press.

Owen, Richard. 1841. "Report on British Fossil Reptiles. Part II." *Report of the British Association for the Advancement of Science* 10: 60–204. London: John Murray.

———. 1843. *Lectures on the Comparative Anatomy and Physiology of the Invertebrate Animals*. London: Longman, Brown, Green and Longman.

———. 1844. "Considérations sur le plan organique et le mode de développement des animaux." *Annales des Science Naturelles* 2 (3): 162–68.

———. 1848. *On the Archetype and Homologies of the Vertebrate Skeleton*. London: John van Voorst.

———. 1849a. *On the Nature of Limbs*. London: John van Voorst.

———. 1849b. *On Parthenogenesis, or the Successive Production of Procreating Individuals from a Single Ovum*. London: John van Voorst.

———. 1851a. "Lyell – On Life and Its Successive Development." *Quarterly Review* 89: 412–51.

———. 1851b. "On Metamorphosis and Metagenesis." *Notices of the Meetings of the Royal Institution*, 7, February, 9–16.

———. 1855. *Lectures on the Comparative Anatomy and Physiology of the Invertebrate Animals*. 2nd edn. London: Longman, Brown, Green and Longman.

———. 1858. "Presidential Address." *Report of the 28th Meeting of the British Association for the Advancement of Science*, xlix–cx. London: John Murray.

———. 1860. "Darwin on the Origin of Species." *Edinburgh Review* 111: 487–532.

———. 1866. "On the Aye-Aye." *Transactions of the Zoological Society* 5: 3–101.

———. 1868. *On the Anatomy of Vertebrates*. 3 vols. London: Longmans.

———. 1894. *The Life of Richard Owen*. 2 vols. London: John Murray.

Porter, Theodore M. 1985. "The Mathematics of Society: Variation and Error in Quetelet's Statistics." *British Journal for the History of Science* 18: 51–69.

Powell, Baden. 1855. *Essays on the Spirit of the Inductive Philosophy, the Unity of Worlds, and the Philosophy of Creation*. London: Longman.

Read, Donald. 1961. *Press and People, 1790–1850*. London: E. Arnold.

Richards, Evelleen. 1976. *The German Romantic Concept of Embryonic Repetition and its Role in Evolutionary Theory in England up to 1859*. PhD diss. University of New South Wales.

———. 1983. "Darwin and the Descent of Woman." In *The Wider Domain of Evolutionary Thought*, edited by David Oldroyd and Ian Langham, 57–111. Dordrecht and Holland: Riedel.

Rudwick, Martin J. S. 1972. *The Meaning of Fossils: Episodes in the History of Palaeontology*. London: Macdonald.

Rupke, Nicolaas. 1985. "Richard Owen's Hunterian Lectures on Comparative Anatomy and Physiology, 1837–1855." *Medical History* 29: 237–58.

Ruse, Michael. 1979. *The Darwinian Revolution: Science Red in Tooth and Claw*. Chicago and London: University of Chicago Press.

Russell, E. S. 1916. *Form and Function: A Contribution to the Study of Animal Morphology*. London: John Murray.

Sedgwick, Adam. 1845. "Vestiges of the Natural History of Creation." *Edinburgh Review* 83: 1–85.

———. 1850. *A Discourse on the Studies of the University of Cambridge*. 5th edn. Cambridge: Cambridge University Press.

Shapin, Steven. 1982. "History of Science and Its Sociological Reconstructions." *History of Science* 20: 157–211.

Temkin, Owsei. 1950. "German Concepts of Ontogeny and History around 1800." *Bulletin of the History of Medicine* 24: 227–46.

Yeo, Richard. 1984. "Science and Intellectual Authority in Mid-Nineteenth-Century Britain: Robert Chambers and *Vestiges of the Natural History of Creation*." *Victorian Studies* 28: 5–31.

Young, Robert M. 1973. "The Historiographic and Ideological Contexts of the Nineteenth-Century Debate on Man's Place in Nature." In *Changing Perspectives in the History of Science*, edited by Mikulas Teich and Robert M. Young, 344–438. London: Heinemann.

———. 1980. "Natural Theology, Victorian Periodicals and the Fragmentation of a Common Context." In *Darwin to Einstein: Historical Studies on Science and Belief*, edited by C. Chant and J. Fauvel, 69–107. Harlow and New York: Longman.

3

THE "MORAL ANATOMY" OF ROBERT KNOX

The interplay between biological and social thought in Victorian scientific naturalism

Could you possibly be afraid of applying the calculation of chances to moral phenomena, and of the afflicting consequences which may be inferred from that inquiry, when it is extended to crimes and to quarters the most disgraceful to society? . . . But is the anatomy of man not a more painful science still? – that science which leads us to dip our hands into the blood of our fellow-beings, to pry with impassible curiosity into parts and organs which once palpitated with life? And yet who dreams at this day of raising his voice against the study? Who does not applaud, on the contrary, the numerous advantages which it has conferred on humanity? The time is come for studying the moral anatomy of man also, and for uncovering its most afflicting aspects, with the view of providing remedies.

– L. A. J. Quetelet (1842, viii)

In 1842, William and Robert Chambers of Edinburgh issued an English translation of Quetelet's *Sur l'homme*. Quetelet himself provided a new preface for the English edition, in which he evoked a "moral anatomy" that would subject social and political phenomena to the scalpel of the social dissector and thus to the rule of natural law. This paper explores the ways in which Quetelet's call for a "moral anatomy" was taken up and developed by his translator Robert Knox, Edinburgh anatomist and ethnologist – or, more properly speaking, "anthropologist."

Knox, long a favourite subject of medical historians and playwrights because of his tragic involvement in the Burke and Hare affair, which led to the passage of the Anatomy Act of 1832, has more recently come under the scrutiny of historians of biology for his leading role in the teaching and dissemination of transcendental anatomy in early nineteenth-century British biology.[1] As well, historians of anthropology and social theory are now agreed that his (previously underestimated) role in the development of scientific racism was crucial. Philip D. Curtin

---

1 Richards 1976, Ch. 6; Desmond 1984a; Jacyna 1983b; Rehbock 1983, 31–55.

has described Knox as "the real founder of British racism and one of the key figures in the general Western movement towards a dogmatic pseudoscientific racism".[2]

From these differing disciplinary perspectives, the historical Knox has been fragmented and the remarkable consistency and coherence of his anatomical, anthropological, and political views have been lost. My object is to reintegrate the whole man and his views within the context of Victorian scientific naturalism. In particular, I intend to explain the relation of Knox's biology to his racism and his politics, and, through such contextual analysis, to clarify his evolutionism.

Knox's views on organic development have always been contentious. Baden Powell, writing in 1855, described him as "one of the most zealous supporters of the principle of transmutation in this country" (Powell 1855, 395). But few of his contemporaries seem to have shared Powell's opinion. Although Baden Powell was himself given due recognition in the "Historical Sketch" that Charles Darwin appended to the *Origin of Species*, Knox did not rate a mention.[3] The recent revival of interest in Knox's transcendental anatomy has been attended by a similar conflict of opinion. Adrian Desmond has described Knox as a materialist and transmutationist (Desmond 1984a, 159, 1984b, 397), whereas Philip F. Rehbock has depicted him as an idealist and antitransmutationist who was "not thinking in terms of an evolutionary process" and did not hypothesize a "theory of universal descent" (Rehbock 1983, 50–1).

Historians of anthropology, on the other hand, readily identify Knox as an "evolutionist," and a number have pointed to the congruence of his and Darwin's views on human evolution.[4] Here, however, another difference of interpretation centres on Knox's racism and his political views. Like Curtin, those historians who have discussed Knox's anthropological views have tended to see his politics of race as reactionary, or at least conservative – to collapse them into the discriminatory ambience of late Victorian scientific racism. However, a recent analysis of Knox's race science by Michael Biddiss offered a different interpretation: Biddiss stated that Knox subscribed to a "very peculiar political radicalism" and that consequently he propounded a "rare and rather paradoxical . . . racism with substantial traces of benevolence." Biddiss also emphasized the high degree

---

2 Curtin 1965, 377; see also Harris 1968, 99–101; Burrow 1966, 124, 130; Stocking 1971, 374; Biddiss 1975; Rainger 1978; Stepan 1982, 41–6.

3 Darwin's omission of Knox is all the more curious in that Powell's work (cited by Darwin) contained an extract from one of Knox's works in demonstration of his "transmutationism" (Powell 1855, 399–400). Moreover, Darwin had read this very work of Knox's, for he cited it in *The Descent of Man* (2nd. ed. 1889, 17, 21, nn.). Both these citations refer to Knox's *Great Artists and Great Anatomists* (1852), and there is a further reference to Knox's *The Races of Men: A Fragment* (1850a) in a footnote (Darwin 1889, 168, n.). From the details given in these references, Darwin evidently read both works with some care. Both works, as I shall demonstrate, contain numerous statements consistent with Darwin's requirement of his "anticipators" that they believed "existing forms of life have descended by true generation from pre-existing forms" (Darwin 1959, 59).

4 Curtin 1965, 377–81; Harris 1968, 99–100; Stepan 1982, Chs. 2, 3.

of systematization of Knox's racism and its close ties with his transcendental anatomy (Biddiss 1976).

By developing Biddiss's suggestive analysis, and focusing on the interplay of biological and social thought in the production, acceptance, and applications of Knox's race science or "moral anatomy," I am able to offer a reconstruction that goes some way toward resolving the above contradictions. My interpretation is in line with recent work in the social history of evolutionary biology that has led to a major revision of the positivist historiography forged by the dominant Darwinians of the late nineteenth century, who undervalued and distorted the role of transcendental anatomy. The role of institutional and social factors is crucial to this revised historiography. Thus, in a series of compelling studies, Adrian Desmond has demonstrated how transcendental conceptions of nature were deployed by two leading British comparative anatomists for different institutional and social purposes: Richard Owen, doyen of the Royal College of Surgeons, harnessed transcendental anatomy to an anti-Lamarckian paleontology and biology to meet conservative institutional and social needs, while his professional rival Robert Grant of University College made it do service to reformist institutional and radical-democratic political interests by casting it in a progressivist transformist mould.[5] My analysis of Knox demonstrates yet another deployment of transcendental anatomy in a nonprogressivist theory of development. In brief, Knox's biology, which was developed largely outside any institutional context, was consistent with his racist radicalism, but, after his death, it was pressed into institutional service by James Hunt, the racist founder of the Anthropological Society of London. Their adoption of Knoxian biology and anthropology not only underpinned their racism and reactionary politics, but gave Hunt and his followers the intellectual and institutional strength to resist incorporation into the Darwinian anthropological model proffered by Huxley, and to offer considerable professional opposition to the takeover of London science by the Darwinian "new guard." The struggle between these rival bodies for scientific and ideological hegemony shaped the "new" anthropology of the 1870s, and certain of Knox's views were thereby perpetuated in late Victorian scientific racism.

My paper is organized in two parts. Part 1 analyses the relationship between Knox's anatomical, anthropological, and political views; Part 2 deals with the appropriation, retooling, and institutionalization of Knox's "moral anatomy."

Part 1. Knox the "moral anatomist"

Any attempt to come to grips with the historical Knox has to confront the mythology that has accreted around his disastrous association with the Burke and Hare murders. This is the stuff of drama, and not only has it dominated most historical accounts of Knox and his work, it has also offered a ready-made explanation for

5 Desmond 1982, 1984a, 1984b, 1984c, 1985.

his professional failure and lack of scientific recognition. The attempt to pry Knox apart from his established image as medical martyr is complicated by the paucity of source material on him. Knox himself destroyed most of his correspondence and his manuscripts shortly before his death. Those letters and materials which his former pupil and partner Henry Lonsdale managed to acquire from family sources and from James Hunt when he was preparing his biography of Knox, cannot be traced. Even Knox's official correspondence with bodies such as the Edinburgh Royal College of Surgeons has largely disappeared.[6]

The historian is forced back onto problematic contemporary accounts and Knox's published writings. These latter, although voluminous, present further difficulties of interpretation and historical evaluation. Knox only began to elaborate his theoretical views after 1842 – the year in which he finally relinquished all hope of professional employment in Edinburgh and left his native city, supporting himself thereafter mostly by hack journalism and public lecturing, and settling finally in London where he died in 1862. These writings were perforce commodities. Through their sale Knox managed to keep one step ahead of pauperism, and their production was tailored to his pressing financial needs. They were written in haste, badly organized, and rarely revised. They ranged over the topics that a fickle public might find of interest, and where Knox wrote within his own major areas of interest he subordinated the elaboration of his theoretical views to his efforts to reach and retain a wider popular audience. In short, he did not present his "moral anatomy" systematically or comprehensively but scattered it through his journal articles and popular books, and it has to be reconstituted from these many and varied sources. To add to the problem of reconstruction, Knox wrote as a propagandist and polemicist on his obsessive topics of race and transcendental anatomy, consistently sacrificing clarity and precision to vehemence and rhetoric. Those qualities which made him so brilliant and stimulating a lecturer do not translate so well to paper. Nevertheless, it is possible to piece together a conception of life and the universe that is surprisingly self-consistent and truly synthetic in scope. In fact, it is not inappropriate to suggest that Herbert Spencer's "Synthetic Philosophy" had an earlier analogue in Knox's "moral anatomy," and some obvious parallels between their works will emerge in the course of this study.[7]

The standard version of Knox's involvement in the Burke and Hare affair is a familiar and quickly told story. Once the most popular extramural teacher of

6 Lonsdale 1870, viii–lx; Rae 1964, 142–6; Bettany 1892. Lonsdale's biography remains the best source on Knox's life and work. Rae's account suffers from its almost exclusive focus on the Burke and Hare affair and from poor comprehension of Knox's biology and anthropology.

7 Spencer's early developmentalism was shaped by transcendental thought. His "Synthetic Philosophy" (from about 1851 on) was an attempt to apply a formula of evolution, centered on von Baer's embryological law of increasing divergence, to every kind of phenomenon throughout the universe, including biology, ethics, sociology, politics, and art. He subsequently gave this a Darwinian gloss, and extrapolated his Social Darwinism from it (Burrow 1966, 179–227; Young 1967; Canguilhem et al. 1960).

anatomy in Edinburgh, whose classes had reached the unprecedented number of over five hundred students, and who had seemed destined for a brilliant career as an anatomist, Knox was professionally ruined and reduced to penury by this most celebrated medical scandal of the nineteenth century. During the years 1827 and 1828, William Burke and William Hare collaborated to murder around sixteen people for gain, selling their bodies to Knox's medical school (at ten pounds apiece) for dissection by his students. When the murders where finally uncovered, the public, long incensed by the activities of the busy Edinburgh "resurrection men," was particularly outraged by these circumstances, and Knox shared in the opprobrium cast upon the murderers. He was hanged and burnt in effigy, vilified in the popular press of the day, and for some time went in fear for his life at the hands of the Edinburgh mob. Although he was officially exonerated of all complicity in the murders, the episode haunted Knox for the rest of his life. According to one account, he never again dissected a human body and he gave up his researches on human anatomy (Ross and Taylor 1955). Others represent him as a scapegoat for the contemporary body-snatching proclivities of the medical fraternity at large, who was hounded and ostracized by society for the rest of his life, and was denied the scientific recognition that was his due (Currie 1933; Comrie 1932; Rae 1964).

While Knox certainly became a social outcast and there is some substance to these interpretations, they overlook the role played by Knox's conception of anatomy and his radical political views in the conservative and highly competitive Edinburgh medical context. And Knox's lack of scientific recognition by his contemporaries possibly had more to do with the subsequent appropriation of his views by that latter-day "resurrection man," James Hunt (see Part II, below), than with his association with the murderous activities of Burke and Hare.

Unlike most of his colleagues, and as Rehbock (1983) has recently emphasized, Knox was no mere anatomist, but a "philosophical anatomist". In his own words: "Anatomy is not a science, but merely a mechanical art, a means to an end. It is pursued by the physician and surgeon for the detection of disease, and the performance of operations; by both to discover the functions of the organs; and by the philosopher with the hope of detecting the laws of organic life, the origin of living beings, and the transcendental laws regulating the living world in time and space" (Knox 1852, 141–2). Clearly, Knox was not interested in the accumulation of dry anatomical facts, nor simply in the pure morphological search for organic homologies and the establishment of the underlying unity of the diversity of nature, but in the problem of diversity itself – with the origin and laws of life. Moreover, for Knox, as we shall see, the "philosopher's" concern for the elucidation of nature's laws was necessarily bound up with a concern for their social and political implications. He was, in this sense, closer to Quetelet's invocation of "moral" anatomist, than to Rehbock's narrower ascription of "philosophical" anatomist.

Knox had graduated in medicine at Edinburgh University in 1814. After a period of duty as an army surgeon at Waterloo and then at the Cape, he studied in Paris. Here, according to his own account and that of his biographer and pupil Henry Lonsdale, his anatomical knowledge and his fluency in the language put him on familiar terms with the foremost comparative anatomists of the day – the functionalist teleologist Georges Cuvier and the transcendental morphologist Éti-enne Geoffroy St Hilaire.[8] For Knox, conservative Cuvierian function soon lost out to Geoffroyan form and its contingent radical political ties,[9] and he returned to Edinburgh a convert to a radical and heterodox transcendental anatomy. From about 1826, when he took control of John Barclay's private anatomy school, he began to preach the new doctrine to his students and in his famous Saturday morn-ing public lectures, to which his eloquence and brilliance, and perhaps the novelty and social implications of his views, attracted the Edinburgh intelligentsia. Even after the Burke and Hare affair, Knox could still attract large and enthusiastic crowds to these lectures on "Comparative and General Anatomy and Ethnology." Reference has been made to his role in the dissemination of transcendental anat-omy in British biology, and there seems little reason to doubt his own estimation of his significance in this. As Rehbock has documented, his students included a number of subsequently prominent anatomists and naturalists who became, in turn, exponents of transcendental natural history.[10]

Initially Knox enjoyed considerable professional success. He quickly built Bar-clay's old school into the largest and most popular anatomy school in Edinburgh. As well, he was instrumental in setting up the Comparative Anatomy section of the Museum of the Edinburgh Royal College of Surgeons and, in recognition of his efforts, was elected founding conservator of the Museum in 1826. He was elected to the Royal Society of Edinburgh, was on the councils of the Plinian Society and the Wernerian Natural History Society, and was a noted contributor to the Medico-Chirurgical Society. He was the valued associate of Robert Jameson, professor of natural history and founder of the *Edinburgh Philosophical Jour-nal*. During the late 1820s, undoubtedly through Knox's influence and that of his fellow transcendentalist, Robert Grant, Jameson's *Journal* made Geoffroy's transcendental views available to a wider reading audience (Geoffroy St.-Hilaire 1829, 1830). As Desmond has noted, during the 1820s it was possible for Knox, Grant, and others to express their heterodox views in the various forums available to them – notably the Plinian Society, which had a pronounced materialist bent (Desmond 1984a, 199–200).

At the same time, however, Knox's very success as a lecturer, his uncompro-mising and unorthodox anti-teleological stance, his all-too-evident disdain for

8 Knox 1850a, 141–2; Lonsdale 1870, 17, 18, 21.

9 Geoffroy's morphology (which emphasized serial development, recapitulation, transformism, and unity of composition, and was founded on the sovereignty of material laws), was directed against Cuvier and aimed at the young medical reformers and republicans of Paris, and it appealed equally to visiting British radicals like Knox and Grant (Desmond 1989; Appel 1987).

10 Knox's pupils included John Goodsir, Edward Forbes, and possibly Richard Owen (Rehbock 1983, 56–1 1 4; Knox 1850a, 73, 211–12, 1855c). Both Knox and Rehbock overlook the contribu-tions of Robert Grant as teacher and disseminator of transcendental anatomy – but, as Desmond had demonstrated in his studies of Grant, he was probably as influential as Knox in this respect (Desmond 1984a, 1984b, 1984c, 1985).

organized religion, and his radical politics bred professional enmity. The Burke and Hare affair was opportunistically manipulated by his direct competitors in the anatomy marketplace to discredit and undermine him. The cutthroat entrepreneurial competition for students that had always characterized higher learning in Edinburgh grew more intense as the medical school began to lose its hegemony and the numbers of students attracted to Edinburgh declined.[11] Knox, an outspoken advocate of university reform, became caught up in the incessant internecine warfare between the university, the Royal College of Surgeons, and the extramural anatomy schools, and by 1842 he had been effectively excluded from all of them. In 1831 he was harried into resigning his position as conservator of the Museum of the Edinburgh Royal College of Surgeons – the museum that his own enthusiasm, initiative, and industry had helped to establish. Knox's students, who had initially stood by him, were eventually alienated by his increasingly bitter and aggressive attacks on professional rivals and colleagues and his growing sarcasm and cynicism. His classes dwindled.[12] In the mid-thirties, his financial problems were exacerbated by the decision of the university to introduce its own compulsory intramural courses on anatomy, and Knox pulled no punches in belabouring this "scandalous monopoly" and the poor quality and expense of the teaching within the university medical school. Knox was an unsuccessful (and hardly a serious) candidate for the University Chair of Pathology in 1837. This was the very chair whose foundation in 1832 he had denounced as a Whig conspiracy, a charge he reiterated in his application, describing it as a "political job of the very worst description."[13] He was passed over as an applicant for the Chair of

11 By the time of the Burke and Hare affair, Knox had two-thirds of the whole Edinburgh medical school in his classrooms, and Lonsdale gives a vivid and partisan account of the machinations of Knox's "jealous rivals" (Lonsdale 1870, 81–91, 113–14, 130, 190–3). According to Knox, by 1841 the chairs of the university had "fallen much below the income of a steady-going retail grocery or bakery," and between 1838 and 1841 the number of medical students had declined from 556 to 356. Knox attributed this decline to three causes: "the overloading of the curriculum, the absence from the University of all men of originality and of European reputation, and the baneful effects of a monopoly exercised by the University, whose sure result, like all other monopolies, is first to ruin itself and afterwards its neighbours" (Lonsdale 1870, 261–4, 196). For an analysis of the traditional laissez-faire and entrepreneurial nature of the university teaching, see Morrell 1972.

12 Even Lonsdale found inexcusable Knox's plagiarism of some anatomical discoveries by his former pupil and partner John Reid in 1840, and their public controversy further diminished Knox's Edinburgh reputation. By 1842, as the final humiliation, Knox was unable to get up a class in his own anatomy school (Lonsdale 1870, 219–20, 257; letters from John Reid to William Sharpey, 18 October 1840; 15 December 1842, Wellcome Institute for the History of Medicine, Library, MS 69099; Creswell 1926, 77–84, 240–50; Rae 1964, 105–26).

13 Robert Knox, "Letter to the Right Honourable the Lord Provost and Town Council of Edinburgh," 6 July 1837, p. 6; "Second Letter to the Right Honourable the Lord Provost and Town-Council of Edinburgh," 15 July 1837 (Archives, Royal College of Surgeons, London, Tr. 1160 [17 and 18]). Knox was widely suspected of being the author of the highly critical pamphlet *An Examination into the Causes of the Declining Reputation of the Medical Faculty of the University of Edinburgh* (Edinburgh: Burgess, 1834), and from its tone and contents this seems quite likely; see Rae 1964, 117. Note also Knox's comments in his preface to the second edition of his translation of Cloquet

Physiology in 1841; in the same year, he did not manage to secure even one vote in the election for the post of lecturer in anatomy to the art students of the Scottish Academy – a position for which, with his romantic enthusiasm for art, he would have been admirably suited (Lonsdale 1870, 261–4).

Apart from Knox's aggressive involvement in institutional politics, his political convictions and his lack of "Calvinistic credentials" (Lonsdale 1870, 264) marked him out as a dangerous and subversive radical. In the politically and socially unsettled post-Reform Bill period of the thirties and forties, attitudes toward heterodoxy hardened. There was little chance of professional advancement in Edinburgh for a lecturer capable of holding up a cranium before his students and provocatively declaiming: "Are we to be told that the Caffre of this cerebral stamp is a savage because he lives in the 'wilde,' and that John Bull is the happy creature of civilization because he wears breeches, learns catechisms, and does his best to cheat his neighbours – always, of course, on Christian principles!" (Lonsdale 1870, 149).

The opportunity to superintend and edit the English translation of Quetelet's work during 1841, when he was so financially hard-pressed, must have been greatly welcomed by Knox. Possibly it was Knox himself who suggested the project of a "People's Edition" to the Chambers brothers. He was already familiar with this "admirable work" of the "illustrious Quetelet," which had been published in French in 1835, and he extolled its "leading idea" as "'that bright and original conception of a great mind."[14] Quetelet's work was also of considerable interest to Robert Chambers, at that stage secretly working on the evolutionary work *Vestiges of the Natural History of Creation*, which he would publish anonymously in 1844. Chambers invoked Quetelet's statistical regularities to support his thesis that the natural world, including humanity and mind, had evolved by law.[15] Knox was also strongly influenced by this naturalistic assumption and he too put forward a theory of organic development, but it differed in important respects from Chambers's, as did his conception of natural law. Their differences may be best summed up in political terms: where Chambers was a liberal reformer and a "progressive" in both the social and biological senses, Knox was, as his obituarist in the *Medical Times* put it, a well-known "savage radical."[16] Knox's major criticisms of *Vestiges* were that its "development hypothesis" was located within a teleological framework, and that it subscribed to the ideology of progress (Knox

---

(Knox 1831, vii); and see Anon 1923, 5. Knox's role in Edinburgh institutional politics and medical reform warrants closer examination than this study can provide.

14 Knox, "Translator's Appendix," (Quetelet 1842, 119). In 1837 Knox had included Quetelet's statistics in a paper on the diurnal changes in the pulse, and he devoted a portion of his "Appendix" to this same topic; see also, Knox 1837a; Lonsdale 1870, 257.

15 Chambers 1844, 328–32; Millhauser 1959. On Quetelet and his influence on British science and social thought, see Merz 1965, 2: 577–87; Burrow 1966, 108, 253; Schweber 1977; Solomon Diamond, "Introduction," in Quetelet [1842] 1969, v–xii.

16 "The Late Dr. Knox," *Med. Times Gaz.* (27 December 1862), p. 684; see Desmond 1984a, 198.

Figure 3.1 Robert Knox, outspoken and charismatic anatomist, in lecturing mode. At the height of his fame, Knox, with his one eye and a dandy's taste for fine linen, high collars and gold jewellery, cut a mesmerising figure. Line engraving, courtesy of Wellcome Collection, London.

1850a, 27).[17] His own transcendental anatomy was grounded in materialism and was thoroughly nonprogressive, and Knox's highly idiosyncratic developmental views can only be interpreted by reference to his materialist ideology and his peculiar brand of political radicalism.

Knox's anomalous radical materialism

Just how "savage" Knox's radicalism was in his early Edinburgh days is not clear, but it seems that he tempered it during his brief period of success and only became more outspoken and satirical with his professional decline.[18] He inherited his radicalism from his schoolmaster father, who had been an admirer of the French Revolution and (until its suppression) a member of the Jacobin inspired "Friends of the People" (Lonsdale 1870, 3). Knox's liberal education and his own Edinburgh experiences probably further radicalized him. Throughout his life, he believed the greatest curses of humanity to be "Kingcraft" and "Priestcraft," and in his writings he often evoked the rights of man: "that inestimable treasure beyond all price or value, freedom of speech, thought and action" (Knox 1862, 564, 1852, 75–6). This was not mere sloganeering. Knox was a vehement, indeed a "savage" critic of British colonial policy and its debasement and oppression of the "coloured races" of India, Africa, Australia, and New Zealand. He abhorred slavery. He denounced the Americans for their hypocrisy in refusing to extend the rights of man to the Negro: "The rights of men is a phrase forever in their mouths; by men we now know they mean white men" (Knox 1862, 552). His radicalism even led him to argue against the legislation of prostitution on the grounds that such legislation would infringe upon the rights of men "amongst whom we are bound to include women. . . . Able writers . . . have forgotten to take into consideration the inherent and innate right which every woman has in her own person. Society has legislated only against woman, ignoring her rights innate and external, to use her person as she may think fit, in so long as she commits no outrage on society."[19]

Knox urged the ineffectiveness of piecemeal philanthropy and legislation directed to the relief of poverty and unemployment and the reform of working

---

17 The influence of transcendental conceptions on Chambers's "development hypothesis" is well established, and the links between Chambers and Knox warrant closer investigation (see Hodge 1972).

18 It is unlikely, for all his anatomical ability, that Knox would have been taken into partnership by Barclay in the first instance, had he made his radical materialism public. Barclay was intolerant of "sceptics" and a devout teleologist (Barclay 1827, 126–32). Lonsdale makes the point that up until the time of the Burke and Hare repercussions, Knox devoted himself to science and kept aloof even from institutional politics (Lonsdale 1870, 91).

19 Knox 1857b, 142, 197. Lonsdale attributed this work to Knox (Lonsdale 1870, 370–1), and the bracketed portions are undoubtedly his. Note the Quetelet influence on the title and the contents. Typically, Knox thought that the majority of women were forced to prostitution by unemployment and want, but that the tendency to licentiousness was innate in the female character, notably the French! (Knox 1857b, 49–50).

conditions, stating bluntly: "Against competition for work there can be but one remedy – combination [i.e., unionism]" ([Knox] 1857b, 56). He had a fundamental radical objection to the augmentation of state powers and intrusiveness on individual rights and liberties, and his targets ranged from the state regulation of salmon fishing[20] to the "Sanatory [sic] Movement." According to his analysis of the latter, published in a London radical weekly, Edwin Chadwick and his "rich and crafty" aristocratic supporters were perpetrating the "gigantic fraud" of sanitary reform in order to accrue new sources of "patronage, place, power and wealth to the few" and to find work for the "most dangerous 'of all the classes,' the brutal, savage, but shrewd and powerful navvy," whose current unemployed state, caused by the "calming of the railway mania," was a threat to the "oligarchy."[21] This piece of class analysis aside, Knox's criticisms of some of the effects of sanitary reform, such as the pollution of the Thames through the discharge of sewage, and the hardships it imposed on the poor, have been reiterated by modern analysts (Smith 1979).

In spite of the Jacobin origins of his radicalism, Knox did not subscribe to an Enlightenment egalitarianism and environmentalism. He was uncompromisingly a man of the nineteenth century in his insistence on the universality and inevitability of natural law, and a rigid determinist in his views on social organization and the essential inequality of humanity, which from an early period he linked to race and grounded in materialism. For Knox, the human mind and conscience were as much subject to natural law as the human body and the rest of nature:

> [A]s man merely forms a portion of the material world, he must of necessity be subject to all the physiological and physical laws affecting life on the globe. His pretensions to place himself above nature's laws, assume a variety of shapes: sometimes he affects mystery; at other times he is grandly mechanical. Now, all is to be done through the workshop; in a little while, the ultimatum . . . is to be gained through religion: and thus man frets his hour upon the stage of life, fancying himself something whilst he is absolutely nothing.
>
> (Knox 1850a, 479)

Although he explicitly rejected the doctrine of human perfectibility, Knox, in common with other nineteenth-century "progressives," was clearly influenced by secular naturalism. Like them, he was concerned with subjecting the whole of

---

20 Knox 1837b, 7–8, 1854, 68.
21 [Knox], "The Sanatory Movement," *Empire,* 1 September 1855, p. 633; see also [Knox], "A Plea for the Thames," ibid., 25 August 1855, p. 617; [Knox], "The Jobs of the Sanatory Reformers," ibid., 8 September 1855, pp. 648–9. Lonsdale attributed these leaders to Knox (Lonsdale 1870, 382). Indubitably, the radical aims of the *Empire* would have struck a responsive chord in Knox: "Freedom in Commerce, Equality in Religion, Impartiality in Representation, and Justice to Man, as Man, all over the World." These articles suggest that Knox had some radical contacts in London.

nature and society to the sway of natural law and opposing such naturalistic or "scientific" explanations to traditional theological modes of explanation.[22]

The appeal of Quetelet's "moral anatomy" for Knox and other scientific naturalists of the day lay in its application of statistical method to the measurement of man's "moral faculties." Quetelet's "moral statistics" on the incidence of marriages, suicides, illegitimate births, murders, and so forth, demonstrated the regularity of such statistics. These "moral" events, which were conventionally considered to be wilful actions, were actually recurrent and predictable. Human behaviour, intelligence, and morality were not arbitrary and capricious, not subject to supernatural interference, but regulated by fixed and immutable laws.

There were, however, significant differences between Knox's and Quetelet's interpretations of human nature. For Quetelet, the regularity and predictability of human actions pointed to the dominant influence of underlying social forces, and he looked to social reform as the "remedy" for crime and immorality. The democratic concept of the "average man" was central to his analysis and he stressed the common factors between men. For Knox, however, the "average man" was an illusion – merely the statistical analogue of the natural type, and, like it, having no real existence. Real men, "natural" men, varied around this statistical abstraction, and their individual characters and moralities were as distinct as their physical differences, and referable to them.[23] Where Quetelet looked to natural law and social reform, Knox looked entirely to natural law and rejected the meliorative power of reform. He had a darker view of human history, and although he championed revolution, he was thoroughly pessimistic of its outcome. The revolutionary ideals of liberty, equality, and fraternity were circumscribed by human nature; these abstractions, while glorious and admirable and the inspiration of all thinking men, were as incapable of realization as the models or archetypes that informed transcendental anatomy. "Civilization" had failed, as inevitably it must, to "better man's condition on the globe" (Knox 1850a, 478). Human ideals and aspirations – in effect, all hope of human progress – were brought up short against the ironclad laws of human nature. For all his radicalism, Knox's was essentially a doctrine of despair – of political nihilism – and this anomalous radicalism (which demarcates him from other, more reform-oriented, radicals of the period) structured his interrelated political and biological views.[24]

22 On Victorian scientific naturalism see Turner 1974, 1–37; Barnes and Shapin 1979, 93–186; Young 1973. Roger Cooter argues for a generally earlier date than is usually accepted for the establishment of scientific naturalism among British phrenologists and other marginal men, and Desmond stresses the connection of the new naturalistic sciences of the Reform Bill period of the 1830s with the radical Dissenting campaigns against ToryAnglican privilege (Cooter 1984; Desmond 1989).
23 Quetelet 1842, 1848; Porter 1985; Knox 1848, 98, 242.
24 Although he paraded his lack of "Calvinistic credentials", the fatalism and rigid determinism that pervaded Knox's materialistic and anti-Providential ideology is suggestive of a kind of deconsecrated Calvinism that may be attributed to his Edinburgh background. Desmond has pointed to a similar Calvinistic fatalism in Robert E. Grant's later views on organic development: Desmond 1984a. I have discussed Knox's religious beliefs in note 47 below.

Yet, this important distinction aside, Knox's subsequent "moral anatomy" shows the powerful impress of Quetelet's views. In spite of his emphasis on the dominance of social forces, Quetelet was also very aware of the biological aspects of human "nature," as his anatomical terminology indicates. He was not only concerned with calibrating moral statistics, but with anthropometric measures such as height and weight, and with anthropological questions. He made a number of statistical analyses of human physiognomy, and these studies instigated the subsequent large-scale anthropometric investigations of the varieties of man that were the stock-in-trade of the late Victorian physical anthropologists and race scientists (see Part II). In his preface to the *Treatise*, Quetelet expressed interest in phrenologist George Combe's suggestion of correlating man's moral and physical statistics, especially cranial measurements.[25] An insistence on such a correlation was central to Knox's later writings on race. However, Knox's debt to Quetelet was more specific than this. In his *Treatise*, Quetelet had argued that the type of each race could be statistically correlated with its climate and environment. Knox took up and extended this view in the appendix he attached to his translation of the *Treatise:* here he presented further data stressing the specific and immutable adaptation of each race to its particular climate, and argued the inability of the Celts and Saxons to maintain themselves in tropical countries.[26] This was a theme that was to be elaborated and reiterated in all his subsequent writings on race, and, as I shall show, it was crucial to his conception of organic development.

In addition, in a larger and more diffuse sense, Knox's mature "moral anatomy" suggests the crystallizing impact of Quetelet's *Treatise*. As Quetelet had stated, a comprehensive and scientific "moral anatomy" should encompass three major interrelated inquiries: "1. What are the laws of human reproduction, growth, and physical force – growth of his intellectual powers . . . the laws regulating his passions and tastes . . . the laws of human mortality. . . . 2. What influence has nature over man. . . . 3. Finally, can human forces compromise the stability of the social system?"[27] Over the next few years, Knox provided his own answers to each of these fundamental questions. Around them he constructed a unique "moral anatomy," compounded of his transcendentalism and his biological determinism, and pervaded by his radical, but nihilistic, materialism. With his own bitter experiences behind him, Knox could hardly have failed to respond to Quetelet's powerful evocation of the "painful science" of anatomy. Perhaps he derived some personal absolution from Quetelet's demonstration of the regularity and predictability of murder. But in any case, as the appendix he attached to the *Treatise on*

---

25 Quetelet 1842, vi–vii; see also 94–100. On the influence of Quetelet's anthropometry on nineteenth-century race science and anthropology see Haller 1971, 21–34.

26 Knox, "Translator's Appendix," in Quetelet 1842, 122–3.

27 Quetelet 1842, 8–9. Lonsdale states that Knox "indoctrinated the majority of his friends with his more advanced views' on race "after 1834" (Lonsdale 1870, 295), and if this is so, it adds weight to my suggestion of the crystallizing impact of Quetelet's *Sur l'homme*, which was first published in 1835.

Man suggests, Knox was already moving in the direction pointed by Quetelet. And if, unlike Quetelet, he had little in the way of "remedies" to provide, neither did he have any fear of the "afflicting consequences" of his researches.

Rehbock has posed the problem of why Knox waited until the last phase of his career before publishing his transcendental philosophy of natural history (Rehbock 1983, 55). The answer would seem to be obvious: his transcendental philosophy was bound up with his radical materialism and only became publishable when it could no longer affect his professional and social aspirations – when he was, in effect, an institutional and social outcast. Save for one brief and unsuccessful attempt in 1844, Knox never again practiced his profession of anatomy lecturer. In 1847, following yet another scandal (involving the wrongful certification of the class attendance of one of his supposed earlier Edinburgh pupils, a John Osborne), the Royal College of Surgeons withdrew his teaching qualification (Rae 1964, 131–46). When he left Edinburgh at the end of 1842, Knox was forty-nine years old, an angry, disillusioned man, motivated by poverty and his radical convictions, and with little to lose by speaking his mind.[28] And speak it he did over the next twenty years, from the lecturing platforms of English provincial towns, and in the numerous journal articles and several books he industriously churned out. These were on such diverse topics as ethnology, zoology, fishing, and prostitution. But whatever his topic, the message was always the same: nature, including human nature, was only to be understood through the laws of transcendental anatomy. They were the key that would provide the solutions to all the diverse problems of life and of human behaviour, however recondite. Properly understood, they made man's morality as accessible to the scientist as his anatomy. The diversity and complexity of social phenomena could be scientifically explained and rendered utterly predictable through the rigorous application of these fundamental laws.

As is clear, Knox applied an extremely blunt scalpel to the task of social dissection. But he was not alone in his overweening nineteenth-century confidence in the "certainties" of science, nor in his extrapolation of biological method to society. Robert M. Young has argued persuasively for a "common context" of biological and social thought in the first half of the century, and Gay Weber, in her analysis of nineteenth-century anthropology, has highlighted the interplay of biological and social thought. Knox was merely one of many nineteenth-century

---

28 Knox's wife of seventeen years had died of puerperal fever in 1841, and this was followed shortly by the death of his four-year-old son; Lonsdale has documented his distress and despair at this personal loss (Lonsdale 1870, 241–2). He was forced to leave his surviving children in Edinburgh in the care of his nephew and oldest daughter in impoverished circumstances, while he tried to turn his public lecturing and journalism to his and their financial support. Knox had married "a person of inferior rank," and Lonsdale suggested that this also created social and professional difficulties for Knox. According to Lonsdale, Knox attempted to overcome these by keeping the marriage secret, and by maintaining two households, one for domestic life with his wife and five children, and one "acknowledged" residence (where his sister was hostess) for social purposes (ibid., 36, 222–4). However, Rae discounts this (Rae 1964, 48–9), and indeed it does not square with Knox's image as a devoted husband and father, nor with his radical convictions.

biologists and social theorists who insisted on the "homology between nature and culture"; as Weber has pointed out, he simply pushed the principle to its logical extreme, and his racism must be interpreted in this context (Weber 1974, 268; Young 1969, 109–45).

Knox's racial determinism

Some four years after his translation of Quetelet's *Treatise* (i.e., around 1846), Knox began to elaborate publicly his view of the social and political implications of his biology. In brief, his particular "moral anatomy" reduced all social and political phenomena to the basic biological category of "race." Knox summed up this principle in the preface to his major ethnological work, *The Races of Men*, which ran through two editions in his lifetime and was based on his earlier public lectures: "That race is in human affairs everything, is simply a fact, the most remarkable, the most comprehensive, which philosophy has ever announced. Race is everything: literature, science, art – in a word, civilization, depends on it" (Knox 1850a, v).[29] What he meant by this was that human history could only be studied through the application of biological method. He explicitly rejected the dominant "Prichardian" environmental approach to the study of man (see Part II) and argued that the only certain knowledge of human history was that which could be ascertained through the biological study of the existing human races: "The basis of the view I take of man is his Physical structure; if I may say so, his Zoological history. To know this must be the first step in all inquiries into man's history . . ." (Knox 1850a, 1).

Knox's interest in racial questions dates back to his South African period, when he had experienced at first hand the bloody struggle between British and Boer colonists and the dispossessed indigenous races. He started his collection of crania at this stage, and he seems never to have doubted that racial traits were biologically based.[30] As early as 1823, he had presented a paper to the Wernerian Society that

---

29 Knox's earlier public lectures were also published in the *Medical Times* (Knox 1848). Lonsdale states that Knox's public lectures in Manchester and other provincial towns "caused a sensation by their novelty, and led to much talk out of doors; and no small amount of controversy in the press" (Lonsdale 1870, 295).

30 Knox's early Edinburgh environment would have been conducive to such a belief. He must have had some contact with the Edinburgh phrenologists with their naturalistic ideology and reformist platform; see Shapin 1975, 1979). But it seems almost certain, as Lonsdale suggests, that Knox was more directly influenced by the writings of the Edinburgh anatomist and physiognomist Alexander Walker (Lonsdale 1870, 294–5). Walker, better known for his acrimonious dispute with Charles Bell over the functions of the roots of the spinal nerves, wrote a series of popular works in which he located the supposed mental and moral differences between the sexes and races in their anatomical and physiognomical differences. Just before he died, Knox was considering reediting Walker's *Intermarriage,* and had made some notes on this project (Lonsdale 1870, 383). Walker and Knox were acquainted, for when Walker appealed to Sir Robert Peel in 1849 for a government pension in recognition of his contribution to physiology, he enclosed a supporting letter from Knox in which Knox professed the highest esteem for Walker's work: "No one has thought more clearly

stressed the "characteristic differences" of the native races of South Africa (Lonsdale 1870, 24–5). From Lonsdale's account, Knox gave as much emphasis to the inculcation in his students of the concept of biological race as to the precepts of transcendental anatomy – and, indeed, as he later elaborated them, they were interrelated concepts. If Knox's significance in the teaching and dissemination of transcendental anatomy is acknowledged, his equally important role in the dissemination of the concept of biological race must also be conceded. According to Lonsdale:

> Knox could not glance at a cranium for the common descriptive anatomy without speaking of its ethnological bearings; it was the same with the external features and form of man . . . Knox seemed to the manner born to investigate distinctive anatomical characters: even when walking along the streets, thronged with men and women, he was always on the *qui vive* for Race features. He could see at a glance what ordinary men could hardly distinguish at their leisure. . . . Previous to his time, little or nothing was heard about Race in the medical schools: he changed all this by his Saturday's lectures, and Race became as familiar as household words to his students, through whom some of his novel ideas became disseminated far and wide, both at home and abroad.[31]

Knox's later writings were simply a more systematic statement and elaboration of his earlier assumption that the mental differences between the human races were as pronounced and self-evident as their physical differences, and that both sets of phenomena were reducible to the same causes or "laws." In this sense, he was profoundly racist.[32] To Knox the human races were so different and distinct that they were "entitled to the name of species" (Knox 1862, 591). Any anthropological or political theory that did not take this fundamental principle of innate and ineradicable racial differences into account was "unscientific," and doomed to failure: "Wild, visionary and pitiable theories have been offered respecting the colour of the black man, as if he differed only in colour from the white races; but

---

on the great physiological questions than you have" (copy of letter from Dr. Knox to Alexander Walker, August, 1848; letter from Walker to Sir Robert Peel, 22 February 1849, Peel Papers 40601, fols. 50, 51, British Library, Manuscript Room). See also Walker 1841, 1834.

31 Lonsdale 1870, 292–3, 330; see also Lonsdale 1868, I, 27.
32 Following Biddiss, I am here employing the term *racism* to signify something narrower than prejudiced attitudes and discriminatory actions, i.e., "certain relatively systematic attempts at using race as the primary or even sole means of explaining the workings of society or politics, the course of history, the development of culture and civilization, even the nature of morality itself" (Biddiss 1976, 245). Biddiss represents Knox as one of a group of mutually independent pioneers of such racist theory, which included Gustav Klemm and Karl Gustav Carus in Germany and Arthur de Gobineau in France. There are some superficial similarities between Knox's and Gobineau's pessimistic schemas of racial history, but Knox's published work predates Gobineau's (Biddiss 1970).

he differs in everything as much as in colour. He is no more a white man than an ass is a horse or a zebra . . ." (Knox 1850a, 245).

As this quotation suggests, Knox differed from the majority of his ethnological contemporaries in placing far more emphasis on the "moral" than on the physical differences between the races.[33] This emphasis was a corollary of his materialism: "The mind of the race, instinctive and reasoning, naturally differs in correspondence with the organization." Human character, intellect, and morality were neither divinely induced nor environmentally produced, but were rooted in the "all-pervading, unalterable, physical character of race" (Knox 1850a, 2–3, 21). By maximizing these supposed mental differences between the races, Knox was able to construct an elaborate racial history of Europe and her colonies. According to Knox, what had previously been interpreted as nationalism and national conflict was better understood as racial conflict, as each race sought to dominate its own geographic locale and erect its own government and civilization in keeping with its own distinctive nature. Innate racial differences and antipathies inevitably overrode Christian morality, demonstrating its irrelevance to natural law:

> The doctrine which teaches us to love our neighbours as ourselves is admirable, no doubt; but a difficulty lies somehow or other in the way. What is that difficulty, which all seem to know and feel, yet do not like to avow? It is the difficulty of race. Ask the Dutch Boer whence comes his contempt and inward dislike to the Hottentot, the Negro, the Caffre; ask him for his warrant to reduce these unhappy races to bondage and to slavery; to rob them of their lands, and to enslave their children; to deny them the inalienable right of man to a portion of the earth on which he was born? If he be an honest and straightforward man, he will point to the firearms suspended over the mantlepiece – "There is my right!" The statesmen of modern Europe manage such matters differently; they arrive, it is true, at the same result – robbery, plunder, seizure of the lands of others – but they do it by treaties, protocols, alliances, and first principles.
>
> (Knox 1850a, 43–4)

Nevertheless, such measures could only temporarily repress the inevitable struggle of race against race: "The eternal laws of nature must prevail over protocols and dynasties: fraud, – that is, the law; and brute force – that is, the bayonet, may effect much; have effected much; but they cannot alter nature" (Knox 1850a, 8).

Until statesmen, scholars, and revolutionaries came to terms with these inexorable laws of racial antagonism and subordination, they could not hope to explain or control events. Knox himself offered his audience a detailed racial analysis of the contemporary world situation. Among other events, he could "scientifically"

---

33 "Men differ more in their intelligence than in their *physique*. . . . These intellectual qualities are equally fixed, permanent, and unalterable, and are much more important than the physical characters of the race" (Knox 1863a, 257–8).

explain the inability of the Celtic Irish to endure Saxon government and Saxon laws (Knox 1848b, 1850a, 15). He claimed to have predicted the 1848 revolution as an irresistible European racial convulsion in which the various tyrannized races struggled to throw off their alien rulers and reconstruct their own government and laws in accordance with their innate racial predilections (Knox 1848a, 97, 1850a, 22). Lonsdale, who subscribed to much of the Knoxian analysis, vividly depicted Knox's deterministic schema of political events: "The actions of men . . . were to Knox like a game of chess: here were kings and pawns on the board, and castles behind which were sheltered statecraft and priestcraft; the knights might be military, diplomatic, or revolutionary, but ever sought to top over the pawns or to crush the people; and all the movements obtained direction from Race".[34]

As Biddiss has noted, such a conception of human history clearly necessitated the stability and immutability of races and racial traits, at least for the duration of recorded history, and Knox's biology was consistent with this. He totally excluded the possibility of environmentally induced change and rejected the concept of transmutation. No race was convertible into another "by any contrivance whatever" (Knox 1850a, 8, 100–1; Biddis 1976, 248). Nor could races, being the equivalent of species, alter their structure through hybridization: "Nature produces no mules: no hybrids, neither in man nor animals. When they accidentally appear they soon cease to be, for they are either non-productive, or one or other of the pure breeds speedily predominates, and the weaker disappears" (Knox 1850a, 65–6). Associated with this natural barrier to racial hybridity was the further principle that each race was adapted to its own specific geographical region and climate, and could not long survive its transposition to another. The future of Europeans in the tropical world was in doubt, and Knox confidently predicted their ultimate defeat by the tropically adapted and fierce Negro: "From St. Domingo he drove out the Celt; from Jamaica he will expel the Saxon; and the expulsion of the Lusitanian from Brazil, by the Negro, is merely a matter of time"(Knox 1855a, 243–4, 256; Curtin 1965, 379–80).

Yet, paradoxical as it may seem, the Knox who insisted on the human races – or, rather, "species" – as immutable biological entities, also subscribed to a theory of organic development – specifically, of saltatory descent. And it is here that we may see most clearly the ways in which his unique blend of political radicalism and racism shaped his biology.

Knox's theory of organic development

Knox's early speculations on organic development were probably inspired by those of his mentor Geoffroy St Hilaire, who in the late 1820s put forward a theory of transmutation – that is, of progressive unilineal species change.[35] But Knox,

34 Lonsdale 1870, 291; see Biddis 1976, 249.
35 Geoffroy St.-Hilaire 1828. A shortened and loose translation of this paper was published in Jameson's *Journal* for 1829, presumably through Knox's influence (Geoffroy St Hilaire 1829). On Geoffroy's theory of transmutation see Russell 1916, Ch. 5; Appel 1987, Ch. 5; Gould 1977, 49–52.

for what I would argue were primarily ideological reasons, rejected Geoffroy's concept of transmutation, and by the 1850s he had elaborated his own distinctive version of development. Like Geoffroy's, it was based on a fundamental concept of transcendental anatomy, the idea that the embryo in its development repeats or mirrors the universal development or *Entwicklung*.[36] In Knox's version, the embryo represented not only all past and existing species, but all possible future species as well, and it was in this way that he allowed for the introduction of new species.

The evidence for Knox's premise rested on his extensive studies of the family Salmonidae, dating back to the early thirties. It is notable that he applied Quetelet's statistical analysis to the salmon in determining colour, weight, proportions, etc., to arrive at the notion of the "average" salmon or "type" of each species.[37] Such external characters, he emphasized, are more significant in the distinction of species than are the internal structures, which, being organized on the one basic "generic" plan, are too homologous or similar to serve as specific characters. This was the same taxonomic method he applied to the distinction of the various races or "species" of the human family. In the case of the salmon, Knox asserted that the young or embryonic members of the family, no matter to what species they belonged, were all essentially similar. This was, of course, by no means an original observation: Karl Ernst von Baer had stated this generalization in his great embryological treatise of 1828, and if Knox was unfamiliar with this source, he must have been aware of Martin Barry's exposition of von Baer's embryology in the *Edinburgh New Philosophical Journal* for 1836–1837.[38] In any case, Knox departed significantly from von Baer's interpretation. He assumed the embryo to have a more complex structure than the adult, and that it is

> chiefly by laying aside some of the characters present in all the young that the adult comes afterwards to be recognized. . . . In the young of the true salmon, I found the specific characters of all the sub-families of

---

36 Knox 1850a, 29–30; Temkin 1950; Gode-von Aesch 1966, 120.

37 Knox 1831–32, 1832–33, 1833, 1855d, esp. 4789.

38 Baer 1828, I, 221–3; Barry 1836–37a, 1836–37b. On the influence of von Baer's embryology on British paleobiology, see Ospovat 1976; Richards 1987. Knox's views on embryogenesis bear some relation to the Kantian concept of "generic preformationism" adopted by von Baer and others of the German teleomechanist school; see Lenoir 1982, 81–95. But Knox's version of this concept seems to me to be closer to that of Carl Vogt, the political radical and "scientific materialist," as described by Lenoir (ibid., 134–40). Like Vogt, Knox rejected spontaneous generation and insisted on the "simultaneous linkage of the phenomena of life to the pre-existence of structure rather than to hypothetical potencies" (ibid., 136). Hence, both identified the embryonic potencies of Kant, von Baer, etc. with material structures capable of direct observation. So Knox claimed to be able to observe all the specific characters of the different species of the salmon genus in the young salmon. Vogt also based his theoretical arguments largely on his study of salmon embryology, and his *Histoire naturelle des poissons de l'eau douce* (1838–42) must surely have been read by Knox. Vogt went on to support Darwinism, but like Knox he rejected the view that chance variation and natural selection could explain the generation of form, and again like Knox he insisted on the fixity and persistence of human racial differences. See Part II.

the genus present; that is, red spots, dark spots of several kinds, silvery scales, proportions and a dentition identical. The young fish before me was, in fact, a generic animal, including within it the specific characters of all the species composing the natural family. To connect this generic animal with any species, you have but to imagine the disappearance of certain characters then and there present. Nothing requires to be added.[39]

Thus, for Knox, all species were originally "generic," and he saw in the generic character of the young the real affiliation that species have to each other: "If this view be correct, it places zoology upon a scientific basis, and explains why one form of life prevailed at one time, and afterwards another; it provides for the extinction of one species and the appearance of another, differing, it is true, from the extinct, but generically the same . . . thus is secured the perpetuity of animal life under different forms, each in unison with the existing order of things" (Knox 1855a, 358).

Knox concluded from this that the successive appearance of new forms or species is "no new creation, but merely the development of forms already existing in every natural family. . . . To institute a species all that is required is to omit or cause to disappear, or cease to grow some parts of the organ or apparatus already existing in the generic being" (Knox 1855b, 627). Humanity was, of course, subject to the same laws, the human embryo containing within itself all the species or races (extinct, extant, and future) of mankind (Knox 1850a, 444, 1862, 503). While Knox usually limited his speculations to consideration of the development of new species within a given family or genus, and his "generic embryo" obviously only allows of a limited development within a particular genus, his belief in the unity and "consanguinité" of all life was fundamental to his materialism: "for life, being a property inherent in matter, must at its origin have been one" (Knox 1850a, 444, 1862, 507, 509). Man's embryonic changes shadow forth all other forms, "worm, mollusc and fish," and he is linked by consanguinity to all other animals that have lived or may live: "A last question remains – the origin of natural families: Have they been distinct from all times? I think not. . . . [T]he law of unity . . . extends to all" (Knox 1855c, 218).[40]

---

39 Knox 1855a, 358, 1855b, 627, 1855c. These papers of 1855 comprise Knox's most comprehensive presentation of his developmental views.

40 My interpretation differs from that of Rehbock, who argues that Knox believed in a community of hereditary descent only among the species of a particular genus and that this genetic connection did not extend to different genera, which, according to Knox, were permanent and distinct and had been separately created (Rehbock 1983, 50). However, in my opinion this is a misinterpretation of Knox's meaning and bears out my emphasis on the need to relate Knox's biology to his radical materialism. Rehbock tends to collapse Knox's views into those of his one-time pupil, the idealist Edward Forbes, who believed that the genus was the "permanent and original" idea (ibid., 73). But Knox did not accept Forbes's conception of the creation of genera and the radiation of species from such "centres of creation" (Knox 1855c, 45), nor his belief in the supremacy of ideas: "The idea of new creations, or of any creation saving that of living matter, is wholly inadmissible.

That Knox held to a theory of organic descent is beyond question, as the following statement makes explicit: "I believe all animals to be descended from primitive forms of life, forming an integral part of the globe itself" (Knox 1852, 109).[41] Yet he as explicitly denied "any transmutation of species, the one into another" (Knox 1855c, 45), or that species were the "direct descendants of each other" (ibid., 46). It is only when this last statement is coupled with his further one that the "law of generation" or "descent" is "generic," not "specific" (ibid., 217; Knox 1855b, 627), that his meaning become clear. For Knox, new species were not produced by change in the mature animal or "species" – that is, by "transmutation" – but by embryonic or "generic" change. It was in keeping with this that he rejected the Lamarckian inheritance of acquired characters, which implied form change in the mature organism: according to Knox, this was not possible, for the species were fixed for all time in the "generic embryo" (Knox 1850a, 100–1). Species are immutable; it is the embryo that is "generically perfect, pliable, adaptive – above all, including within it all the forms which the natural family is destined to assume when developed and specialized in time and space" (Knox 1855a, 359). The embryo contains all possible specific forms for that genus. As conditions in the external world change, species become extinct and are replaced by others from embryonic forms existing in all the species of that genus. So long as one species survives, so does the genus and all other possible species, ready to come into existence when the "order of things" is appropriate: "and thus the law of generation being *generic*, and not *specific*, marks the extent of the natural family, its unity in time and space, the fixity of its species, the destruction of some and the appearance of others being but the history, not of successive creations, but of one development, extending through millions of years, countless as the stars of the firmament" (Knox 1855b, 627).

In other words, Knox held to a theory of saltatory descent – of gross embryonic change – with persistence of species over countless generations[42] – not one of gradual, progressive, or unilineal species change or "transmutation." It is this distinction that has created so much confusion over his developmentalism, his antitransmutationist statements being interpreted as antievolutionary. To a certain extent, the distinction that Knox drew between "generic" and "species" change was more a matter of semantics than biology: the "generic animal" or embryo must itself be the generative product of a species, and in this sense, "generic"

The world is composed of matter, not of mind" (Knox 1850a, 444). Their differences may be best understood by reference to Jacyna's excellent analysis of the early nineteenth century conflict between immanentist (Knox) and transcendentalist (Forbes) cosmologies: Jacyna 1983b. Apart from explicit statements such as the one I have quoted in the text, Knox made it clear that his focus on the relation of species to genera was but the obvious and first step to the "more difficult" question of the development of genera (Knox 1855b, 627, 1855c, 71, 162).

41 This was the passage cited by Baden Powell as evidence of Knox's "transmutationism"; see note 3 above.

42 Not unlike the theory of "punctuated equilibria" of some modern evolutionists.

change may be assimilated to species change or transmutation. Knox himself conceded this, but insisted on maintaining the distinction:

> My immediate object is to prove the existence of a *generic animal*, the product, no doubt, of hereditary descent from a species, but in itself including the characteristics of all the species belonging to that natural family: or, in other terms, proving hereditary descent to have a relation primarily to genus or natural family. . . . My ultimate aim is to offer a scientific explanation of the appearance, from time to time, of seemingly new species on the earth, and of the extinction of others, thus restoring to legitimate science that branch of philosophy which the theory of successive creations, invented by Cuvier and still maintained by his followers, had clearly removed from it.
>
> (Knox 1855c, 4841–2)

Knox's developmentalism clearly served anticreationist and naturalistic purposes, but more than this, his insistence on "generic" change and his associated rejection of transmutation brought his biology into line with his racism and his radical ideology. By rejecting the possibility of transmutation, he was able to affirm the permanency of race and its fixed and unalterable role in determining the character and behaviour of the different races. The human races were unchanged and unchangeable. He could even, on occasion, deny the consanguinity of races, in the sense that they were only related at the generic, rather than the species level (Knox 1862, 507). The concept of racial permanency, or "specificity" of races, enabled him to argue against miscegenation, on the grounds of the necessary sterility of hybrids, and to explain the profound and "natural" antipathy of one race for the other and their inevitable antagonism and conflict. Nevertheless, while they were of distinct species, the races of men belonged to the same natural family or "genus," and shared a common heredity and humanity. All races were thus "naturally" entitled to the rights of man, and the inevitable efforts of black slaves to free themselves must be applauded. At the same time, Knox's rejection of unilineal transmutation allowed him to override the conventional ranking of races, with whites at the top of the scale and blacks at the bottom, and to rebut the charge that he meant to "disparage" any race: "The white races are not the more fully developed, and the negro the more imperfectly developed, species of one common natural family. The development of each is perfect in its way – equally so."[43]

This anti-transmutationism also permitted Knox to set "natural" limits to colonialism. Each race had been unalterably shaped to its particular locale and climate, to the "existing order of things." European colonists could thus survive for only a limited time in tropical countries, and only by dint of enslaving or oppressing

---

43 Knox 1848b, 242, 1855c, 26. At the same time, although he refused to rank races, Knox, in common with most of his contemporaries, assumed the biocultural inferiority of the "dark races" who were everywhere losing ground to colonial expansionism: Knox 1850a, 215–317.

the indigenous races (with whom they could not interbreed), and by constantly replenishing the European stock by immigration. But sooner or later, natural law must inevitably assert its effects, and the oppressors would be eliminated through their inability to adapt to their "unnatural" environment or through "natural" and inevitable racial conflict. Imperialist expansionism was thus curtailed by natural law, and Knox derived some gloomy satisfaction from assigning its natural limits:

> A new crusade has been formed, the banners of which are a cross sur-
> mounting a bale of cotton; Oxford and Manchester combine to push for-
> ward the good work, which, aided by the Armstrong gun, cannot fail to
> reduce Africa to the condition we now so much admire in the United
> States of America, Australia, India, etc. – the native races exterminated,
> or ground to the earth in the most abject condition humanity can assume.
> All this endures for a time. At last nature resumes her course, and the
> intrusive race disappears.
>
> (Knox 1862, 576 and passim)

As the above quotation indicates, Knox's anti-transmutationism was also consistent with his radical rejection of the ideology of progress: "One thing is certain,
the development of new species has no relation to any kind of successive perfectibility" (Knox 1855b, 627). In keeping with its transcendental origins, his own
conception of organic development was a dialectical process, not a mere progress
from the simple to the complex. According to Knox, every organism is influenced
in its development by two antithetical principles: one is the law of unity of organization or of deformation, which is "ever ready to retain the embryonic form"; it
is opposed by the law of specialization or of formation, which leads to the formation of the individual. Where the law of unity of organization is dominant, a
deformation results – that is, the embryonic form is retained. Thus, development
of the individual may be either progressive or retrogressive. Knox made it clear
that it was by retrogressive development – that is, by a return to the embryonic
form – that new species were generated: "By progressive development, I mean
that which tends towards the highest specialization of the individual; by retrogressive development is meant, the development of forms other than those of the species to which the individual belongs" (Knox 1855c, 45, 1850a, 35, 1852, 60–3).
New species originate as the "combined result of these [inherent] laws and the
external circumstances in which they are placed" (Knox 1855c, 218). What Knox
seems to have meant by this was that "deformations" are constantly generated;
those which are not "viable" are unable to survive and reproduce themselves,
while those which are compatible with existing geographical and geological conditions reproduce and increase in number, and so a new species is established.[44] In

---

44 Knox 1850a, 445–6, 1855c, 45, 1862, 503. Knox was clearly not content to leave the expression
   of his "law of generation" in the metaphysics of polarity: ". . . these varieties [of man] must have
   a producing cause, and that cause must be physical. Nothing metaphysical can exist, and it is an

this sense, he could state: "Species is the product of external circumstances, acting through millions of years. When produced they continue until extinguished by external circumstances . . ." (Knox 1855c, 70). As the "material conditions of the external world" change, so the species "disappears," but it may be reestablished by "generic descent" if the appropriate conditions return.[45]

Knox brought paleontological and teratological evidence in support of this non-linear conception of development. The fossil record, he argued, does not illustrate a progression, because some of the extinct animals were equal, if not superior, to existing species (Knox 1850a, 28). And the evidence from teratology was, for Knox, even more compelling. He took issue with the transcendentalist explanation of foetal abnormalities as arrests of development because of its progressionist implications, preferring an explanation consistent with his interpretation of abnormalities as "retrogressive development[s]" toward other forms (Knox 1852, 63, 1855c, 26; Lonsdale 1870, 249–53). The human races were the result of such "deformations," and Knox cited the cuticular fold in the corner of the Eskimo's eye as an example (Knox 1850a, 278). On the ideological level, he linked his antiprogressionism to his antiteleological stance and his reiterated contempt for William Paley and the "Bilgewater" Treatises (Knox 1843, 530, 1850a, 34, 420, 1856; Carter Blake 1870, 334). He was, if anything, even more scathing of those "low transcendentalists" (such as Richard Owen) who had managed to reconcile their transcendentalism with the exigencies of natural theology (Knox 1850a, 28, 437–8; Desmond 1984a, 198). But at the bottom of his rejection of progressive species change, lay his profoundly pessimistic vision of human history. The idea of human progress was "Utopian" and contradicted by the reality of worldwide racial conflict. Human nature, "race," was immutable and ineradicable, and this

---

outrage on common sense to give the nonentity a corporeal existence" (Knox 1863a, 256). At the same time, he was insistent that these physical causes "must have a direct relation to the existing order of things" (Rehbock 1983, 50). However, his ideological exclusion of environmentalism (because of his racial determinism) meant that he was clearly at a loss for any other materialistic explanation of species generation, although he readily invoked environmental agencies for the extinction of species. This in my view accounts for the equivocation in his writings detected by Rehbock (ibid.), and for his falling back on a demystified version of the "law of deformation" in combination with the indirect action of the environment. Cf. Desmond 1984a, 198.

45 Knox 1855a, 358. It must be acknowledged that for all his antiprogressionism Knox had an underlying romantic commitment to the great chain of being and the associated principle of continuity whereby species merge into one another and have no separate reality (Rehbock 1983, 49–52). He could therefore invoke the *"serial unity* of all that lives, or has lived, or may here-after," and through his concept of "generic descent" explain the apparent gaps in the fossil and taxonomic series – especially the gulf he insisted on between the apes and humans. According to Knox, a "class or natural family between man and animals is wanting, or they never have appeared"; either fossil evidence of "anthropomorphous apes or pithecian men" would be uncovered, or such affiliating representatives would be generated sometime in the remote future in accordance with "Nature's great plan or scheme" of unity of organization (Knox 1852, 63, 1855b, 627). Knox had a romantic – but not, it should be stressed, theological – aversion to bestialism (Lonsdale 1870, 255–6).

profound and irresistible biological truth must inevitably conflict with all attempts at social change, whether by reform or by revolution.

Knox's nonprogressivist version of saltatory descent cannot be dissociated from his radical politics, his racial determinism, and his materialist ideology, and attempts to interpret his biology without reference to these latter have led to contradiction. In order to appreciate to the full the need to view Knox's biology in political terms, it is instructive to compare his version of organic development with that of his fellow transcendentalist Robert Edmond Grant, as reconstructed by Desmond. From this it is clear that both Knox and Grant adapted Geoffroy's theory of transformism in ways consistent with their different politico-institutional positions, and that both tried to produce self-consistent materialistic theories of life. Grant, the radical democrat of the University of London, committed to social and institutional reform, "blended Geoffroy's views with a powerful historical Lamarckism."[46] His theory of serial development emphasized gradual, continuous organic change, and related this progressive development to directional temperature change; his causal mechanism for the "direct generation" of new species was therefore environmental in nature. Desmond has stressed the compatibility of Grant's environmentalism and serial developmentalism with his reformist platform, and has shown how he structured his paleobiology in conformity with this socially derived perspective: "Grant needed an undeviating Lamarckian ascent to establish the operation of materialistic laws; like later reformers and 'evolutionists' . . . he would have welcomed an inexorable lawful ascent as a weapon against aristocratic resistance to social melioration and continued political progress" (Desmond 1984c, 10).

Although Knox was involved in attempts at institutional reform during his Edinburgh period, during his most intellectually productive period (from 1842 to around 1855) he was an "outsider" with no institutional affiliations. His peculiar brand of political radicalism was fundamentally anti-reformist and nihilistic, and he rejected the meliorative power of reform. He was antiprogressive, antienvironmentalist, and a racial determinist, and he adapted Geoffroy's transformism to these ideological requirements. He rejected a reforming and improving

---

46 Desmond 1984a, 198; see also Desmond 1984a, 1984b, 1984c. Desmond, in his brief references to Knox's "transmutationism," does not take sufficient cognizance of the ideological differences between Knox and Grant. My interpretation explains why Knox "leaned more toward a demystified *Naturphilosophie*" than Grant. It is difficult to form any concrete opinion of the relations between Knox and Grant. Neither ever referred to the other, or to the other's views, in his published writings, so far as I have been able to ascertain. Lonsdale scarcely refers to Grant (who left Edinburgh in 1827), but it would seem that he and Knox were rivals in the Edinburgh context. In 1826 Grant was supported by Knox's enemies for the position of curator of the Museum of the Royal College of Surgeons (Rae 1964, 36). However, Grant subsequently supported Knox during the Burke and Hare scandal; see Godlee 1921 (I am grateful to Adrian Desmond for this reference). There is some evidence that Knox and Grant moved in the same London reformist circles, in that when an attempt was made to found the Royal Free Medical School in 1853, both Knox and Grant were advertised as lecturers (Rae 1964, 152–4).

Lamarckism for a theory emphasizing radical nondirectional change – the abrupt nonlinear embryogenesis of new species – with unchanging persistence of species over long steady-state geological intervals. His ideological exclusion of environmentalism and progressionism led him to invoke a demystified version of the transcendental laws of polarity in combination with the indirect action of the environment as a causal mechanism for the introduction of new species. In contrast to Grant, his conception of geological change was nondirectional and more cyclical in nature, and he conceived life as "coeval with the globe" rather than spontaneously generated from inorganic matter. Nevertheless his conception of life was also thoroughly naturalistic and materialistic, and he excluded any possibility of a remote or intrusive supernatural power transcending or directing organisms. For Knox, the "living zoological world" was a "selfcreated, self-creating world," and human morality, intellect, and social organization were grounded in the material laws of life.[47]

If, as Desmond emphasizes, Grant's transmutationism was actually a "*constitutive* part of the ideology of radical reform," we must also view Knox's theory of "generic descent" as constitutive of his anomalous ideology of radical racism.[48]

Part 2. Knox and the "resurrection men" of the Anthropological Society of London: the institutionalization of Knox's moral anatomy

Racism, as Biddiss emphasized, is not easily combined with Knox's brand of radicalism: its "benevolent implications" usually lost out to his conviction that the innate hostility between races would always make it "politically unrealistic to preach Equality and Fraternity across their boundaries" (Biddis 1976, 250). When Knox's racism began to gain popularity during the fifties, it was almost inevitably dissociated from its radical roots and harnessed to conservative political ends. Benjamin Disraeli, in 1852, argued in Knoxian terms against the emancipation of West Indian slaves: "In the structure, the decay, and the development of the various families of man, the vicissitudes of history find their main solution. All is

---

47 Knox 1855c, 218. Knox has been represented as a deist, but he seems more of a pantheist to me. Although he sometimes referred to "secondary causes," there is little implication in his cosmology of a remote deity; rather, we find anthropomorphic references to "Nature's great plan" and a good deal of romantic nature worship. Perhaps, like the *Naturphilosophen*, he conceived of a God somehow immanent in the unfolding of nature's plan, and his cosmology was thoroughly deterministic in true *Naturphilosophie* fashion. But his insistence on material causality demarcates him from the idealistic *Naturphilosophen*. What Desmond has said of Grant is equally true of Knox: "his problem was not theology . . . but the production of a self-consistent materialistic theory of life" (Desmond 1984a, 208 n74).
48 Desmond 1984c, 9–10. Grant's environmental determinism and Lamarckian transmutationism are assimilable to the more popular evolutionism of the artisan radicals of the thirties and forties, whereas Knox's developmental views do not fit easily into this more "orthodox" radical framework; see Desmond 1987.

race."[49] An American edition of Knox's *Races of Men* had been issued in 1850, and the Knoxian laws of race antagonism and subordination quickly found their way into some of the more notorious and influential American proslavery texts on race and were reimported into England in this form. In 1856, the *Westminster Review* favourably reviewed Josiah Nott and George Gliddon's racist *Types of Mankind* and noted:

> One of the earliest to apply the doctrine of the essential diversity of human races, so fertile of results, to historical, political, and other problems, was, we believe, Dr. Robert Knox, in his singular work . . . "The Races of Men". . . . This view explains much heretofore most obscure. One term of sacred import, Civilization, receives from it a limitation of application which the benevolent spirit is disposed to brook ill. We are generously inclined to desire for all whom we include as fellows in humanity, the same privileges, rich and expansive blessings, as those we enjoy ourselves. . . . Stern experience, however, teaches that in its wide application to the family of man, it must be often modified, and sometimes restrained within very narrow limits indeed. The capacity to receive the boon of civilization is very different in the different races of men. Some, we are constrained to admit, are so low in the scale of improveability that they are totally incapacitated for its reception. . . . And, amongst those fitted to receive it, there are so many shades and degrees of capacity, limiting and defining their progressive advancement, that nothing less than an extended acquaintance with human races can preside over the proper administration to their wants.[50]

Knox himself stated that his views on race as the key to social, political, and historical explanation, did not generate much public interest until the 1848 revolution. But following on his successful prediction of this cataclysmic event his ideas were taken up by the press (so much so that Knox accused the editor of the *Times* of consistent plagiarism), and became widely known.[51] He attributed the growing influence of racist theory to the predictive power and "truth" of his science, but

---

49 Quoted in Curtin 1965, 381. Other reviewers discussing colonial policy and racial issues began to employ Knoxian arguments to promote conservative opinion (ibid.).

50 "Types of Mankind," *Westminster Rev., n.s.,* 9 (1856), 378–9; see also Knox 1850b; Nott and Gliddon 1854, 53. John Campbell also cited Knox in his notorious *Negro-Mania;* see Curtin 1965, 372.

51 Knox 1850a, 23, 1862, 565. See also "Races v. Nations," *Medical Times and Gazette,* 11 (1862), 226–7: "Dr. Knox, who has laboured all his life to establish the influence of *race* in the destinies of *nations,* is well avenged by finding that those who once denied, finish by proclaiming his theories as if discoveries of their own, or else *adopting* them – of course without acknowledgement." Knox himself felt constrained to enter a caveat upon the overly enthusiastic applications of some of his "plagiarists" and followers: "Day by day the opposition weakens; the great questions of race are discussed in a calmer and more philosophic tone, and there is every danger of their running to the other extreme, and undervaluing those acquired and artificial qualities strictly the result of national

we may detect other forces at play. With his emphasis on race as the "overweening determinant of character and culture, of individual and collective behaviour," Knox defused the environmental and social explanations of the reformers and radicals and catered to the increasingly negative evaluation of the cultural worth of non-European peoples that accompanied the economic expansion of Britain (Biddis 1976, 250). His radicalism and racial determinism were in essential tension within his system, but his audience and followers were less interested in Knox's "moral anatomy" as a self-consistent synthesis, than in his emphasis on biological race and the (usually conservative) political implications they might draw from it.

To a certain extent, Knox himself was responsible for this. He made little attempt to present his views in any systematic form. His disorganized, vehement, and scattered writings and lectures were not conducive to a general comprehension of the more recondite theoretical aspects of his work. The *Medical Times* made the point that his *Races of Men* was a work that Knox's "acrimony, scepticism, want of proper arrangement, carelessness, and repetition would have damned, had it not been for its truth."[52] Given such a presentation, it is not surprising that his contemporaries found his conception of "generic descent" difficult to grasp. Luke Burke (another early British exponent of the "Science of Race"), in reviewing Knox's early "Lectures" on race, endorsed his emphasis on the "permanence of all the characteristics of race," but could not come to grips with his developmentalism. He wrote that the "two propositions involved [racial permanence and organic change] . . . are mutually destructive. One may be true, but both cannot be so."[53]

But Knox was virtually forced to popular exposition with all its attendant dangers of oversimplification and distortion. The real point at issue is that he had no scientific audience and no institutional forum for his views. From all contemporary accounts, his London life was overshadowed by his notoriety and questionable "morality" (i.e., his radicalism), and restricted by his poverty. His participation in established London medical and scientific circles and institutions was more or less limited to the articles he managed to have published in journals such as the *Medical Times*, the *Lancet*, and the *Zoologist*, and his translations of anatomical and zoological texts. His London medical reputation was further compromised when the *Lancet* publicized the "Osborne scandal" as part of its campaign for medical reform.[54] Nor did he have an established ethnological audience. His "ethnology" was more in tune with, and was certainly better appreciated by,

---

influences" (Knox 1862, 566, 596). Note also his statements at the conclusion of his *Man, His Structure and Physiology, Popularly Explained and Demonstrated* (Knox 1857a, 170–1).

52 "The Late Dr. Knox," *Medical Times and Gazette* (27 December 1862), 684. The *Lancet* made the same point: "The Late Dr. Knox," *Lancet* (3 January 1863), 19–20, esp. 20.

53 Burke 1848, 94. Other reviewers of the same period failed to perceive Knox's developmentalism at all: "If we understand Dr. Knox's theory, it is that men were originally created of different races, like the wild animals . . ." ("Human Progress," *Westminster Review*, 52 [1850], 2).

54 See *Lancet*, 1847, 1: 565–71, 630, 653–4, 685; Rae 1964, 134–61; Lonsdale 1870, 343–94; Blake 1870. All accounts of Knox's London period are very sketchy and incomplete. Knox does seem to have had some contact with London medical reformist circles (see note 47 above), and he

French physical anthropologists,[55] than by the members of the conservative and religion-oriented Ethnological Society of London, who did their best to exclude Knox from their ranks.

In any case, by the close of the fifties, Knox's developmental views were overtaken by the Darwinian theory of natural selection. Not that Knox himself was overly impressed by the *Origin of Species:* "Darwin's work," he wrote to James Hunt, "leaves the question precisely where it was left by Goethe, Oken, and Geoffroy St Hilaire."[56] Whatever the intrinsic merits of the Darwin/Wallace theory, it is instructive to compare Knox's lack of scientific credentials and of an effective institutional power base with those Darwin had already established by the time he published his evolutionary views (Herbert 1977, 156–7; Rudwick 1982).

The institutionalization of Knox's moral anatomy

However, toward the end of his life Knox briefly found an institutional niche in the Ethnological Society, and his anthropological doctrines inspired the breakaway Anthropological Society that was founded shortly after his death. It was, in fact, only posthumously that he achieved the essential institutional backing at the hands of the above-mentioned Hunt, who in 1863 led the secession of the Anthropologicals from the Ethnological Society. Hunt, as George W. Stocking has stated, "in a paradoxical and antithetical way was one of the most influential figures in English anthropology in the 1860's" (Stocking 1971, 376). Knox had first met Hunt around 1855, and it was under Hunt's aegis that he eventually gained entry into the Ethnological Society (Hunt 1868a, 432).

The Ethnologists, who had their roots in Quaker and Evangelical philanthropy, conducted their inquiries within a framework of religious assumptions that provided the ethnological problem of accounting for racial variety in terms consistent with the biblical account. Like their mentor James Cowles Prichard, they were primarily "monogenists," who accepted some modification over time as races had diverged from their original unity of type. Their methodology was historical,

---

eventually found employment in 1856 as pathological anatomist to the Cancer Hospital, founded by the reformer William Marsden.

55 The French physical anthropologist Paul Broca, founder of the Société d'Anthropologie de Paris, incorporated Knox's arguments on the infertility of racial hybrids into his polygenist writings, and in 1861 Knox was elected the first foreign corresponding member of the Société. See Lonsdale 1870, 385; Hunt 1864b; Broca 1864, 61–71.

56 Quoted in Lonsdale 1870, 368. See also Knox 1862, 570, 589, 594; Knox 1863b, esp. 267. Knox dealt very peremptorily and dismissively with the *Origin,* but it is tempting to speculate that Darwin's utilitarian Malthusian mechanism of natural selection was unacceptable to the radical and anti-Malthusian Knox; see Knox 1862, 580. As well, Knox, like the other transcendentalist-influenced critics of Darwin, would have found the chance element of natural selection incompatible with his deterministic schema of development; see Russell 1916, 241–5. His dismissive attitude toward the *Origin* possibly accounts for Knox's failure to make any bids for acknowledgment as a "precursor," unlike Grant who gloried in the connection: see Desmond 1984a, 191–2.

based largely on linguistic analysis that demonstrated the unity of humanity, and they emphasized environmental factors in the formation of race (Stocking 1971, 1973). Hunt, who served as secretary to the society for a number of years, later characterized it as dominated by a Quaker clique and made moribund by religion (Hunt 1868a, 432).

According to Hunt, Knox first attempted to join the Ethnological Society in 1855, but was black-balled by the Quakers (Hunt 1868a, 432). In the Council Minute Book of the society there is note of a "letter read from Dr. Knox" for the meeting of February 7, 1855, and undoubtedly Knox's pungent views, marginal status, and radical reputation would have been meat too strong for Quaker stomachs.[57] But over the next few years the structure and orientation of the Ethnological Society underwent some changes, and a number of physical anthropologists, including Hunt, joined the society. In 1860 Hunt became joint secretary and the polygenist John Crawfurd was elected president. Following on this, Knox was finally elected an honorary fellow "to the horror and indignation of the Quakers,"[58] and from then until his death two years later he was a prominent member of the society.

On the sparse evidence of the minutes, Hunt unabashedly used his powers as secretary to promote Knox, ensuring that his papers were given priority for reading at the society's meetings and were selected for publication in the *Transactions*. However, he did not have matters all his own way in this, and there is some evidence that Knox's reputation was a continuing source of conflict within the society. There is a revealing entry in the minutes for the meeting of May 20, 1862. Here resolution no. 7 reads: "That Dr. Knox M. D. be appointed 'Honorary Curator' to the Society," but a heavy line has been drawn through the whole entry. The events behind this deletion can only be conjectured, but it seems fairly clear that the secretary had exceeded his authority on this occasion and was called to account by the Quaker opposition. It took Hunt another two meetings before he was able to organize the resolution back onto the books and triumphantly record Knox's appointment, and this time the entry was allowed to stand.[59] The triumph was short-lived, however; within a few months Knox was dead, and Hunt was

57 Ethnological Society of London (ESL) Minutes, 7 February 1855, "Council Minute Book, 1844–1869," Archives, Royal Anthropological Institute. The society had previously purchased a copy of Knox's *Races of Men* as part of its library collection: ESL Minutes, 11 June 1851.

58 Hunt was elected a fellow in 1856; Knox was elected honorary fellow on 27 November 1860: ESL Minutes; Hunt 1868a, 432. Hunt gives the date of Knox's election as 1858, but according to the Minutes this is incorrect.

59 ESL Minutes. Knox was formally appointed honorary curator on 17 June 1862. Over the two years of his membership he read six papers in all to the society, of which three were published in the society's *Transactions* – a record exceeded only by Crawfurd, the president; see Bloxam 1893. This same period was a stormy one, with conflict over the issue of the admission of women to the Ethnological Society's meetings (forcefully opposed by Hunt, who resigned as secretary at one point, ostensibly on health grounds, but was persuaded to withdraw his resignation; ESL Minutes for 27 November 1860; 6 February, 20 February 1861). Hunt later represented this issue as one of the major reasons for his secession from the Society (Hunt, "Dedication to Broca," in Vogt 1864, viii–ix; see Richards 1989).

engaged in organizing the rival Anthropological Society of London, which met for the first time on January 6, 1863.

It is difficult to determine Knox's exact role in the formation of the new society. Hunt subsequently represented Knox, Richard Burton, and himself as the real founders of the Anthropological Society, and considering the close communication between Knox and Hunt over the period 1860–1862, and the similarity of their anthropological views (which Knox himself endorsed), Hunt's claim seems plausible. Moreover, it was Knox who introduced Hunt to the French physical anthropologist Paul Broca, and Hunt modelled his society on Broca's Société d'Anthropologie de Paris.[60] However, Knox's poor health during this period (he died of a long-standing heart condition on December 20, 1862 – just two weeks before the inaugural meeting of the society) makes his active organizational involvement questionable. But irrespective of his organizational role, he was indubitably the Anthropological Society's intellectual mentor. Hunt later claimed to have "imbibed" his views from Knox, and all the available evidence supports this claim.[61] When he first met Knox, the much younger Hunt had been something of a marginal medical man also. He had inherited a practice in the treatment of stammering from his father, and had published a treatise on his father's system that was primarily concerned with defending it (and Hunt) from the charge of quackery. Although Hunt later represented this work as the basis of his interest in race, investigation reveals almost nothing on this "great question"; what little there is suggests a Prichardian environmentalism and monogenism: "Savages do not stammer; in them the human animal remains unchanged. In the civilized world, on the contrary, refinement has materially altered the physical man. Robustness yields to delicacy, and the very structure of organs undergoes metamorphosis."[62] Within a few years, under Knox's tutelage, Hunt had shed his early environmentalism and become a demagogic "anthropologist."

Several studies of the history of the Anthropological Society have now been published, and there is general agreement that Hunt conceived it as a platform for his anthropological-cum political opinions and that it was his racism that precipitated the break with the Ethnologists. As Stocking and others have represented them, the Anthropologicals were primarily "polygenists" who advocated the ultimate diversity of human races, and took issue with the theological concern of the ethnologists to derive all races from the one stock. They were generally men with a background in medicine or biology, and their method was that of the continental

---

60 Hunt, "Dedication," in Vogt 1864, vii–viii; Hunt 1868a, 432–4; Lonsdale 1870, viii, 384–7; Knox 1862, 600. Cf. Rainger 1978, 56–7.

61 Hunt 1866b, esp. 336. Hunt was collecting material for a biography of Knox, and had advertised to this effect in the *Anthropological Review,* when Lonsdale made known his proposed biography; Lonsdale 1870, viii.

62 Hunt 1854, 25. This work does indicate Hunt's early preference for naturalistic explanations: ibid., 12. Little is known of Hunt's early career. See "James Hunt," *Dictionary of National Biography, 28:* 266–7; also Stocking 1971, 376; Rainger 1978, 52.

physical anthropologists. They placed great emphasis on describing, measuring, and classifying the physical "types" of humanity, forming rigid categories that maximized racial differences and justified the polygenist emphasis on essential human diversity and inequality.[63] Like Knox, they were inflexibly determinist, seeing race as the *cause* of civilizational achievement rather than a product of cultural experience. But where Knox had tempered his racism with his humanitarianism, Hunt and his followers advocated an extreme racism, underpinning it with a bastardized Knoxian anthropology and biology, and promoting a range of reactionary policies that were at odds with Knox's radicalism. Knox became the figurehead behind which the Anthropological Society, with Hunt at the helm, steered full tilt against the tide of liberalism, personified by John Stuart Mill.

From its inception, Hunt made it clear that he and his fellow Anthropologicals were founding not merely a new society, but a "new science," and that the overwhelming significance of the new science devolved on its political implications:

> It is frequently the habit of scientific men to exaggerate the importance of their own special study to the detriment of other branches of knowledge; but do I exaggerate when I say that the fate of nations depends on a true appreciation of the science of anthropology? . . . Does not the success of our colonization depend on the deductions of our science? . . . Is not the wicked war now going on in America caused by an ignorance of our science? These and a host of other questions must ultimately be resolved by inductive science.
>
> (Hunt 1864a, lxxxi, xcii)

In Hunt's hands, Knox's "moral anatomy" was to lay the foundations of a new applied moral science – "Anthropology." Race was for Hunt, as it had been for Knox, the key to "scientific" political legislation and social procedure. But he enunciated this much more explicitly than had Knox, shored it up with positivist rhetoric, and deployed it specifically to preclude the objectionable "unnatural" notion of equal human rights:

> [T]he science of political economy must be based simply and solely on the facts discovered by the anthropologists. . . . Now a social science cannot be based on mere philanthropic theories. In other words, social science must be based on the facts of human nature as it is, not as we would wish it to be. . . . We are the students and interpreters of nature's laws, and it is our duty carefully to ascertain what those laws are, and not attempt to raise up in the name of "social science" a code of morals based upon an assumption of human equality and consequently equal human

---

63 See Stocking 1971; Rainger 1978; Burrow 1963–64). Stocking's is by far the best and most detailed analysis.

rights, because we know that human equality is a mere dream and all systems based on it are mere chimeras.

(Hunt 1867a, lxi–lxii; cf. Rainger 1978, 61)

During the first half of the nineteenth century, Mill, with his thesis that human "nature" was primarily socially determined, was the greatest liberal defender of racial egalitarianism and, as such, Hunt's prime target. To Hunt, Mill's claim for black suffrage was a scientific absurdity, contradicted by the "facts of human nature," which, according to Hunt, were best served in those parts of the world in which "the Negro [is] in his natural subordination to the European."[64] To the left of Mill stood the socialists and the communists who adhered to a more radical environmentalism, and, not surprisingly, their claims too were completely routed by "anthropology": "the theories of socialism, communism, and republicanism find not a fact in anthropological science to support such chimeras."[65] As Marvin Harris has pointed out, it was Mill and those who suffered from what Hunt termed "the rights-of-man mania" who were the objects of Hunt's most vitriolic attacks, rather than the rival Ethnologists, who were merely subject to "religious mania" (Hunt 1866a, 1867a, lix; Harris 1968, 101).

In fact, the positions of the Ethnologists and Anthropologicals were not as irreconcilable as might appear, and after the death of Hunt the two societies were reunited in 1871 (largely through the efforts of Thomas Henry Huxley) to form the Anthropological Institute of Great Britain and Ireland. As a number of historians have stressed, the "new" Darwinian anthropology that emerged may be viewed as the logical and historical synthesis of the preceding two major models of anthropological enquiry – ethnology and physical anthropology.[66] The concept of race was central to both models, and both were preoccupied with the problem of racial diversity and subscribed to a naturalistic conception of this diversity. Thus the Ethnologist Prichard had argued that civilization for humans was like domestication for animals, and consequently that the physical features of the superior races had been "improved by civilization"; while the Anthropologicals, by extending their biological model to human nature, could explain cultural and social differences in terms of anatomical and physical differences. Although the Ethnologists opposed the more extreme racial views of the Anthropologicals, both were agreed

---

64 "On the Negro's Place in Nature", Hunt 1863, 51–2. According to Hunt, this paper was initially presented at the Newcastle meeting of the British Association, where it was hissed by the audience. He subsequently read it to the Anthropological Society, where he received "the cordial and earnest support of our scientific brethren" (ibid., vi). This paper contains a number of references to Knox's anthropology in support of Hunt's views (ibid., 13, 17). As Stocking has noted, Hunt's defence of slavery was well timed to coincide with the American Civil War: Stocking 1971, 376.

65 Hunt 1867a, lx. Here Hunt expressed his own preference for a "well-selected hereditary aristocracy" as being "more in accordance with nature's laws than those glittering trivialities respecting human rights which now form the stock-in-trade of some of our professors of political economy, and many of our politicians" (ibid., lxi).

66 Notably, Stocking 1971, 384–6; Weber 1974, 269–72.

on a causal relationship between race and civilization and both factions assumed the biocultural inferiority of non-Caucasoids.

The extreme antiquity of man was established by the 1850s, and this, together with mounting pressure from biblical criticism and anticlericalism and the increasingly overt racism that went hand in hand with British expansionism, undermined the ethnological position. In this context, the Anthropologicals, like the contemporaneous "young guard" Darwinians, promoted their position as the more scientific one, as unhampered by Christian apologetics, based on a tried and tested scientific method, and consistent with natural facts (Stocking 1971, 385; Turner 1978). Given their similar naturalistic and anticlerical orientation and their ultimate synthesis, it might have been expected that the Anthropologicals and the Darwinians would have made common cause against the more conservative and religiously oriented Ethnologists – but such was not the case. For most of its brief history, the Anthropological Society was explicitly and vehemently "anti-Darwinian," and Hunt and Huxley were in overt conflict. The leading Darwinians, such as Huxley, A. R. Wallace, Lane Fox, Francis Galton, George Busk, John Lubbock, and Edward Burnett Tylor, were all members of the Ethnological Society and had little to do with, or were actively hostile toward, the Anthropological Society. It is by focusing on their conflict with the Darwinians that the significance of Knox's views for the Anthropologicals, and for Hunt in particular, may be best understood. This analysis will also clarify the relation of Knox's anthropology and biology to late Victorian scientific racism.

The institutional and ideological conflict between the Darwinians and Anthropologicals

Most accounts of the conflict between the Anthropologicals and the Darwinians have stressed their basic intellectual incompatibility.[67] This interpretation hinges on the polygenism and racism of the Anthropologicals, which, it is argued, they perceived as under threat from the Darwinian thesis of the common descent of the human races; it was because of the antipathy of Hunt and his Anthropological Society to theories of development, and to Darwinism in particular, that evolutionists chose to join forces with the Ethnologicals – who, in spite of their tradition of religious orthodoxy (or perhaps because of it), were more oriented toward theories of human change over time. There are some problems, however, with this interpretation. It is true that Hunt made it clear that what he chose to construe as the reaffirmation of monogenetic doctrine by such leading Darwinians as Wallace and Huxley, constituted the major objection of the Anthropologicals to evolution by natural selection.[68] But the Anthropologicals were well aware that Darwinism

---

67 Burrow 1963–64; Stocking 1971, 378; Rainger 1978, 58–9.
68 Notably: "I cannot think that any advance can be made in the application of the Darwinian principles to anthropology until we can free the subject from the unity hypothesis which has been identified with it, especially by the influence of Professor Huxley" (Hunt 1866b, 339).

was not incongruent with their polygenism, and the Darwinians were instrumental in bringing this more forcefully to their notice.

It was to a meeting of the Anthropological Society in 1864 that Wallace addressed his first paper on man, in which he demonstrated how monogenism and polygenism might be reconciled in evolutionary biology. According to Wallace, all races were derived by natural selection from a single, originally homogeneous stock, but racial traits, once developed, were fixed and very ancient. Their common ancestry lay so far in the past that it might fairly be said "that there were many originally distinct races of men." With a conflation of Knoxian race laws and natural selection, Wallace confidently predicted the extinction of the "inferior" races from the "inevitable effects of an unequal mental and physical struggle".[69] His audience might have been more receptive to this polygenist compromise had Wallace not gone on to draw a splendid utopian vision of an earth ultimately peopled by a superior race of perfectly equal beings – an "eloquent dream" from the incipient socialist that outraged Hunt's belief in essential human inequality.[70] His audience was also unreceptive to Wallace's thesis that man's moral faculties could evolve without concomitant physical changes. As Luke Burke pointed out, this thesis contradicted the fundamental anthropological correlation of mental differences with physical differences: "It divorces our power of judging of the mind from the body".[71] If they found Wallace's interpretation ideologically unacceptable, in the same year the Anthropological Society published a translation of the *Lectures on Man* by the German-born polygenist and Darwinian Carl Vogt (translated and edited by Hunt), which offered the more congenial "anthropological" picture of an ever-increasing gulf between the races that was virtually unbridgeable. Hunt indicated, in his preface to this work, that it was Vogt's advocacy of polygenism within a Darwinian framework that made his *Lectures* so valuable to the members of the society:

> Prof. Vogt acknowledges that, to a great extent, he is willing to accept the conclusions of England's great modern naturalist, Charles Darwin; but, unlike many of that profound observer's followers in this country, he entirely repudiates the opinions respecting man's unity of origin which a section of the Darwinians in this country are now endeavouring to promulgate. The author's views on this point I hold, in the present state of science to be especially sound and philosophical: and I hope that this work may help to counteract the inconsistent and antiquated doctrines now being taught by one of our government Professors respecting the small distinction which exists between the members of the genus *Homo*.[72]

69 Wallace 1864; Stepan 1982, 68–70; Schwartz 1984, 272–5.
70 Hunt 1867b, esp. 113; Hunt also objected to Wallace's exemption of man from natural law (Wallace 1864, clxxx).
71 Burke in Wallace 1864, clxx.
72 Hunt, "Preface" in Vogt 1864, xv. See also Hunt's endorsement of Vogt's interpretation in Hunt 1867b, 114, 118.

Hunt and his followers were not opposed to the Darwinian thesis per se, but rather to Huxley's deployment of it. It is quite misleading to state, as does Ronald Rainger, that Hunt's society had an "antagonism to theories of development" (Rainger 1978, 58). As we might expect from a self-proclaimed disciple of Knox, Hunt consistently made it clear in his major writings on the matter that he "accept[ed] the great principle of natural development to explain man's origin" (Hunt 1866b, 340, see also Hunt 1867b). Hunt was not forthcoming on his own theoretical views; for the most part, he confined himself to "hints," formally eschewing "speculation" on the subject. On Knox's authority, evolution by natural selection was merely one of a number of speculations by "popular writers" adopted from the philosophy of Goethe and the morphological speculations of Geoffroy. While such Goethean developmentalism was probably correct, "the really scientific men do not as yet look to the theory as established on a strictly scientific basis."[73] Such vaguely defined developmentalism (or "continuity," as he came to term it) sufficed to provide Hunt with the essential naturalism, and the even more essential naturalistic proscription of the revolutionary ideas of the "so-called rights of man." He approvingly quoted William R. Grove, president of the British Association, to this effect: "Our language, our social institutions, our laws, the constitution of which we are proud, are the growth of time, the product of slow adaptations, resulting from continuous struggles. Happily in this country . . . practical experience has taught us to improve rather than to remedy; we follow the law of nature and avoid cataclysms" (Hunt 1867a, lx).

I suggest that Hunt's failure to delineate his own views, or to ally his society with the Darwinian or any other specific developmental model was largely strategic. In the context of the Darwinian disputes of the sixties, a degree of ambiguity was manipulable. While Darwinism remained a controversial doctrine, particularly in its application to humanity, it was politically expedient for the Anthropologicals to maintain their "positivist" independence of it.[74] At the same time, their commitment to a vaguely defined naturalistic developmentalism could be deployed against outmoded and "unscientific" theological explanations of human origin. It is consistent with this interpretation that Carter Blake – Hunt's closest colleague, and secretary of the Anthropological Society (and also a Knoxian) – could maintain, in the midst of his highly critical review of Huxley's *Man's Place in Nature:* "The day is long gone by when the probability of transmutation could

---

73 Hunt 1866b, 326. Hunt did make some statements that are suggestive of a Knoxian/Vogtian embryological model of "natural development" in his unsigned "Race in Legislation" (Hunt 1866a, 120, 129; see Vogt 1864, 183–92). I have pointed to the affinities between the views of Knox and Vogt (see above, note 38). However, Hunt was undoubtedly far more interested in the political applications of Knox's views than in the biological details of his developmentalism, which Knox did not present very coherently in his major anthropological writings. Also, Hunt made very clear his preference for an interpretation of development that did not promote revolutionary change (Hunt 1867b, 119, 120).

74 It is possible that Hunt modelled this strategy on Broca's in the "parent" Société d'Anthropologie de Paris. See Harvey 1983.

be sneered down as the phantasm of a dreamer, or the product of the scepticism of an infidel. The possibility, nay, even the extreme likelihood of such a law being eventually established is now rapidly becoming a tolerated doctrine in the creed of deep thinking scientific men" (Blake 1863a, 161, 1863b). It is also in keeping with such a strategy that Hunt and his followers went to some lengths to cultivate the most prominent of the anti-Darwinian developmentalists, Richard Owen.[75]

That the Anthropologists attempted initially to put a conscious strategy of neutrality into practice is indicated by their choice of the first five honorary fellows to be elected to the newly formed society: they comprised the polygenist Crawfurd, three Darwinians (Darwin, Huxley, and Lyell), and Richard Owen (who by this stage had emerged publicly as an advocate of "continuous creation").[76] But this carefully staged neutrality was fragmented when, within a few months, in reaction to Carter Blake's abovementioned "coarse attack" on *Man's Place in Nature*, Huxley resigned his diploma as honorary member of the Anthropological Society and joined the rival Ethnological Society.[77] This same meeting of the Ethnological Society witnessed Hunt's resignation on racist grounds and a takeover of the society by the leading Darwinians: Crawfurd was deposed to vice president and Lubbock elected president in his stead; Galton replaced Hunt as honorary secretary; and Huxley and Busk were made members of the Council.[78] From this point on, relations between the two groups were characterized by "recurring conflict, bitterness, recriminations, and the failure of several attempts at reconciliation" – and an increasingly anti-Darwinian edge to Hunt's rhetoric. When Huxley became president of the Ethnological Society in 1868, this, in the view of the Anthropologicals, consolidated the transformation of that society into "little more than a sort of Darwinian Club" (Stocking 1971, 381, 377).

75 Carter Blake to Owen, 22 December 1863; 5 September 1865; 14 August 1868; 29 August 1873, British Museum (Natural History), Owen Collection, *4*, fols. 202, 204, 209, 211. See also the invitation to Owen to attend the Anthropological Society meeting of 6 December 1864, to comment on a collection of human remains: ibid., *8*, fol. 343a. Note also Hunt's remarks re Owen in Hunt 1867b, 117. Owen was another of those who based his developmentalism on embryogenesis; see Richards 1987.

76 ASL Council Minutes, 18 February 1863.

77 Ibid., 12 May 1863; ESL Minutes, 5 May 1863; Huxley to Carter Blake, 2 and 5 May 1863, Huxley Papers, *V. XI*, fols. 17–20, Imperial College Archives (hereinafter cited as Huxley Papers); Rolleston to Huxley, ibid., *XXV*, fol. 165. Huxley's resignation was ostensibly over Carter Blake's mauling of Rolleston; however, as Desmond has noted, the article was a general attack on *Man's Place*, so Huxley had a more personal reason for resigning. See Desmond 1982, 223 n51; Huxley to Lubbock, 3 May 1863, Avebury Papers, Correspondence of Sir John Lubbock, *III*, 49640, fol. 53, British Library (hereinafter cited as Lubbock Correspondence). Matters were not improved when Hunt entitled his proslavery paper "On the Negro's Place in Nature," in obvious paraphrase of Huxley; Stocking 1971, 379.

78 ESL Minutes, 5 May 1863. The previous two years had seen an influx of Darwinians into the ESL: Charles Darwin had been elected an honorary fellow on 14 May 1861; Francis Galton became a member on 1 March 1862, and Erasmus Darwin (who served on the Council for a time) on 18 March 1862. On the reasons for Hunt's resignation, see Stocking, 1971, 376.

If it was largely for personal reasons that Huxley broke with the Anthropological Society and threw in his lot with the Ethnologists, it soon became apparent that the differences between the Ethnological-based Darwinians and the Anthropologicals were more political in nature than personal or theoretical. Their opposing institutional locations and affiliations, rather than their theoretical differences, set them apart; as I will argue, they were not so much locked in theoretical conflict, as in ideological and professional competition with one another. The charge of "monogenism" became a convenient peg on which Hunt could hang their differences and thus demarcate the Anthropologicals from the "Darwinian Club." In a number of crucial areas, Huxley's anthropological position was congruent with that of the Anthropologicals, and he soon set about the pressing political task of liberating Darwinism from the "monogenism" with which Hunt and his cohorts persisted in identifying it.

The urgency of the task was occasioned by the phenomenal growth of the Anthropologicals and the formidable professional competition they offered the Ethnologicals. No longer could they be sneeringly dismissed by Huxley as a bunch of "quacks" and a "nest of imposters" and left to their own devices (Stocking 1971; Desmond 1982, 81). By mid-1865, they had over five hundred members (about twice the membership of the Ethnological Society) and were on the point of establishing provincial branches (one of which was actually established in Manchester in 1866). The society was involved in an active publication program, including its own *Memoirs*, a series of translations of foreign anthropological works, the *Anthropological Review* (owned and controlled by Hunt), and the society's *Journal*. Moreover, the society had made a formal assault on the scientific establishment with a series of determined attempts to have their "new science" officially recognized by the British Association alongside the traditional ethnology. Although Hunt and his followers were consistently rebuffed by the Association's conservative leaders, who rallied to the support of the established Ethnologicals, Huxley was well aware that not even the powers of the scientific establishment could withstand for long the sheer force of Anthropological numbers. In the mid-sixties the Darwinians were still seeking to establish themselves scientifically, and Huxley saw the tactical need to support the Anthropologicals' claims to recognition by the Association. From this point on, he committed himself to the strategy of reconciling and uniting the rival societies. As he put the case to Lubbock, by contrast with the Ethnological Society, the Anthropological Society was "certainly alive and vigorous and under proper direction may become a very valuable organization."[79]

There was another dimension to this strategy in that the Darwinian disputes of the sixties centred on the highly contentious issue of "man's place in nature"; as

79 Huxley to Lubbock, 18 October 1867, Lubbock Correspondence, *V,* 49642, fol. 63; Stocking 1971, 377, 381–2. Stocking does not attribute such an overtly manipulative role to Huxley, but see Turner (1978) on the takeover of London science by the Darwinian "young guard"; see also Desmond 1982, 110–12, 158–64.

the self-constituted chief spokesman of the Darwinians on this "question of questions" (Huxley [1865] 1908, 39), Huxley must have been very conscious of the problems presented by the professional schism within the discipline most closely focused on the study of man. With the leading Darwinians clustered in the minority faction, the obvious solution was conciliation and unification with the all-too-successful and vociferous Anthropologicals, who might then be kept in order and given a "proper direction" by the Darwinians.

Huxley's major anthropological production of 1865, "On the Methods and Results of Ethnology," was designed not only to promote Darwinism as the key to the scientific study of man, but also to bridge the theoretical and, he hoped, the institutional gap between the two societies. He followed Wallace's lead by endorsing Darwinism for its potential of "reconciling and combining all that is good in the Monogenistic and Polygenistic schools," and he made a number of important concessions to the Anthropologicals (Huxley [1865] 1908, 121). In spite of the ethnological emphasis of his title, Huxley made it clear that insofar as method went he supported the new anatomical method of the physical anthropologists against the traditional linguistic approach of the ethnologists. On the basis of skin colour, hair type, and skull shape, he identified eleven racial types, which he designated "persistent modifications" or semi permanent stocks. Again, while Huxley claimed that the anatomical evidence was against the specific differences asserted by the polygenists and overwhelmingly in favour of the unity of the origin of mankind, he stressed that a belief in the diversity of human species did not necessitate diversity of origin. He excluded direct environmental influences on race as rigidly as any Anthropological, arguing that the races had arisen singly or appeared in a number of contemporaneous examples in some remote epoch and had evolved by natural selection. He suggested that the distinct racial types might have evolved so far as to prevent fully fertile crosses, and although he did not support the polygenist evidence for this, he would be "*A priori . . .* disposed to expect a certain amount of infertility between some of the extreme modifications of mankind; and still more between the offsprings of their intermixture." In conclusion, Huxley even suggested that satisfactory evidence of such infertility might well provide the crucial proof of Darwin's theory of evolution.[80]

Although Huxley's efforts to have the "new science" recognized by the establishment were applauded by the Anthropologicals, his attempts at institutional reconciliation were rejected and his polygenistic overtures met with the ritual incantation of "monogenism."[81] The point was that at this stage the Anthropologicals

---

80 Huxley 1908, 118, 123. See also Stepan 1982, 78–9. On Huxley's "Persistence," see Desmond 1982, 84–112.
81 Hunt 1866b, 320. Hunt actually wrote to Huxley in acknowledgment of Huxley's conciliating role, expressing the willingness of the Anthropologicals to consider amalgamation under Huxley's presidency. Whatever the reality of this offer (the Council of the ASL refused to even consider Huxley's candidacy for honorary fellow), a few weeks later Hunt, on hearing that Huxley had joined the Jamaica Committee, publicly derided Huxley for his recent attack of "negromania": Hunt to

simply saw no need for the Darwinian bridge that Huxley was intent upon constructing. For one thing, in the context of the ongoing controversy over man's place, Hunt could still make some capital out of demarcating the Anthropologicals from the Darwinians. When the Anthropologicals were forced to hold their own impromptu conference at Dundee in 1867 after the proposed anthropological section of the British Association meeting was cancelled at the last minute for fear of local religious reaction, Hunt could vehemently assure the conservative opposition of the anti-Darwinian stance of his society:

> I will invite those who will persist in attacking us, and endeavouring to raise a feeling of disgust against us, because of our adherence to Darwinism – to earnestly look at the real facts. It they will do so, they will find that if there be one society or one body of men who have more earnestly, more continually, persisted in attacking and endeavouring to refute the doctrines respecting man's origin by Mr. Darwin, or either of his disciples, that body is composed of men calling themselves Anthropologists.[82]

Over and above such rhetoric, Hunt's Knoxian biology and anthropology provided a theoretical basis that not only was more congenial to his racism and polygenism, but also rendered the Darwinian anthropological model proffered by Huxley redundant. Knox could be cited chapter and verse in opposition to Huxley's rejection of specific difference between the races, and invoked to support Hunt's contention that the mental differences between the races were crucial to such determinations (Hunt 1866b, 322, 325–6). As for Huxley's risible suggestion that evidence of human hybrid infertility would establish the truth of the Darwinian hypothesis, a Knoxian was even more disposed to expect such evidence a priori (and, moreover, convinced that its existence was proven), but hardly required the Darwinian hypothesis in order to account for it. Nor was a Knoxian in need of Huxley's Darwinian exclusion of environmentalism. The maxim that each race was specific to its particular locale precluded the possibility of human physical or physiological adaptation to new environments, either in the remote past or in the present, and Hunt had "never yet seen any reason to change my views, which I imbibed from the late Dr. Knox." In view of the contradictions between the

---

Huxley, 6, 12, and 18 October 1866, Huxley Papers, *V. XVIII,* fols. 334–57; Hodgson to Huxley, 3 November 1866, ibid., fol. 201. Hunt subsequently lambasted Huxley in print as "for five years . . . our most deadly, and sometimes even our most bitter, foe" (Hunt 1868b, 77). As Stocking has noted, Hunt was capable of some duplicity in his dealings with Huxley, and on one occasion even apologized to Huxley for lampoons that had appeared in the *Anthropological Review* "at the caprice of the Editor" – i.e., Hunt himself, as subsequent enquiries revealed (Stocking 1971, 382).

82 Hunt 1868b, 77. It is significant that this setback to his British Association aspirations provoked Hunt's most vehement denunciations of Darwinism, and it is noteworthy that the *Dundee Courier,* in reporting Hunt's speech, voiced the "faint suspicion" that Hunt's disavowal of Darwinism had been written "with just a tinge of a desire to suit the latitude and longitude of Dundee" (quoted in Hunt 1868b, 83).

interpretations of Darwin's various disciples, it was Hunt's expressed wish that "Mr. Darwin himself may be induced to come forward" and apply his own theory to the origin and future of mankind (Hunt 1866b, 336, 341).

If the Darwinians could not come to some consensus on these matters, Hunt's own views were settled, and he got on with the more important anthropological task of putting them into practice. Armed with Knox's *Races of Men* ("a little mine of suggestive and interesting thought") and the anthropologist's brief to deliver his expert opinion on the practical applications of his science, Hunt engaged the Anthropological Society in a number of topical political controversies.[83] For above all, the Darwinian competitive struggle for existence between races was superfluous (and hardly new and original) for those who accepted the "inexorable" Knoxian laws of race antagonism and race subordination; with their discovery, Knox ("this great practical anthropologist") had brought the science of man down out of the clouds to its "intimate relations with humanity in religion, politics, government, national conduct, and every department of human action" (Hunt 1868c, 276, 278). The sociopolitical implications of these laws were immense, and Hunt could recognize their clear manifestation in a whole range of contemporary social and political issues. Knox was invoked to support the society's anthropological endorsement of Governor Eyre's bloody repression of black revolt in Jamaica:

> Upwards of fifteen years ago, one of the most eminent anthropologists of the country, declared that there would be a Negro revolt in Jamaica. I quote Dr. Knox's own words. . . . These words appear to the mind of the vulgar prophetic; but they were based on sound theories, ignored by nearly all our then statesmen. . . . The merest novice in the study of race-characteristics ought to know that we English can only successfully rule either Jamaica, New Zealand, the Cape, China, or India, by men such as Governor Eyre.[84]

That perennial political problem, the Irish question, could be expertly settled through the application of Knoxian race laws:

> Ireland has been politically sick, and a number of doctors are fighting and squabbling about the efficacy of their respective drugs while the

---

83 Hunt 1866c, 24, 1866d, 1. Hunt's short-lived *Popular Magazine* lasted only from January to October 1866. Hunt was not only owner and editor, but author of this venture, which was largely devoted to a defence of Eyre on "anthropological" grounds; see Rainger 1978, 62–3.

84 Hunt 1866e, lxxviii. The ASL organized a public meeting in defence of Eyre, at which Captain Bedford Pim (who had been hastily admitted to the society for the purpose) delivered a racist diatribe on "The Negro and Jamaica" to the loud cheers of his audience and their unanimous vote of thanks (Stocking 1971, 379). Huxley, of course, was a noted leader in the liberal attack on Eyre. Some members of the Anthropological Society tendered their resignations over the affair; letters from Bainsford (9 March, 16 April 1866) and Buxton (6, 26, and 29 February 1866), ASL, Letters to the Society, 1865–66, Archives, Royal Anthropological Institute.

patient is dying. – When anyone ventures to hint that the patient is of a different *race*, and that the medical treatment which exactly suits the constitution of Britannia may be most detrimental to Erin, they unite in laughing the suggestion to scorn. When will our medicine men perceive that what suits Saxon England will not suit Celtic Ireland? Let us call in an anthropological doctor. Let Dr. Knox instruct us from his grave.

(Hunt 1868d, 190–1)

Knox was in fact regularly exhumed by that arch-resurrectionman Hunt, and over the years of the Anthropological Society's existence, Knox the erstwhile "savage radical" not only posthumously supported the infamous Eyre and rejected Home Rule for Ireland, but also endorsed British imperialism, became an apologist for slavery, and opposed the extension of the franchise to women and blacks (Hunt 1863, 1866a).

Although the Anthropologicals did their best to monopolize them, the politics of race were not their exclusive preserve. Here too, Huxley made some Darwinian bids for the Anthropological brief. In his 1865 "ethnological" essay, he had made his own tribute to white supremacy,[85] and in the same year he spelled out the racial implications of Darwinism more explicitly in his well known address, "Emancipation Black and White." Here Huxley's ethical "oughts" clashed with the biological "ises" of natural selection. While on the one hand he supported the abolition of slavery on liberal democratic grounds, on the other hand he could reassuringly assert that the innate inferiority of blacks would never endanger white supremacy:

[I]t is simply incredible that, when all his disabilities are removed, and our prognathous relative has a fair field and no favour, as well as no oppressor, he will be able to compete successfully with his bigger-brained and smaller-jawed rival, in a contest which is carried on by thoughts and not by bites. The highest places in the hierarchy of civilization will assuredly not be within the reach of our dusky cousins, though it is by no means necessary that they should be restricted to the lowest.[86]

For all their differing political positions, Hunt and Huxley were in fundamental agreement on a "natural" hierarchy of race, and both put their respective anthropologies to sociopolitical use. While the leading Darwinians, such as Huxley, Galton, and Lubbock, did not engage in the provocative political polemics of the Anthropologicals, they were as prone to offer biologically based moral and

85 To wit "With [the white races] has originated everything that is highest in science, in art, in law, in politics, and in mechanical inventions. In their hands, at the present moment, lies the order of the social world, and to them its progress is committed" (Huxley, "On the Methods and Results of Ethnology," in Huxley [1865] 1908, 104–23, quotation on 114).
86 Huxley (1865) 1870, 25–30; Stepan 1982, 79–80. Huxley made the same point with respect to the higher education of women; see Richards 1983, 92–3.

social guidance. What Weber has called a "moralizing naturalism" characterized the dominant Darwinian tradition from the 1870s on, and she suggests that it was the Darwinians who actually realized Hunt's project of raising anthropology to the level of an applied moral science (Weber 1974, 280).

In the mid-sixties, however, Hunt was still in control of his project, and Darwinism had not achieved dominance. On the basis of Weber's persuasive analysis, I suggest that what was really at stake in the negotiations and confrontations between the "new guard" Darwinians and the Anthropologicals during this period was the struggle for hegemony of the ideological role of anthropology in Victorian society. In a period when traditional theological modes of explanation were giving way before a secular redefinition of the world, the Anthropologicals and the Darwinians offered two competing versions of a legitimating scientific naturalism. Hunt's overt introjection of Knoxian biology and anthropology into politics and social legislation impinged on the Darwinian program, as described by Frank M. Turner, of "relat[ing] the advance of science and its practitioners to the physical, economic, and military security of the nation, to the alleviation of social injustice, to the Carlylean injunction of a new aristocracy of merit, and to the cult of the expert" (Turner 1978, 363). These were all factors in the struggle that Huxley and his fellow scientists with a Darwinian axe to grind were waging to establish science as a profession worthy of middleclass status and rewards. Within science, in the mid-sixties, both the Darwinians and the Anthropologicals were still outside the establishment. Over the next decade, as Roy M. Macleod has put it, the Darwinians – from the epicenter of the influential X Club – "increasingly, *were* the Establishment" (Macleod 1970, 1981, 82). In the process, they shaped and used Darwinism to further their interrelated professional and social interests, and their negotiations and struggles with the Anthropologicals during this period left their lasting impression on the Darwinian anthropological model.

By 1868, the continuing schism between the two societies had become a serious obstacle to Darwinian dominance of this key discipline and an embarrassment to their professional aspirations – in Huxley's words, a "scientific scandal."[87] He accepted the presidency of the Ethnological Society on condition that its Council support his efforts toward amalgamation (Stocking 1971, 382). In the meantime, Hunt and the Anthropologicals had become more amenable to amalgamation by virtue of a decline in their numbers and serious financial difficulties. However, once again the negotiations broke down in bitterness. Over the next few years, Huxley kept pushing the idea of amalgamation in the face of the (not unjustifiable) suspicions of the Anthropologicals that he intended to take advantage of dissension within their society in order to "crush" them.[88] While Hunt's active engagement of the Anthropological Society in topical political controversies

---

87 Huxley to Joseph Hooker, 24 October 1868, Huxley Papers, *V. II*, fol.140; Stocking 1971, 382.
88 Stocking 1971, 383. Peter Martin Duncan had been elected to the Council of the ASL and was actively working on Huxley's behalf to undermine Hunt's authority: Duncan to Huxley, 8 and 25 September 1868; 2 and 16 June 1869, Huxley Papers, *V. XV*, fols. 26–32.

and the provocative free-wheeling discussions of the meetings initially attracted a large and enthusiastic membership, his iconoclasm and the dominance of the society by Hunt and a small inner coterie, the "Cannibal Clique," alienated more conventional members and led to factionalism. From about 1868 on, the society was "plagued by debt, resignations, and internal dissension"; in the same period, the Ethnologicals were reorganized under Huxley's leadership, stepped up their publications, and built up their membership.[89]

As Stocking has pointed out, there was a significant difference in style between the leaderships of the rival societies, which was all to the advantage of the Ethnologicals. While the Darwinians were committed to "one large heterodoxy" and adopted fairly advanced theological positions, they were not inclined to complicate things unnecessarily by flouting the conventions (Stocking 1971, 380–1). The leading Darwinians were solid middle-class Victorians – gentlemen and family men "of complete financial, political and sexual respectability" (Burrow 1968, 4). Their collective respectability was of great advantage in the promotion of unorthodox opinion and their acceptance by the scientific establishment, and Huxley and the entire membership of the X Club capitalized on it. By contrast, the members of Hunt's "Cannibal Clique" – who included the notorious Richard Burton – went out of their way to confront middle-class moral values in their dedicated and fearless pursuit of a science untrammelled by theological or social restraint. Their frank and free discussion of subjects such as phallic symbolism, female circumcision, and the anatomy of the "Hottentot Venus," the ferocity and bad taste of their lampoons of prominent scientists, their anti-missionary crusade (Rainger 1978, 61–2), and their political posturings violated Victorian canons of good taste and propriety. Their unsavoury reputation and the internal dissension within the society contributed to their increasing scientific marginalization.

But perhaps of greater importance than the question of style so much emphasized by Stocking (after all, Huxley could be fairly ferocious and unscrupulous in his rout of "parsondom" and in his dealings with scientific opponents) was the narrower ideological appeal of the Anthropologicals' anthropology. Hunt's head-on confrontation with classical liberalism and political economy, his extreme racism, and his biologization of a range of reactionary political and social positions were out of step with the political and social needs of a rapidly advancing liberal and "progressive" bourgeoisie. He was ideologically outmatched by Huxley's more subtle accommodation of Darwinian anthropology and biology to the contemporary need for a means by which a "fundamentally inegalitarian society based upon a fundamentally egalitarian ideology rationalized its inequalities" (Hobsbawm

89 Huxley to Hooker, 24 January 1868, Huxley Papers, *V. JI,* fol. 140; Stocking 1971, 383. One of Huxley's innovations was to demarcate between special meetings, where "popular" topics could be discussed, and ordinary meetings, which would be for "scientific" discussions, to which "ladies will not be admitted": "Report of the Council," *J. Ethnol. Soc., n. s., I* (1869), viii–xv. He thus demonstrated his concurrence with Hunt on this contentious issue, and removed one of the major obstacles to amalgamation.

1975, 268). As Desmond has noted, 1866–1870 was a time of "growing trade-unionism and demands for reform, suffrage and educational opportunities," and Huxley's "social stratagem" was to bring science, and Darwinism in particular, to the "*stabilization of capitalist society*." Huxley was as aware of the socialist threat as Hunt, and his *Lay Sermons* of this period were designed to naturalize the status quo, to make science the "essential accompaniment of right morality and civil order".[90]

Hunt's sudden death in mid-1869 cleared the way for more harmonious relations between the two societies, but Huxley did not succeed in his objective of amalgamation until the beginning of 1871. Even in their state of decline the Anthropologicals were a thorn in the Darwinians' side. While they held out for greater recognition of their science in any proposed amalgamation, they continued with their mix of politics and anthropology, engaging Huxley in a spirited controversy over the Irish problem, which they persisted in interpreting in terms of Knoxian racial history.[91] Among other diversionary tactics, they disrupted a British Association meeting chaired by Huxley by according a tremendous ovation to his old enemy Richard Owen, merely in order to score points against a furious Huxley.[92] Finally, faced with the inevitable amalgamation, they refused to accept Huxley as president, and Lubbock was installed as a compromise candidate. The leading Anthropological dissidents were not ultimately quelled until several years after amalgamation, but by 1873 the Darwinians were firmly in control of the newly formed Anthropological Institute of Great Britain and Ireland, and Huxley's active involvement in racial matters was over. He turned to other issues, while for the next two decades the Anthropological Institute was led by one after another of the Darwinian "ethnologists": Lubbock, Busk, Lane Fox, John Evans, Tylor, William Flower, and Galton. Significantly, it was Huxley who came up with the name that, as Stocking so succinctly stated, "recognized the science but not the Society" of the Anthropologicals (Stocking 1971, 383; Beddoe 1910, 215–16).

Conclusion

Historians are now generally agreed that the Darwinian recognition and institutionalization of the polygenist position was more than merely nominal.[93] Wallace, Vogt, and Huxley had led the way, and we may add Galton (1869) to the list of those leading Darwinians who incorporated a good deal of polygenist thinking into their interpretations of human history and racial differences (Stepan 1982, 126–8). Eventually "Mr. Darwin himself," as Hunt had suggested he might, consolidated

---

90 Desmond 1982, 158–64. Helfand's rereading of Huxley's later famous Romanes Lecture supports Desmond's interpretation: Helfand 1977. See also Stepan 1982, 80–2.

91 See "Professor Huxley on Political Ethnology," *Anthrop. Rev., 8* (1870), 197–216; Rainger 1978, 64–5.

92 "Owen threw wide the door and entered with nods and wreathed smiles, while his great adversary scowled as if he could kill him" (Beddoe 1910, 212–13).

93 Stepan 1982, 83–110; Stocking 1971, 384–6, 1973, lxx, 1968.

the Darwinian endorsement of many features of polygenism. Darwin's *Descent of Man* was published in the same year that the Anthropological Institute was founded, and it was no coincidence that it was broadly congruent with Knoxian/ Anthropological race science. Recent scholarship has stressed the derivative character of the *Descent*, and Darwin's views on race were clearly influenced by the earlier interpretations of the above-cited Darwinians.[94]

However, although the *Descent* was written in the light of the anthropological struggles of the 1860s, it is essential to acknowledge its origins in Darwin's notebooks of the late 1830s and early 1840s. A good deal of the congruence between Darwinian and Knoxian conceptions of race may be traced back to these early notebook constructions. As these document, Darwin, like Knox, brought to his very earliest conceptions of human evolution a "commitment to the idea of human races as discrete biological units with distinct moral and mental traits."[95] The young Darwin had been concerned with the same sorts of questions on racial biological and cultural differences that preoccupied Knox around the same time, and he was committed to as ruthless a naturalism. Apart from their individual and independent debts to Quetelet's "moral statistics," both Darwin and Knox drew heavily on the general themes of struggle and adaptation in the contemporary "common context" of biological and social thought.[96] Given their common context, the broad general similarities between the Knoxian laws of race antagonism and subordination and the Darwinian struggle for existence between races need occasion no strained historical explanation of direct influence.[97]

Nevertheless, in more explicit ways, the *Descent* does show the conflation of Knoxian/Anthropological and Darwinian racial views, and Darwin located his discussion of these issues squarely within the dispute "of late years" between polygenists and monogenists.[98] His mature views on race were shaped by the contemporaneous confrontations and negotiations between the Darwinians and the Anthropologicals. It is within this context that the minor historical puzzle of Darwin's failure to acknowledge Knox's "generic descent" may be explained. Apart from the difficulties of integration and interpretation of his scattered theoretical

94 Greene 1977; Jones 1978; Durant 1985.
95 Stepan 1982, 51; see also Schweber 1977; Herbert 1977. Darwin's notebooks are transcribed in: Darwin 1974; De Beer 1960–61, 1967.
96 Young 1969; Gale 1972; Shapin and Barnes 1979, 125–42; Greene 1959. For Darwin's debt to Quetelet, see Schweber 1977, 287–93, 1980.
97 Although the connection is tenuous, the young Darwin possibly attended some of Knox's famous Saturday lectures on ethnology while he was studying medicine in Edinburgh in the year from 1826 to 1827, when Knox was at the height of his fame as a lecturer and just before the Burke and Hare affair. Darwin's exposure to transcendental and Lamarckian views via his association with Grant in this period is well known, and Manier has recently stressed the young Darwin's enthusiastic response to romanticism: Manier 1980. See also Sloan 1985.
98 Darwin 1889, 176, 180. As his references indicate, Darwin had read Knox's major works (see above, note 3) and (in spite of his distaste for their racist ideology) was an assiduous reader of the publications of the defunct Anthropological Society.

writings, Knox, through his adoption by Hunt and the Anthropologicals, became identified with anti-Darwinism and therefore with antievolutionism.[99] Moreover, Knox, the disreputable and marginal "savage radical" and lately resurrected and equally unsavoury "Anthropological," was hardly an acceptable "precursor." Yet, paradoxically, it was via the antithetical medium of the Anthropological platform that Knox's race science made an indirect and unacknowledged, but lasting, impact on the Darwinian anthropological model.

In the *Descent*, Darwin argued that racial traits arose very early in the prehistory of man, were not biologically adaptive, and were therefore relatively fixed in character. By viewing race formation as a distant and closed episode of human history, Darwin endorsed the Knoxian categories of race as fixed and unalterable types. Although he thought it irrelevant whether human races were called species or subspecies, he conceded more to the Knoxian view than Huxley by granting that a naturalist confronted for the first time by specimens of Negro and European man "might feel himself fully justified in ranking the races of man as distinct species."[100] Consistent with the Knoxian interpretation, struggle, competition, and survival occurred between racial units rather than between individuals and, in Darwin's view, accounted for the superiority of the Anglo-Saxon and the inevitable triumph of the more intellectual and moral races over the lower and more degraded ones.

Darwin was as insistent as Knox on the biological basis of intellectual and moral differences, and, through his tendency to reduce social and cultural differences to biology, he maintained the essential Knoxian/Anthropological link between race and culture.[101] For above all, the *Descent* did much more than offer a naturalistic explanation of human evolution: it proffered social interpretation, justification, and prescription, and its timely appearance gave a powerful boost to the "moralizing naturalism" of Huxley and Galton, and to Spencer's "Social Darwinism".[102] We may draw a straight line from Knox's "moral anatomy," through Hunt's "anthropology," and on to "Social Darwinism" and the "social surgeons" of the eugenics movement.

The Darwinians did not, of course, owe their tendency to naturalize existing economic and social relations to Knox or to Hunt and the Anthropologicals – they were simply reflecting the same general intellectual trend that had affected Knox and the Anthropologicals as well. And in the larger context, the forces that had

99 Richard Owen's struggle to have himself included in the pre-Darwinian evolutionary roll call may be recalled in this connection, and the establishment and ultra-respectable Owen had a good deal more going for him as an acceptable "precursor" than did Knox. See Macleod 1965; Richards 1987.

100 Darwin 1889, 166–99, quotation on p. 173. Note the reference to Knox in this connection: Darwin characterized Knox as "another firm believer in the specific distinctness of the races of man" (ibid., 168n 5). See also Stepan's excellent discussion of the *Descent* in Stepan 1982, 52–66.

101 I have discussed Darwin's biological determinism within the context of Victorian scientific naturalism elsewhere: see Richards 1983; note also Stepan 1982, 86.

102 Richards 1983, 87–9; Weber 1974, 280; Young 1985; Moore 1986.

created a climate receptive to Knox's racism had intensified: in the seventies, the need to justify white imperialism and class and racial inequalities was greater than ever. Scientific racism no longer appeared an aberration but the very essence of the scientific study of man, taking on a newfound respectability in the "new" evolutionary anthropology. But in more specific ways, through the struggle between the Darwinians and Anthropologicals for scientific and ideological hegemony, Knox's "moral anatomy" was institutionalized and perpetuated in late Victorian scientific racism. In the process, the delicate balance that Knox had maintained between his radicalism and his racism was outweighed by conservative institutional and social needs, and his "moral anatomy" was retooled – first by Hunt, and then by the Darwinians – to fit those needs.

Acknowledgments

I should like to thank Adrian Desmond, Everett Mendelsohn, Philip Rehbock, John Schuster, Sylvan Schweber, and James Secord for discussions and criticism; and the following institutions and libraries for permission to study manuscript material: the British Library, the British Museum (Natural History), the Wellcome Institute for the History of Medicine, the Royal College of Surgeons of England, the Imperial College of Science and Technology, and the Royal Anthropological Institute.

Bibliography

Anon. 1923. "An Hitherto Unpublished Letter by Dr. Robert Knox." *Glasgow Medical Journal* 100: 5.

Appel, Toby. 1987. *The Cuvier-Geoffroy Debate: French Biology in the Decades Before Darwin.* Oxford: Oxford University Press.

Baer, Karl Ernst von. 1828. *Über Entwickelungsgeschichte der Thiere: Beobachtung und Reflexion.* Königsberg: Bornträger.

Barclay, John. 1827. *Introductory Lectures to a Course of Anatomy.* Edinburgh: Mac-Lachlan and Stewart.

Barnes, Barry, and Steven Shapin. 1979. *Natural Order: Historical Studies of Scientific Culture.* Beverly Hills, CA and London: Sage.

Barry, Martin. 1836–37a. "On the Unity of Structure in the Animal Kingdom." *Edinburgh New Philosophical Journal* 22: 116–41.

———. 1836–37b. "Further Observations on the Unity of Structure in the Animal Kingdom." *Edinburgh New Philosophical Journal* 22: 345–64.

Beddoe, John. 1910. *Memories of Eighty Years.* Bristol: Arrowsmith.

Bettany, G. T. 1892. "Robert Knox." In *Dictionary of National Biography,* vol. 31, 331–3. London: Smith Elder.

Biddiss, Michael D. 1970. *Father of Racist Ideology: The Social and Political Thought of Count Gobineau.* London: Weidenfeld and Nicolson.

———. 1975. "Myths of the Blood: European Racist Ideology 1850–1945." *Patterns of Prejudice* 9 (4): 11–19.

———. 1976. "The Politics of Anatomy: Dr. Robert Knox and Victorian Racism." *Proceedings of the Royal Society of Medicine* 69: 245–50.

Blake, C. Carter. 1863a. "Man and Beast." *Anthropological Review* 1: 161.

———. 1863b. "On the Relations of Man to the Inferior Animals." *Anthropological Review* 1: 107–17.

———. 1870. "The Life of Dr. Knox." *Journal of Anthropology* 1: 332–8.

Bloxam, G. W. 1893. *Index to the Publications of the Anthropological Institute of Great Britain and Ireland, 1843-1891*. London: Anthropological Institute.

Broca, Paul. 1864. *On the Phenomena of Hybridity in the Genus Homo*. London: Anthropological Society.

Burke, Luke. 1848. "Criticism: Lectures on the Races of Men, by Robert Knox, M.D., F.R.S.E." *The Ethnological Journal: A Magazine of Ethnography, Phrenology, and Archaeology, Considered as Elements of the Science of Races: With the Application of this Science to Education, Legislation, and Social Progress* 2: 94.

Burrow, John W. 1963–64. "Evolution and Anthropology in the 1860's: The Anthropological Society of London." *Victorian Studies* 7: 137–54.

———. 1966. *Evolution and Society: A Study in Victorian Social Theory*. Cambridge: Cambridge University Press.

———. 1968. "Introduction to Charles Darwin." In *The Origin of Species*, reprint of 1st edn., edited by J. Burrow. Harmondsworth: Penguin.

Canguilhem, Georges et al. 1960. "Du développement à l'évolution au XIXe siècle." *Thales* 11: 25–9.

Chambers, Robert. [1844] 1969. *Vestiges of the Natural History of Creation*. 1st edn., reprint. New York: Humanities Press.

Comrie, John D. 1932. *History of Scottish Medicine*. London: Balliere, Tyndall and Cox.

Cooter, Roger. 1984. *The Cultural Meaning of Popular Science: Phrenology and the Organization of Consent in Nineteenth-Century Britain*. Cambridge: Cambridge University Press.

Creswell, C. H. 1926. *The Royal College of Surgeons of Edinburgh: Historical Notes from 1505–1905*. Edinburgh and London: Oliver and Boyd.

Currie, A. S. 1933. "Robert Knox, Anatomist Scientist and Martyr." *Proceedings of the Royal Society of Medicine* 26: 39–46.

Curtin, Phillip D. 1965. *The Image of Africa: British Ideas and Action, 1780–1850*. London: Macmillan.

Darwin, Charles. 1889. *The Descent of Man, and Selection in Relation to Sex*. 2nd edn. London: John Murray.

———. [1861] 1959. "Historical Sketch." In *The Origin of Species. A Variorum Text*, edited by Morse Peckham, 59–70. Philadelphia: University of Pennsylvania Press.

———. 1974. "M and N Notebooks and Old and Useless Notes." In *Darwin on Man: A Psychological Study of Scientific Creativity*, edited by H. E. Gruber. New York: E. P. Dutton.

De Beer, Gavin, ed. [1960–61] 1967. "Darwin's Notebooks on Transmutation of Species." *Bulletin of the British Museum (Nat. Hist.) Hist. Ser.*, 2 (2–6); 3 (5).

Desmond, Adrian. 1982. *Archetypes and Ancestors: Palaeontology in Victorian London 1850–1875*. London: Blond and Briggs.

———. 1984a. "Robert E. Grant: The Social Predicament of a Pre-Darwinian Transmutationist." *Journal of the History of Biology* 17: 189–223.

————. 1984b. "Robert E. Grant's Later Views on Organic Development: The Swiney Lectures on 'Palaeozoology' 1853–1857." *Archives of Natural History* 11: 395–413.

————. 1984c. "Interpreting the Origin of Mammals: New Approaches to the History of Palaeontology." *Journal of Linnean Society (Zool.)* 82: 7–16.

————. 1985. "Richard Owen's Reaction to Transmutation in the 1830's." *British Journal for the History of Science* 18: 25–50.

————. 1987. "Artisan Resistance and Evolution in Britain, 1819–1848." *Osiris* 3: 77–110.

————. 1989. *The Politics of Evolution: Morphology, Medicine, and Reform in Radical London.* Chicago: University of Chicago Press.

Diamond, Solomon. 1969. "Introduction." In *Treatise on Man*, edited by Adolphe Quetelet. Facsimile Reproduction of the English Translation of 1842, v–xii. Gainesville: Scholars' Facsimile and Reprints.

Durant, John R. 1985. "The Ascent of Nature in Darwin's *Descent of Man*." In *The Darwinian Heritage*, edited by David Kohn, 283–306. Princeton, NJ: Princeton University Press.

Gale, G. 1972. "Darwin and the Concept of a Struggle for Existence: A Study in the Extra-scientific Origins of Scientific Ideas." *Isis* 63: 321–44.

Galton, Francis. 1869. *Hereditary Genius: An Inquiry into Its Laws and Consequences.* London: Macmillan.

Geoffroy St.-Hilaire, Étienne. 1828. "Mémoire où l'on se propose de rechercher dan quels rapports de structure organique et de parenté sont entre eux !es animaux des ages histor-iques, et vivant actuellement, et les espèces antediluviennes et perdues." *Mémoires du Muséum d'Histoire Naturelle* XVII: 209–29.

————. 1829. "Of the Continuity of the Animal Kingdom by Means of Generation, from the First Ages of the World to the Present Times." *Edinburgh New Philosophical Journal* 7: 152–5.

————. 1830. "On the Philosophy of Nature." *Edinburgh New Philosophical Journal* 8: 152–4.

Gode-von Aesch, Alexander. 1966. *Natural Science in German Romanticism.* New York: AMS Press.

Godlee, Rickman J. 1921. "Thomas Wharton Jones." *British Journal of Ophthalmology* 93: 145–81.

Gould, Stephen Jay. 1977. *Ontogeny and Phylogeny.* Cambridge, MA: Harvard University Press.

Greene, John C. 1959. "Biology and Social Theory in the Nineteenth Century: Auguste Comte and Herbert Spencer." In *Critical Problems in the History of Science*, edited by Marshall Clagett, 419–46. Madison, WI: University of Wisconsin Press.

————. 1977. "Darwin as a Social Evolutionist." *Journal of the History of Biology* 10: 1–27.

Haller, John S. 1971. *Outcasts from Evolution: Scientific Attitudes on Racial Inferiority, 1859–1900.* Urbana, Chicago and London: University of Illinois Press.

Harris, Marvin. 1968. *The Rise of Anthropological Theory.* New York: Thomas Y. Crowell.

Harvey, Joy. 1983. "Evolutionism Transformed: Positivists and Materialists in the *Société d'Anthropologie de Paris* from Second Empire to Third Republic." In *The Wider Domain Of Evolutionary Thought*, edited by David Oldroyd and Ian Langham, 298–310. Dordrecht: Riedel.

Helfand, Michael S. 1977. "T. H. Huxley's 'Evolution and Ethics': The Politics of Evolu-tion and the Evolution of Politics." *Victorian Studies* 20: 159–77.

Herbert, Sandra. 1977. "The Place of Man in the Development of Darwin's Theory of Transmutation. Part 2." *Journal of the History of Biology* 10: 155–227.

Hobsbawm, Eric J. 1975. *The Age of Capital 1848–1875*. London: Weidenfeld and Nicolson.

Hodge, M. J. S. 1972. "The Universal Gestation of Nature: Chambers' *Vestiges* and *Explanations*." *Journal of the History of Biology* 5: 127–51.

Hunt, James. 1854. *A Treatise on the Cure of Stammering*. London: Longman, Brown, Green and Longmans.

———. 1863. "On the Negro's Place in Nature." *Memoirs of the Anthropological Society* 1: 1–64.

———. 1864a. "Anniversary Address to the Anthropological Society of London, January 5, 1864." *Journal of the Anthropological Society* 2: lxxx–xciii.

———. 1864b. "Preface to Carl Vogt." In *Lectures on Man: His Place in Creation, and in the History of the Earth*. London: Anthropological Society.

———. 1866a. "Race in Legislation and Political Economy." *Anthropological Review* 4: 113–35.

———. 1866b. "On the Application of the Principle of Natural Selection to Anthropology, in Reply to Views Advocated by Some of Mr. Darwin's Disciples." *Anthropological Review* 4: 320–40.

———. 1866c. "Race Antagonism." *Popular Magazine of Anthropology* 1: 24.

———. 1866d. "Introduction." *Popular Magazine of Anthropology* 1: 1.

———. 1866e. "Anniversary Address." *Anthropological Review* 4: lxxviii.

———. 1867a. "Anniversary Address, January 1, 1867." *Journal of the Anthropological Society* 5: xliv–lxx.

———. 1867b. "On the Doctrine of Continuity Applied to Anthropology." *Anthropological Review* 5: 110–20.

———. 1868a. "On the Origin of the Anthropological Review and Its Connection with the Anthropological Society." *Anthropological Review* 6: 431–42.

———. 1868b. "President's Address." *Anthropological Review* 6: 77.

———. 1868c. "Knox on the Saxon Race." *Anthropological Review* 6: 257–79.

———. 1868d. "Knox on the Celtic Race." *Anthropological Review* 6: 175–91.

Huxley, Thomas Henry. [1865] 1870. "Emancipation Black and White." In *Lay Sermons, Addresses and Reviews*. London: Macmillan, 25–30.

———. [1865] 1908. "On the Methods and Results of Ethnology." In *Man's Place in Nature and a Supplementary Essay*, 104–23. London: Watts.

Jacyna, L. S. 1983a. "John Goodsir and the Making of Cellular Reality." *Journal of the History of Biology* 16: 75–99.

———. 1983b. "Immanence or Transcendence: Theories of Life and Organization in Britain, 1790–1835." *Isis* 74: 311–29.

Jones, Greta. 1978. "The Social History of Darwin's *Descent of Man*." *Economy and Society* 7: 1–23.

Knox, Robert. 1831. *A System of Human Anatomy: On the Basis of the "Traité d'Anatomie Descriptive" of M. H. Cloquet*. 2nd edn. Edinburgh: MacLachlan and Stewart.

———. 1831–32. "Observations on the Natural History of the Salmon." *Report of the British Association for the Advancement of Science*, 587–9.

———. 1832–33. "On the Natural History of the Salmon." *Edinburgh New Philosophical Journal* 14: 397–400.

———. 1833. "Observations on the Natural History of the Salmon, Herring, and Vendace." *Transactions of the Royal Society of Edinburgh* 12: 462–518.

———. 1837a. "Physiological Observations on the Relations of the Heart, and on its Diurnal Revolution and Excitability." In *Memoirs, Chiefly Anatomical and Physiological, Read at Various Times to the Royal Society in Edinburgh, the Medico-Chirurgical, and other Societies*, edited by R. Knox, 1–19. Edinburgh: P. Rickard.

———. 1837b. *Observations upon a "Report by the Select Committee on Salmon Fisheries, Scotland: Together with the Minutes of Evidence, Appendix, and Index"*. Edinburgh: Adam and Charles Black.

———. 1843. "Contributions to Anatomy and Physiology." *London Medical Gazette* 32: 529–32.

———. 1848a. "Lectures on the Races of Men." *Medical Times* 18: 97–9, 114–15, 117–20, 133–4, 147–8, 163–5, 199–201, 231–3, 263–4, 283–5, 299–301, 315–16, 331–2, 365–6; 19: 1–3, 17–18, 33–4, 49–50, 69–70, 121–3, 175, 191–3, 247–8, 315–16.

———. 1848b. "Dr. Knox on the Intermarriages of Jewish Females." *Medical Times* 18: 242.

———. 1850a. *The Races of Men: A Fragment*. London: Henry Renshaw.

———. 1850b. *The Races of Men, A Fragment*. Philadelphia: Lea and Blanchard.

———. 1852. *Great Artists and Great Anatomists*. London: John van Voorst.

———. 1854. *Fish and Fishing in the Lone Glens of Scotland*. London: G. Routledge.

———. 1855a. "Some Remarks on the Aztecque and Bosjieman Children, Now Being Exhibited in London, and on the Races to Which They Are Presumed to Belong." *Lancet* 1: 357–60.

———. 1855b. "Introduction to Inquiries into the Philosophy of Zoology." *Lancet* 1: 625–27.

———. 1855c. "Contributions to the Philosophy of Zoology, with Special Reference to the Natural History of Man." *Lancet* 2: 24–6, 45–6, 68–71, 162–4, 186–8, 216–18.

———. 1855d. "Inquiries into the Philosophy of Zoology." *Zoologist* 13: 4777–92.

———. 1856. "On Organic Harmonies: Anatomical Co-relations, and Methods of Zoology and Paleontology." *Lancet* 2: 245–7, 270–1, 297–300.

———. 1857a. *Man, His Structure and Physiology, Popularly Explained and Demonstrated*. London: H. Bailliere.

———. 1857b. *The Greatest of our Social Evils: Prostitution, as it Now Exists in London, Liverpool, Manchester, Glasgow, Edinburgh and Dublin: An Enquiry into its Cause and Means of Reformation, based on Statistical Documents, by A Physician*. London: H. Bailliere.

———. 1862. *The Races of Men: A Philosophical Enquiry into the Influence of Race over the Destinies of Nations*. 2nd edn. London: Henry Renshaw.

———. 1863a. "Ethnological Inquiries and Observations." *Anthropological Review* 1: 257–8.

———. 1863b. "On the Application of the Anatomical Method to the Discrimination of Species." *Anthropological Review* 1: 263–70.

Lenoir, Timothy. 1982. *The Strategy of Life: Teleology and Mechanics in Nineteenth-Century German Biology*. Dordrecht: Riedel.

Lonsdale, Henry. 1868. "Biographical Memoir." In *John Goodsir, The Anatomical Memoirs*, edited by William Turner. Edinburgh: Adam and Charles Black.

———.1870. *A Sketch of the Life and Writings of Robert Knox, the Anatomist*. London: Macmillan.

Macleod, Roy. 1965. "Evolutionism and Richard Owen, 1830–1868: An Episode in Darwin's Century." *Isis* 56: 259–80.

————. 1970. "The X-Club: A Social Network of Science in Late-Victorian England." *Notes and Records of the Royal Society* 24: 305–22.

————. 1981. "Introduction: On the Advancement of Science." In *The Parliament of Science: The British Association for the Advancement of Science 1831–1981*, edited by Roy Macleod and Peter Collins. Northwood: Science Reviews.

Manier, Edward. 1980. "History, Philosophy and Sociology of Biology: A Family Romance." *Studies in the History and Philosophy of Science* 2: 1–24.

Merz, John T. 1965. *A History of European Thought in the Nineteenth Century*. 4 vols. New York: Dover Publications.

Millhauser, Milton. 1959. *Just before Darwin: Robert Chambers and Vestiges*. Middletown, CT: Wesleyan University Press.

Moore, James R. 1986. "Socializing Darwinism: Historiography and the Fortunes of a Phrase." In *Science as Politics*, edited by Les Levidow, 38–80. London: Free Association Books.

Morrell, J. B. 1972. "Science and Scottish University Reform: Edinburgh in 1826." *British Journal for the History of Science* 6: 39–56.

Nott, Josiah C., and George R. Gliddon. 1854. *Types of Mankind: Or Ethnological Researches . . .* Philadelphia: Lippincott, Grambo; London: Trubner.

Ospovat, Dov. 1976. "The Influence of Karl Ernst von Baer's Embryology, 1828–1859: A Reappraisal in Light of Richard Owen's and William B. Carpenter's Palaeontological Applications of 'von Baer's Law'." *Journal of the History of Biology* 9: 1–28.

Peckham, Morse, ed. 1959. *The Origin of Species: A Variorum Text*. Philadelphia: University of Pennsylvania Press.

Porter, Theodore M. 1985. "The Mathematics of Society: Variation and Error in Quetelet's Statistics." *British Journal for the History of Science* 18: 51–69.

Powell, Baden. 1855. *Essays on the Spirit of the Inductive Philosophy, the Unity of Worlds, and the Philosophy of Creation*. London: Longmans.

Quetelet, Lambert A. J. 1842. *A Treatise on Man and the Development of His Faculties*. Translated by Robert Knox. Edinburgh: W. and R. Chambers.

————. 1848. *Du système social et des lois qui le régissent*. Paris: Guillaumin.

————. 1849. *Letters Addressed to H. R. H. the Grand Duke of Saxe Coburg and Gotha on the Theory of Probabilities, as Applied to the Moral and Political Sciences*. London: Charles and Edwin Layton.

Rae, Isobel. 1964. *Knox the Anatomist*. Edinburgh and London: Oliver and Boyd.

Rainger, Ronald. 1978. "Race, Politics and Science: The Anthropological Society of London in the 1860s." *Victorian Studies* 22: 51–70.

Rehbock, Philip R. 1983. *The Philosophical Naturalists: Themes in Early Nineteenth-Century British Biology*. Madison, WI and London: The University of Wisconsin Press.

Richards, Evelleen. 1976. *The German Romantic Concept of Embryonic Repetition and Its Role in Evolutionary Theory in England up to 1859*. PhD diss. University of New South Wales.

————. 1983. "Darwin and the Descent of Woman." In *The Wider Domain of Evolutionary Thought*, edited by David Oldroyd and Ian Langham, 57–111. Dordrecht: Riedel.

————. 1987. "A Question of Property Rights: Richard Owen's Evolutionism Reassessed." *British Journal for the History of Science* 20: 129–71.

————. 1989. "Huxley and Woman's Place in Science: The 'Woman Question' and the Control of Victorian Anthropology." In *History, Humanity, and Evolution*, edited by James R. Moore, 253–84. Cambridge: Cambridge University Press.

Ross, James A., and Hugh W. Y. Taylor. 1955. "Robert Knox's Catalogue." *Journal of the History of Medicine and Allied Science* 10: 269–76.

Rudwick, Martin J. 1982. "Charles Darwin in London: The Integration of Public and Private Science." *Isis* 73: 186–206.

Russell, E. S. 1916. *Form and Function: A Contribution to the History of Animal Morphology*. London: Murray.

Schwartz, Joel S. 1984. "Darwin, Wallace, and the *Descent of Man*." *Journal of the History of Biology* 17: 271–89.

Schweber, Sylvan S. 1977. "The Origin of the *Origin* Revisited." *Journal of the History of Biology* 10: 284–93.

———. 1980. "Darwin and the Political Economists: Divergence of Character." *Journal of the History of Biology* 3: 195–289.

Shapin, Steven. 1975. "Phrenological Knowledge and the Social Structure of Early Nineteenth-Century Edinburgh." *Annals of Science* 32: 219–43.

———. 1979. "The Politics of Observation: Cerebral Anatomy and Social Interests in the Edinburgh Phrenology Disputes." In *On the Margins of Science: The Social Construction of Rejected Knowledge*, edited by Roy Willis. *Sociological Review Monograph* 27: 139–78.

Shapin, Steven, and Barry Barnes. 1979. "Darwin and Social Darwinism: Purity and History." In *Natural Order*, edited by Barry Barnes and Steven Shapin, 125–42. Beverly Hills, CA and London: Sage.

Sloan, Phillip R. 1985. "Darwin's Invertebrate Program, 1826–1836: Preconditions for Transformism." In *The Darwinian Heritage*, edited by David Kohn, 71–120. Princeton, NJ: Princeton University Press.

Smith, F. B. 1979. *The People's Health 1830–1910*. Canberra: Australian National University Press.

Stepan, Nancy. 1982. *The Idea of Race in Science: Great Britain 1800–1960*. London and Basingstoke: Macmillan.

Stocking, George W. 1968. "The Persistence of Polygenist Thought in Post-Darwinian Anthropology." In idem, *Race, Culture, and Evolution: Essays in the History of Anthropology*, 42–68. New York: Free Press.

———. 1971. "What's in a Name? The Origins of the Royal Anthropological Institute (1837–71)." *Man* 6: 369–90.

———. 1973. "From Chronology to Ethnology: James Cowles Prichard and British Anthropology 1800–1850." In Prichard, James Cowles [1813] 1973 *Researches into the Physical History of Man*, edited by George W. Stocking, ix–cx. Chicago and London: University of Chicago Press.

Temkin, Owsei. 1950. "German Concepts of Ontogeny and History around 1800." *Bulletin of the History of Medicine* 24: 227–46.

Turner, Frank M. 1974. *Between Science and Religion: The Reaction to Scientific Naturalism in Late Victorian England*. New Haven, CT: Yale University Press.

———. 1978. "The Victorian Conflict between Science and Religion: A Professional Dimension." *Isis* 69: 356–76.

Vogt, Carl. 1864. *Lectures on Man: His Place in Creation, and in the History of the Earth*. London: Anthropological Society.

Walker, Alexander. 1834. *Physiognomy Founded on Physiology and Applied to Various Countries, Professions, and Individuals*. London: Smith, Elder.

————. 1841. *Intermarriage; or the Natural Laws by Which Beauty, Health and Intellect, Result from Certain Unions, and Deformity, Disease and Insanity, from Others*. London: John Churchill.

Wallace, Alfred Russel. 1864. "The Origin of Human Races and the Antiquity of Man Deduced from the Theory of Natural Selection." *Journal of the Anthropological Society* 2: clviii–clxxxvii.

Weber, Gay. 1974. "Science and Society in Nineteenth-Century Anthropology." *History of Science* 12: 260–83.

Young, Robert M. 1967. "The Development of Herbert Spencer's Concept of Evolution." *Actes du XIe Congrès International d'Histoire des Sciences* 2: 273–78.

————. 1969. "Malthus and the Evolutionists: The Common Context of Biological and Social Thought." *Past and Present* 43: 109–45.

————. 1973. "The Historiographic and Ideological Contexts of the Nineteenth-Century Debate on Man's Place in Nature." In *Changing Perspectives in the History of Science*, edited by Mikulas Teich and Robert M. Young, 344–438. London: Heinemann.

————. 1985. "Darwinism *is* Social." In *The Darwinian Heritage*, edited by David Kohn, 609–38. Princeton, NJ: Princeton University Press.

4

A POLITICAL ANATOMY OF MONSTERS, HOPEFUL AND OTHERWISE

Teratogeny, transcendentalism, and evolutionary theorising

Monsters have always challenged the boundaries of human identity. Typically, they denote physically or morally deviant states of nature. Monsters are outcasts, "the embodiment of that which is exiled from the self." In both popular folklore and traditional medicine, human monstrosities "grounded discourse on the natural and supernatural, medical and legal, portents and diseases."[1] In fiction, beginning with Mary Shelley's *Frankenstein*, the monster has become a rich and fertile source of images and meanings, a metaphor for the exploration of difference, marginality, and alienation, of the intersection of the terrible with the pitiable and the biological with the technological, of the dualism of nature and culture in Western thought, of alternative possibilities. More recently, feminist theoreticians, notably Donna Haraway, have reinvented the monster as a subversive and potent symbol for a feminist future, an image that may be exploited for its destabilizing and transformative promise, "a regenerative politics for inappropriate/d others." In a related development, John Law, Susan Leigh Star, and others have promoted a "sociology of monsters" as a site for the analysis and contestation of sociotechnical relations and their overlapping dimensions of power.[2]

"Hopeful monsters," it seems, are thriving in the multilayered, boundary-blurred, discursive worlds of feminist postmodernism (or, rather, as Haraway insists, amodernism) and actor/network theory. The biological referent from which the term originated is less viable. Fictional monsters have engendered a large critical literature; biological ones have been less explored by historians and sociologists of science.[3]

---

1 Star 1991, 54; Haraway 1987, 35. See also Park and Daston 1981; Huet 1993. As Haraway reminds us, "*monsters* have the same root as *to demonstrate;* monsters signify"; Haraway 1992, 333, n.16.
2 Haraway 1992, 1987, 1991. On the monster as metaphor in fiction see, e.g., Kranzler 1988/1989; Shaw 1993. On the sociology of monsters see John Law, "Introduction," in Law 1991a, 1–23.
3 It is not without irony that Law lifts the term from a fictional source, Nicolas Mosley's recent prizewinning novel *Hopeful Monsters*. "They are," as one of the characters explains, "things born

As a biological term, *hopeful monster* came to prominence in the writings of the saltationist evolutionist Richard Goldschmidt in the 1930s and 1940s, to be retrieved by Stephen Jay Gould in what Michael Ruse describes as the "second" or "most discontinuous" phase of the theory of punctuated equilibria in the late 1970s and early 1980s (Goldschmidt 1933, 1940; Gould 1980, 1983; Ruse 1993a). The evolutionary legitimacy of monstrous changes or macromutations as discontinuous speciation events became one of the more contentious issues in the ensuing debate between punctuationalists and gradualists. Under attack, Gould pulled back to the "third phase" of punctuated equilibria, where he dissociated "illegitimate" forms of macromutations, or the "sudden origin of new species with all their multifarious adaptations intact *ab initio*," from the theory (Gould 1982, 88–9; Ruse 1993a, 122–3). Hopeful monsters remain, but in a more carefully delineated and circumscribed "legitimate" form.

Ruse (along with others among Gould's critics) has no hesitation in assigning Gould's advocacy of punctuated equilibria theory and the support it has received in large part to politics, both internal and external to science. He identifies an internal political struggle among palaeontologists and other evolutionists, especially geneticists, for cognitive standing in evolutionary theorizing. Palaeontologists, and Gould in particular, want to be seen as making a more useful – perhaps even vitally important – contribution to evolutionary theory than their professional rivals have been wont to concede to them. But the larger and more significant struggle over punctuated equilibria relates to the politics of the external world. Gould's advocacy of punctuated equilibria theory, Ruse claims, is connected with its congruency with the Marxist ideology of dialectical materialism. Ruse goes on to link this claim with Gould's leading role in the contemporaneous battle over sociobiology, which Gould and other critics associated with capitalist libertarianism. The punctuated equilibria theory, Ruse points out, if it is true for humans, "blows holes in any kind of extreme sociobiological thesis".[4]

Nor is this the full story. Underlying these political concerns are the "really deep reasons" for the support of punctuated equilibria theory – its "metaphysical attractions." Punctuated equilibria theory "downplays the significance of natural selection and (especially) adaptation." Their emphasis on functional adaptation is the "key point" for understanding the "thread of continuity" from Darwin to present-day sociobiologists. Gould has been a persistent critic of adaptationism, and, indeed, he belongs to a biological tradition different from the utilitarian adaptationism of Darwin and his fellow thinkers. This is the tradition of German

---

perhaps slightly before their time; when it's not known if the environment is quite ready for them": Mosley 1991, 70, cited in Law 1991a, 1. Haraway rejects the "postmodern" label, agreeing with Bruno Latour that, historically, the "modern," in the sense of the rational and objective pursuit of knowledge, never existed. She prefers the "amodern" label, which implies an active movement away from the "modern," rather than Latour's "premodern" or "non modern." See Haraway 1992. I am grateful to Viviane Morrigan for drawing my attention to this point.

4 Ruse 1993a, 129–31, 143–4, quotation on 131.

transcendentalism, with which Marxism also has intellectual connections. Transcendentalism emphasizes form rather than function, and its biological adherents have included not only Germans like Goethe and Lorenz Oken, but also the French anatomist Étienne Geoffroy St Hilaire and British followers like Richard Owen and Robert Chambers (author of the notorious evolutionary work *Vestiges of the Natural History of Creation*). A number of them were saltationists. Punctuated equilibria theory is situated well within this competing evolutionary tradition or paradigm (Ruse 1993a, 132–9).

My point in dwelling on Ruse's arguments is not necessarily to support the particular claim that there is a Marxist component in Gould's evolutionary theorizing but, rather, to support and extend the larger claims about the transcendental tradition within which saltationists and other hopeful monsters are to be located and about the contingent political ties, both intra- and extrascientific, of evolutionary theorizing in all its forms.[5] More particularly, I want to apply some of the sociological tools and strategies advocated by Law, Star, and Haraway to certain historical monsters, deviant anatomical structures that embodied deviant evolutionary interpretations and deviant politics. My larger historiographic purpose is to draw out the implications of this political anatomy of monsters for both the Darwinian Revolution and the Darwin Industry.

Geoffroy's teratogenic theory of evolution and the politics of the Cuvier-Geoffroy debate

It was the French transcendental anatomists, Étienne Geoffroy St Hilaire and his disciple Étienne Serres, who during the 1820s turned the study of malformations or monsters as anatomical curiosities or *lusus naturae* into the science of teratology. Together they applied the general morphological laws of their *anatomie transcendante* (which had strong affinities with the contemporaneous German *Naturphilosophie*) to teratology, insisting that the formation of monstrosities followed certain fixed and invariable laws. Their guiding principle in teratology was the theory of arrests of development, a concept that had been given wide currency in German Romantic biology and was made anatomically explicit by the anatomist Johann Friedrich Meckel. According to this view, which was identified originally with the embryological law of parallelism (sometimes referred to

---

5 This is, perhaps, a generous interpretation of Ruse's position. It should be noted that Gould himself, in a letter to *Nature* (1981, 289: 742), rejected the charge of developing the theory "as part of a sinister plot to foment world revolution" and represented it as "an attempt to resolve the oldest empirical dilemma impeding an integration of palaeontology into modern evolutionary thought: the phenomena of stasis within successful fossil species, and abrupt replacement by descendants." Nevertheless, he did not deny the "congeniality" of punctuation change and Marxist thought. Note also Ruse's comments on this "retreat," which he attributes to the exigencies of Gould's professional and social commitments at the time and the harm that the "taint" of "controversial ideology" might inflict on them: Ruse 1993a, 143–4. It is worth noting that in the current world political context, the charge of Marxism has lost much of its power either to harm or to promote scientific theorizing.

as the Meckel-Serres law), abnormalities in higher animals represented states of organization that were permanent in lower ones. The majority of abnormalities or monstrosities were thus to be attributed to an arrest of development, so that the organism (strictly speaking, the organ or part) remained fixed at one of those stages through which it ordinarily passed in the normal course of development. In its original form the theory of arrests of development was predicated on the assumption of a linear embryological development that paralleled or recapitulated the uniserial taxonomic series of adult organisms.[6] Geoffroy made this thesis, which he attempted to substantiate via a series of teratological experiments, the backbone of his evolutionary speculations.

In essence, Geoffroy argued by analogy from the production of monsters to the origin of species. He was convinced that he had provoked the hatching of monstrosities from hens' eggs by artificially varying the conditions of incubation and that he could actually relate a particular kind of malformation to a specific disruption in the normal course of development. This meant, according to Geoffroy, that foetal malformations were the result of chance exogenetic or mechanical circumstances and did not, as the conventional view of development would have it, pre-exist in the germ.[7] He was further convinced that he had experimentally demonstrated the way new species arose in nature, and he offered a materialistic account of this process. By extending this experimental model of exogenetic causation of foetal abnormalities to species change, Geoffroy could attribute hereditary changes in the organism to the direct action of the environment. In his view, mechanical and chemical changes in the environment (especially in the respiratory milieu) induced changes in the organism during the embryonic stage that were akin to monstrous development. Through their propagation by inheritance, these embryonic changes brought about the transmutation of species.[8]

In Geoffroy's schema, then, monsters took on an enhanced significance as the harbingers of evolutionary change, and in his writings he referred to them as *êtres ébauchés*, preparatory or precursory beings. Geoffroy also attempted to classify monstrosities according to the supposed cause of malformation and the homologies he identified between malformed structures in higher animals and normal structures in lower ones; he categorized some thirty genera of monsters. Most types of malformations he attributed to foetal adhesions to the placenta which, he hypothesized, would disrupt the blood supply to particular organs or parts, so that their development would be arrested at some stage peculiar to the lower

---

6 As they were conceived in German and French transcendental biology, there was no necessary connection between the theory of arrests of development or the embryological law of parallelism and evolutionary theory. Russell [1916] 1972; Gould 1976.

7 Geoffroy was a committed epigenesist, and the original purpose of his teratological experiments seems to have been to disprove the preformation theory, according to which all the parts of the embryo are preformed or pre-exist in the original primordium or germ.

8 Geoffroy's major evolutionary writings are contained in two memoirs: Geoffroy St Hilaire 1828, 1833. See also Appel 1987, 125–36.

animals. However, the major category of "double monsters" (which included the original Siamese twins, who visited France in the late 1820s) could not be readily compared with lower animals, and for these Geoffroy proposed a different explanation. He pointed out that the union of such double malformations always takes place between homologous parts, and he attributed this to the law of attraction of similar parts, *d'affinité de soi pour soi* (Geoffroy 1833, 85, 1827, 139–41; Appel 1987, 128; Gedda 1961, 125–36).

Geoffroy did not limit the law of attraction of like for like to teratology or embryology but, in Romantic fashion, generalized it into a universal law governing all the phenomena, organic and inorganic, of natural philosophy. Thus the teratological law of *soi pour soi* was universalized into the innate property of all matter; in Geoffroy's mature system, it provided a self-sufficient materialistic explanation for the formation of the solar system and the origin of life (Appel 1987, 181–8).

As Toby Appel has shown in her detailed reconstruction of Geoffroy's famous conflict with his anatomical rival Georges Cuvier, the teleologist and creationist, Geoffroy founded his morphology (which emphasized serial development, parallelism, transmutation, and unity of plan or composition) on the sovereignty of material laws; he directed it against the conservative and elitist Cuvier and toward the young medical reformers and republicans of Paris. Their rancorous debate in the Académie des Sciences was played out against the background of the 1830 revolution, which resulted in the fall of Charles X from the French throne, and the anatomical content of the debate cannot be prised apart from the ideological and political context in which it took place. Cuvier singled out Geoffroy's monster-making experiments for particular attack, deriding the possibility of the creation of a new species by these means while denouncing them for their potential to destabilize not only the science of zoology but also the social order.[9] Geoffroy's contemporaries saw his morphology as contingent upon the progressivist liberal ideology of the 1830 revolution, and their perceptions of the ideological and political dimensions of the conflict shaped their responses to both the man and the issues.

This was no less the case for British anatomists than for the French naturalists who were caught in the middle of the battle. As Adrian Desmond's extensive researches have documented, from the late 1820s on Geoffroy's transmutationist and materialist interpretations of the Romantic law of parallelism and the theory of arrests of development were picked up and widely promoted in the context of a ragbag of radical doctrines that included democratic politics, laissez-faire demands, and doctrines of self-development. This radical version of Geoffroyan anatomy was particularly espoused by the London medical reformers. The environmentalism and the progressive, serial developmentalism of Geoffroy's

---

9 Appel 1987; see also Desmond 1989, 42–56. On Cuvier's concerns about destabilization see Corsi 1988, 249–65.

transmutationism were compatible with their progressive political platform of social and institutional reform. From their institutional bases of London University and the private anatomy schools, the reformist anatomists aggressively deployed transcendental anatomical concepts and Geoffroyan transmutationism against the Tory-Anglican privilege and conservatism of the Royal College of Surgeons and the Royal College of Physicians, seeking to undermine the scientific and moral credibility of their conservative opponents.[10]

While Desmond has identified and analysed the institutional and social threat of radical "monster makers," he has not entered into the specifics of their teratology. His focus is on the paleontological and anatomical conflicts of the rival anatomists and medical men. But Geoffroy's teratological speculations were crucial to his transmutationism and to the radical and reformist goals of his British followers. In the 1840s they were to become even more widely known among the British middle class in the form promoted in the anonymous evolutionary best seller *Vestiges of the Natural History of Creation.* Teratology, as much as palaeontology and comparative anatomy, was a critical site in the contestations among British anatomists in the first half of the nineteenth century. I take as my exemplars the monstrous constructions of two anatomists whose evolutionary theorizing I have studied previously: the radical extramural anatomist Robert Knox; and Richard Owen, doyen of the Hunterian Museum of the Royal College of Surgeons (Richards 1987, 1989). The anatomical careers of both anatomists were profoundly affected by their early exposure to the Cuvier-Geoffroy conflict and by their very good understanding of its ideological and political implications. While both Knox and Owen adapted Geoffroy's teratology to theories of nonlinear saltatory descent, their theories differed significantly in ways consistent with their different politico-institutional positions, and each anatomist reconstituted discrete anatomical abnormalities or monstrosities in conformity with his interrelated theoretical and political needs. In one critical instance the two anatomists constructed divergent representations of the same monsters, the so-called "Aztecque" children who were exhibited in London during the 1850s; these conflicting political anatomies are of particular value to my analysis.

Knox, the "tiger arm," and "savage radicalism"

Robert Knox, who had graduated in medicine from Edinburgh University, studied pathological anatomy in Paris from 1821 to 1822 during the early phases of the Cuvier-Geoffroy conflict. Knox was already an established comparative anatomist with a number of published papers to his credit, and his anatomical understanding and fluency in French put him on familiar terms with both Cuvier and

---

10 Desmond 1987, 1989. Many early nineteenth-century middle-class radicals and reformers objected to the augmentation of state powers and intrusion on individual rights and liberties and espoused laissez-faire and individualist doctrines or what would later be identified as essentially liberal or capitalist politics. See Royle and Walvin 1982.

Geoffroy (Lonsdale 1870; Rehbock 1983, 56–114; Knox 1852). For Knox, life-long opponent of "Kingcraft" and "Priestcraft" and champion of the rights of man, conservative Cuvierian functionalism soon lost out to Geoffroyan morphology and its more congenial political implications.

Knox returned to his native Edinburgh a convert to Geoffroyan transcendental-ism. From around 1826, when he took control of the private anatomy school that he made into the largest and most successful of the Edinburgh extramural schools, Knox preached the new doctrine to his students, who included a number of sub-sequently prominent anatomists and, possibly, Richard Owen himself. By 1842, however, Knox's involvement in the infamous Burke and Hare affair, his aggres-sive participation in the incessant institutional warfare between the university, the Edinburgh Royal College of Surgeons, and the extramural anatomy schools, and, above all, his radical political convictions and lack of "Calvinistic credentials" had combined to bring about his professional and social ruin. He was forced to give up his dwindling and unprofitable anatomy school and to leave Edinburgh, where he could no longer find professional employment. He spent the last twenty years of his life as an institutional and social outcast, earning a precarious living from hack journalism and public lecturing before dying in poverty in London in 1862. During this latter period, Knox elaborated and published what I have termed his "moral anatomy" or race science, which incorporated his teratology and his related evolutionary views.[11]

Knox's moral anatomy was one of the earliest and most comprehensive of the nineteenth-century attempts to biologize social relations, and his popular exposi-tions on race as the key to scientific political theory and social practice made a considerable impact on his contemporaries. He has been described as the "real founder of British racism," and most historians have tended to collapse Knox's politics of race into the discriminatory ambience of late Victorian scientific rac-ism. This is a serious misinterpretation that, as I have argued elsewhere, results from the failure to refer Knox's biology and anthropology to his radical politics and his materialist ideology. Far from deploying his moral anatomy for discrim-inatory purposes, Knox (who was described as a well-known "savage radical" in his obituary in the *Medical Times)* used it to attack colonialism and slavery, denouncing the economic and political repression of non-Europeans and con-demning the Americans for their hypocrisy in refusing to extend the rights of man to the Negro.[12]

11 Richards 1989. Between 1827 and 1828, the grave robbers William Burke and William Hare mur-dered a total of about sixteen people, selling their bodies to Knox's anatomy school. When the murders were uncovered, Knox was implicated in the scandal and was a target of the public dem-onstrations that followed. For an excellent account of the Burke and Hare affair and its ramifica-tions in nineteenth-century medicine and social legislation see Richardson 1987. On the possibility that Owen studied with Knox, see Lonsdale 1870, 280–1; Rehbock 1983.

12 Curtin 1965, 377; Stepan 1982, 41–6. For the obituary see "The Late Dr. Knox," *Medical Times and Gazette,* 27 December 1862, pp. 683–5. On this misinterpretation of Knox's politics of race see Richards 1989; Biddiss 1976.

The "tiger arm" that Knox discovered and dissected in his Edinburgh anatomy school in 1841 was the crucial piece of anatomical evidence for his peculiar and idiosyncratic "savage radicalism" and for his associated conception of "generic descent." Knox's biographer Henry Lonsdale (his former pupil and partner) was present when Knox made his great discovery, and he recorded Knox's excitement and pleasure in his anatomical find. Knox also published his own account of this discovery in the *Edinburgh Medical and Surgical Journal* for 1841 and hinted at its significance for his larger views (Lonsdale 1870, 249–53; Knox 1841). It was all the more significant because, according to both accounts, Knox – reasoning by analogy from the permanent condition of the forelimb of the carnivore – had earlier predicted the existence of this anomalous structure in humans.

Knox had been presented with a "very fine specimen" of a jaguar or American panther that had died suddenly while being exhibited in Edinburgh. During the course of his dissection of the jaguar, Knox had drawn the attention of his students to the presence in the humerus of a foramen or hole, characteristic of the carnivora, through which the median nerve and humoral artery pass on their way to the elbow. Knox was aware that this foramen had been reported as occurring in "certain Bosjemen [Bushmen] skeletons"; he presumed that it would be found to be formed by a supracondyloid process (a hook-like process of bone completed by a band of ligament near the lower end of the humerus) and an arterial and nervous arrangement "similar to what prevails so often in the carnivora." Soon after his dissection of the jaguar, Knox discussed this possibility in one of his lectures on race delivered before the Anatomical and Physiological Society, and he showed the members some specimens of human humeri that he and Lonsdale thought to have vestigial supracondyloid processes. "At the time I made this statement," wrote Knox in 1841,

> I was quite aware that it was opposed to what we know of the usual course of the artery and nerve; but I felt assured, that either *the analogy* drawn by most German anatomists between this rudimentary process in man and the much larger one found in the carnivora, was well founded or it was not; and if the former, that the course of the nerve and artery would be accommodated to this new osseous arrangement and follow the law of *deformation* as strictly as our structures obey usually the law of formation. Now, it happened by a singular coincidence, that within two days after this communication was made to the society, a test case actually occurred to bring the opinion to the test.

The left arm of a stout, muscular middle-aged man was being dissected, and Knox's assistant called his attention to "an anomaly corresponding strictly with what I had predicted two nights previously to the Anatomical Society, viz. the presence of a supra-condyloid process . . . *and the consequent deviation of the main artery and median nerve from their usual course, in order to pass behind this process in the groove, half osseous, half ligamentous, thus formed . . .* in short, the precise anatomy of the arm of the Jaguar."[13]

Knox's "tiger arm," as it was dubbed, went on prominent display in his dissecting rooms early in 1841 and, according to Lonsdale's account, attracted a great deal of professional interest with which Knox was "mightily pleased." For comparative purposes, the dissected carcass of the jaguar, with its perfect supracondyloid foramen, lay on a nearby table, while on yet another table Knox had arranged some specimens of monkeys obtained from a traveling menagerie, "some with and others without, the carnivorous type of humerus." Knox was clearly intent on demonstrating as thoroughly as he could what he regarded as his transcendental triumph over natural theology. The tiger arm, he wrote in his 1841 report, was "startling proof of the strength of the analogies subsisting occasionally between man and the lower animals, and an undeniable argument of the existence of a general scheme or plan embracing the whole range of the animal kingdom, even to a minuteness of detail quite inconceivable by those who adopt the views and physiology of the Bridgewater Treatises" (Lonsdale 1870, 251; Knox 1841, 128).

That particularly British phenomenon, the Bridgewater Treatises, which were issued during the 1830s, embodied the dominant functional teleological view that each anatomical structure had been specially created or designed to fit its particular purpose or function and that, taken together, such purposive structures were cumulative illustrations of the argument from design. Cuvier's work was a major resource for socially and theologically credentialed British anatomists and physiologists such as Sir Charles Bell, who was commissioned to author the treatise that dealt with the anatomical evidence for design. For Cuvier, any resemblances between the structures of higher and lower animals were attributable to the similar functions they performed under their similar "conditions of existence." It was this "Cuvierian mania" to which Knox specifically counterposed his tiger arm. Such anomalous monstrous structures defied Cuvierian functional analysis and were a source of considerable embarrassment to those committed to the notion of perfect adaptation and final cause. Over and over in his writings, Knox brandished his tiger arm in the face of his "Bilgewater" opponents, challenging them to explain this "singular coincidence" in functionalist terms, to invoke their "vile patchwork" of "expedients and contrivances." For Knox, the explanation of this anomaly could reside only in the "fixed, unalterable, eternal laws" of nature, not in the *ad hoc* contrivances of the British Cuvierians. As he explained in greater detail in 1843 in the *London Medical Gazette*, such deviations were to be subsumed under the

---

13 Knox 1841, 125, 126–7 (Knox's emphasis). Knox spent a period of duty as an army surgeon in South Africa before he went to Paris to study. He began his collection of crania at this stage and, given his early and continuing interest in race, eagerly collected any relevant material and information on racial characteristics he came across for his famous Saturday morning public lectures on "Comparative and General Anatomy and Ethnology," which continued to attract large and enthusiastic audiences even after the Burke and Hare affair. See Richards 1989.

"great law" of transcendental anatomy, the law of unity of organization, and were to be accounted for as cases of imperfect development, *arrêt de développement*, and referred to the regular structures of animals lower in the scale: "these are the only views which, in the present state of science, can be adopted, and . . . what has been written against them in France, and more especially in England, is simply, and, to use the mildest phrase, ingenious nonsense; sometimes very pompous and imposing, as in the Bridgewater Treatises, but still downright nonsense, and not meriting the smallest attention from any philosophic mind" (Knox 1843, 501, 1850, 29–35; Ospovat 1978; Yeo 1986; Desmond 1989, 198–9).

Initially, therefore, Knox's tiger arm served overt anti-teleological purposes. As he represented it, this discrete anatomical preparation embodied or concretized his philosophical and ideological objections to the functional Cuvierian explanation of anatomical structure. As Lonsdale's account makes explicit, Knox had exhibited the tiger arm in his dissecting rooms in such a way as to enforce comparisons between this structure and those of lower animals and to point up the inability of the teleologists to account for the occurrence of what was properly an animal structure in a human being. Its uniqueness and significance, both as object and as symbol, were profound. It was for its discoverer the "most remarkable, perhaps, of all deviations in human structure." When Knox left Edinburgh, the tiger arm accompanied him on his peripatetic lecturing circuit of London and the provinces. And, along with its anti-teleological and transcendental significance, Knox impressed the singularity of this prized possession upon his audience: "I do not believe that any similar preparation to the one now before you exists in Britain" (Knox 1843, 529, 1850, 442).

Furthermore, to emphasize the fact that he had actually *predicted* the occasional presence of this anomaly in human beings, reasoning by means of the transcendental laws of unity of organization and arrests of development from the existence of the rudimentary supracondyloid process on the human humerus, Knox had made two detailed drawings that he published with his 1841 report. The first of these depicted the tiger arm, displaying the foramen and the anomalous arrangement of the humoral artery and median nerve to the best advantage by means of a judiciously placed pin; the second showed a humerus with a rudimentary supracondyloid process (Figure 4.1). These published drawings were iconic representations of the predictive power of the new anatomy. They also served the important function of securing a wider professional anatomical audience for this emblematic preparation.

It must be acknowledged, however, that to a certain extent Knox's early deployment of the tiger arm for anti-teleological purposes was so much tilting at windmills. By the time of his momentous discovery such anomalies were in the process of being assimilated to the exigencies of British natural theology via the appropriation of certain transcendental conceptions, notably the notion of an ideal archetype, which permitted less stress on the criterion of perfect adaptation. The transcendental precept of unity of plan was transposed into a natural theological "unity of design," and monstrosities could thus be referred to the general

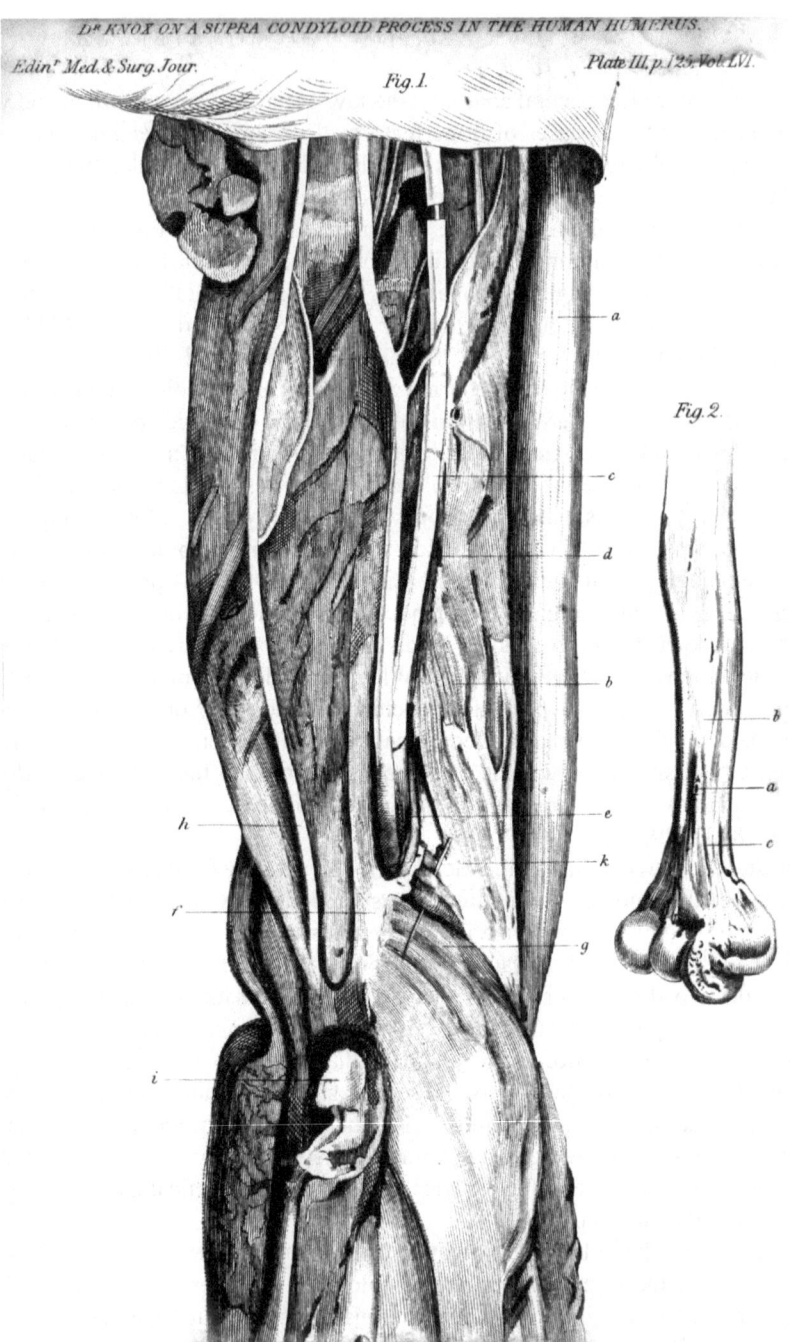

Figure 4.1 Knox's "tiger arm", as depicted in the *Edinburgh Medical and Surgical Journal* (1841). The supracondyloid process is indicated by (f), the foramen or groove by (e), and the consequent deviation of the main artery and nerve by the pin (k). The humerus alongside the main drawing shows a rudimentary supracondyloid process at (a) with a bony groove for the deviated artery and nerve at (c) Knox 1841, 124. University of Sydney Library, Rare books and Special Collections.

vertebrate model or archetype and accounted for through their conformity with the Creator's providential design and continuing stabilizing presence in nature and the social order. Such an interpretation could accommodate the theory of arrests of development and the analogies between human monstrosities and the structures of lower animals. Indeed, by the early 1840s a version of the associated transcendental embryological law of parallelism had become the leading argument for premeditated design among British special creationists or "progressionists" and had even found its way into the Bridgewater Treatises. In the late 1840s this reconciliation of Geoffroyan morphology with Cuvierian functionalism under the auspices of British natural theology was formalized by Richard Owen. The real threat for transcendental anatomists like Knox was not so much that the new anatomy would be suppressed by the dominant teleologists, as that it would be expropriated to an ideologically objectionable teleology. This was a danger of which Knox was well aware, and he poured his contempt upon the "school of *low transcendentalists*" such as Owen, "nibbling, but with great and becoming caution, at the transcendental doctrines," who think that a "portion of [these] views may be admitted without causing scandal, or risking their positions with Orthodoxy and Oxford".[14]

Knox also stood apart from those other British anatomists who, around the same time, were promoting Geoffroy's transcendental and transmutationist views in an overtly materialist form and in a reform-oriented political context (Desmond 1989). While Owen was working to recapture transcendentalism from the reformist anatomists, Knox took it into new radical territory.

Although Knox had been involved in attempts at institutional reform in Edinburgh, during his most intellectually productive period, from 1842 to around 1855, he was an "outsider" with no institutional affiliations. He was a marginal figure whose professional activities were circumscribed by his poverty, the notoriety of his earlier involvement in the Burke and Hare affair, and his unorthodox religious views and radical politics.[15] While he seems to have had some contact with the London medical reformers – and he would certainly have been sympathetic to their materialistic and naturalistic platform – Knox's lack of institutional or professional affiliations meant that he could play little effective part in their campaign. In any case, by this stage Knox had come to reject the meliorative power of reform. There were many radicalisms in the early nineteenth century, some of which were violently in opposition to one another. Knox's own peculiar

14 Knox 1850, 428, 435; see also Desmond 1989, 335–72; Richards 1990; Yeo 1986; Bowler 1976, 47–62.
15 In 1848, following yet another scandal involving the wrongful certification of one of Knox's supposed former pupils at Edinburgh, the Edinburgh Royal College of Surgeons withdrew his teaching qualification and formally notified the Royal College of Surgeons of London of its action. This scandal, and Knox's involvement in it, was publicized by the *Lancet* (the major literary forum of the London medical reformers) as part of the campaign for medical reform. This publicity must have further compromised Knox's professional reputation and his relationship with the London medical reformers. See Rae 1964, 131–61; *Lancet,* 1847, 1: 565–71, 630, 653–4, 685.

political views were fundamentally anti-reformist and nihilistic, and he attacked the sanitation and welfare programs of the reformers (Royle and Walvin 1982, 10; Richards 1989, 384–91). Through his insistence on the universality and inevitability of natural law and his subsequent rigid biological determinism, Knox had become committed to the proselytizing of a radical antienvironmentalist, antiprogressivist ideology. And here the symbolic tiger arm played a crucial role.

In his later writings on race Knox made it clear that the tiger arm embodied and signified much more than his mere endorsement of the transcendental laws of unity of organization and arrests of development. In his 1843 lecture on the tiger arm he had hinted that the doctrine of arrests of development "requires very considerable modification before it can be applied extensively to the history of organization." From the late 1840s on he began to spell out this "considerable modification" in the context of his polemical public lectures and writings on the politics and anatomy of race. Here he indicated his dissent from Geoffroy's "loose views" and from the "formula of Meckel" (i.e., the theory of arrests of development). Knox explained that "Meckel's doctrine" had never satisfied him, even when he was forced to teach it "for want of a higher generalization," and that he was "now prepared to refute it" and replace it with another. It was his discovery of the tiger arm and his failure to relate this "most remarkable" of all human deformities to any consistently present embryological structure in humans that gave him the empirical basis to reject the theory of arrests of development: "As no such structure is constantly observable in the human embryo, it cannot be that such a variety in man is an arrest of development, since such a structure, as a constant law, does not exist." If the tiger arm were really an arrest of development, it should be observable in the normal course of human embryological development. But it was not always present. This meant, according to Knox, that deformities in general were not to be interpreted as arrests of development, but as "retrogressive developments" toward other forms. It was by this means, through the action of the "deformating powers of nature," that Knox accounted for the origin of new species. His "ultimate aim," Knox declared, was to offer a "scientific explanation" of the appearance and extinction of species, "thus restoring to legitimate science that branch of philosophy which the theory of successive creations, invented by Cuvier and still maintained by his followers, had clearly removed from it" (Knox 1843, 501, 1850, 442, 1852, 63, 1855a, 4842, 1855b).

In brief, Knox posited a common material origin of all life and its evolution by a process of saltatory descent. His scheme of "generic descent" supposed the embryos of all members of the same natural family or genus to be essentially similar and to contain within themselves all the "characters" or incipient structures of all the possible different species of that genus. New species arise through the action of the "law of deformation," which opposes the "law of specialization" (Knox's laws of development were a materialistic variant of the transcendental principle of polarity) and therefore suppresses the formation of "some parts of the organ or apparatus already existing in the generic being" or embryo. These "deformations" or monstrosities are constantly generated. Those that are not "viable" perish; those that are compatible with existing geographic and geologic conditions reproduce and increase in number, and so a new species is established. The human races (which Knox regarded as separate species) were the result of such monstrous change.[16]

Although Knox usually confined his evolutionary theorizing to the origin of species from a particular genus or family, he explicitly assumed a common hereditary descent for the different genera, including that of humanity. He viewed the human embryo as a microcosm of all life, "worm, mollusc and fish," and saw humanity as consanguineous with all other animals that have lived or may live (Richards 1989, 398–9). Thus, a structure peculiar to a different natural family, such as the tiger arm, might occasionally appear as a deformity in humans, a testimony to their common descent.

In other words, Knox followed Geoffroy in arguing by analogy from the production of deformations or monstrosities to the production of species. But there were some important distinctions. Knox's rejection of the theory of arrests of development via the all-important tiger arm meant that he rejected Geoffroy's related conception of a unilineal or progressive transmutation of species. Knox was most insistent about this, repeatedly denying "any transmutation of species, the one into another." For Knox the "law of generation" or "descent" of new species was "generic" or embryonic, and he explicitly rejected any notion of form change in the mature species, thereby affirming the "fixity" of species. As I shall show, this was a necessary inference of his "moral anatomy" or race science. Species, in this sense, were immutable, being fixed for all time in the embryo, and they did not emerge in any necessarily progressive way. This accounted for the radical discontinuities in the fossil record and the series of existing animals and was consistent with Knox's radical rejection of the ideology of progress (Knox 1855b, 45; Richards 1989, 404–6).

Knox also departed from Geoffroy's conceptual scheme by totally excluding the possibility of environmentally induced change in the embryo. In Geoffroy's scheme, environmental changes acted directly on the embryo to bring about the transmutation of species. In Knox's scheme of "generic descent," deformations resulted from the action of the innate laws governing embryological development, and the environment could exert its influence only indirectly on these incipient species once they were produced, eliminating the nonviable and permitting the survival and perpetuation of those that were compatible with the "material conditions of the external world" (Knox 1855b, 45, 1855c). So Knox did not relate the occasional appearance of the tiger arm in humans to any changes in external conditions, but explained it in terms of endogenous causation via the materialistic

---

16 Knox 1850, 432–3. Knox supposed all organisms to be influenced in their development by two antithetical principles, one being the law of deformation that tends toward the retention of the embryonic form, and the other the law of specialization or formation that leads to the formation of the individual. According to Knox, new species originate as the combined result of these two inherent laws and the external conditions of existence. See Richards 1989, 396–404.

law of deformation. This exclusion of environmentally induced change was also essential to his moral anatomy.

Knox's developmentalism served overt anti-creationist and naturalistic purposes. But, more than this, his insistence on "generic" change and his associated rejection of environmentally induced progressive or unilineal species change or "transmutation" (for which the tiger arm was his most significant piece of evidence) brought Knox's biology into line with his racism and his radical ideology. His theory of generic descent allowed him to reconcile the seemingly contradictory unity of origin of the human races (and therefore their common humanity and entitlement to the rights of man) and their distinction as separate species (and therefore the permanency of race and its fixed and unalterable role in determining the character and behaviour of the different human races).

Knox's moral anatomy was an attempt to extrapolate anatomical method to society, to reduce all social and political phenomena to the basic biological category of race. According to Knox, the mental and moral differences between the human races were as distinctive and unalterable as their anatomical and physical differences, and both sets of phenomena were to be explained by the same natural causes or laws. These laws rendered human morality as accessible to the scientist as human anatomy. The diversity and complexity of all social phenomena could thus be explained scientifically and made utterly predictable through the rigorous application of these fundamental laws. For Knox, human character, intellect, and morality were neither divinely bestowed nor environmentally produced, but were rooted in the "all-pervading, unalterable, physical character of race" (Knox 1850, 35; Richards 1989, 390–6).

Just as Knox had predicted the occasional existence of the tiger arm in humans, so he could claim to have predicted the 1848 revolution. This he viewed as an inevitable Europe-wide racial struggle, as the various oppressed races fought to throw off their alien tyrants and reinstitute their own governments and laws in accordance with their innate racial predilections. Knox's moral anatomy also enabled him to set natural limits to imperialist expansionism. As each race or species had been unalterably shaped to its particular locale and climate, it followed that European colonists could survive for only a limited time in tropical countries, and only by enslaving or oppressing the indigenous races (with whom, being separate species, they could not hybridize) and by constantly replenishing the European stock through immigration. Sooner or later, natural law must inevitably assert its effects, and the oppressors would be exterminated, either through their inability to adapt to their "unnatural" environment or through "natural" and inevitable racial conflict. Knox could thus confidently (and approvingly) predict the ultimate defeat of the colonizing and oppressive Europeans by the tropically adapted and fierce Negro. Knox's rejection of unilineal transmutation and promotion of "generic descent" also permitted his rejection of the conventional hierarchy of races and enabled him to rebut the charge that he meant to "disparage" any race: "The white races are not the more fully developed, and the negro the more imperfectly developed, species of one common natural family. The development of each is perfect in its own way – equally so."[17]

Knox's biology was constitutive of his anomalous ideology of radical racism, and he deployed it for both naturalistic and radical purposes. His nonprogressivist version of saltatory descent cannot be dissociated from his radical politics, his racial determinism, and his materialist ideology, and his reconstituted tiger arm encapsulated and embodied these ideological needs.

Owen's "conserving reform" and the craniopagus skull

In 1868 the erstwhile "British Cuvier," Richard Owen, was facing his professional eclipse by the triumphant Thomas Henry Huxley. With bitter hindsight and a strong sense of betrayal, he looked back to the Cuvier-Geoffroy conflict and reassessed the foundations of his anatomical career. He conceded the crucial role of Geoffroyan conceptions in his own morphology, and he explicitly invoked the political and ideological considerations that his own earlier writings had helped to obscure and that had initially led him to acclaim Cuvier as his "master." Cuvier, he now finally admitted, in the context of the Darwinian takeover of London science, had had the "advantage of subserving the prepossessions of the 'party of order' and the needs of theology".[18]

The "party of order" was the name the French gave to the union of conservative and formerly moderate forces with the old regimes against the revolutionary forces of 1848 (Hobsbawm 1975, 17). Owen's use of the expression is significant. It indicates his well-developed perception of French politics and, by implication, of English political developments and the relevance of his own work to them. For more than thirty years Owen had moulded his anatomy to similar conservative ends. But by the 1860s his conservative anatomy, and the English "party of order" and natural theology that it subserved, were becoming redundant in the context of a growing secular naturalism and the capitalist ideology of a new party of order. Belatedly, Owen attempted to dissociate himself from the creationist label with which the Darwinians insisted on tagging him and to articulate and find an audience for his long-held views on organic descent – views that had been inspired by Geoffroy's teratological speculations, but that he had earlier muted in response to public and professional criticism.

To the young and ambitious Owen, who first visited Paris in 1831 in the aftermath of the Cuvier-Geoffroy debate and the 1830 revolution, the choice of Cuvier

---

17 Knox 1855b, 26. Although he refused to rank races, Knox, in common with most of his contemporaries, assumed the biocultural inferiority of the "dark races." On inevitable Europe-wide racial struggle see Knox 1850, 22; on the ultimate defeat of colonialism see ibid., 243–4, 456; Richards 1989, 402.

18 Owen 1868, 3: 814. It is significant that Owen depicted Louis Pasteur's triumph over Felix Pouchet in the same terms. See Richards 1987, 156.

as anatomical mentor was ready made. Owen was employed by the conservative Royal College of Surgeons of London (at that stage under siege by the medical reformers) as assistant curator of the Hunterian Collection and, Desmond has argued, was both professionally and ideologically committed to the defence of establishment interests against the social threat posed by the Geoffroyan transmutationism of the medical reformers. During the 1830s Owen promoted himself as Britain's answer to Cuvier, and he structured his anatomy and paleontology in opposition to Geoffroy's uniserial transmutationism, thus securing the patronage of the powerful Oxbridge scientific network and Peelite conservatives and establishing his highly successful anatomical career.[19]

However, Owen was by no means a thoroughgoing teleologist in the traditional Cuvierian sense. From the early 1840s on we may discern in his writings a cautious advocacy of the position that God's design was manifest in nature's uniformity and that creation had to be explained by "natural laws or secondary causes" rather than by divine intervention. As well, Owen had a highly developed critical understanding of the limitations of Cuvierian functionalism in explaining structural similarities or homologies that do not serve similar functions. His well-known response to this teleological problem was to reconcile Geoffroyan transcendental anatomy with Cuvierian typology and teleology, to attempt to dissociate it from its radical, materialist ties and assimilate it to a conservative, non-materialist framework. At the same time, from around 1837 on, Owen was cautiously feeling his way toward a non-materialist theory of saltatory descent in which Geoffroy's teratological speculations were adapted to these same conservative institutional and social interests. But while the "conserving reform" of Owen's revised Geoffroyan morphology was welcomed by his conservative colleagues and influential Oxbridge patrons and his archetype theory was readily subsumed within natural theology, Owen's tentative efforts to give voice to his related views on saltatory descent were censured.[20]

In 1849 Owen's evolutionism, cautious as it was, was brought up short against the larger social forces of the day. The ambiguous evolutionary hints of his recently published *On the Nature of Limbs* were publicly denounced, along with the popular evolutionary work *Vestiges of the Natural History of Creation* and the recent translation of the major transcendental text of the German *Naturphilosoph* Lorenz Oken (which had been published by the Ray Society at Owen's instigation). All were condemned in the pages of the *Manchester Spectator* for promoting a

---

19 Desmond 1985, 1989. Phillip Sloan disputes this interpretation, arguing that Desmond overstates the alleged conflict between Owen and Robert Grant (the leading radical exponent of Geoffroyan transmutationism at the University of London) and is not sufficiently sensitive to the intellectual components of Owen's anatomical theorizing: Sloan 1990, 425–9. See also Sloan 1992. But see Michael Ruse's review of the latter for a rebuttal of Sloan's claims: Ruse 1993b.

20 The expression "conserving reform" comes from Desmond 1985, 50. On Owen's developing sense that creation must be explained naturally, see Owen 1849, 85–6; Ospovat 1978, 36; Desmond 1982, 29–48, 62–4.

"desolating Pantheism" in the form of the "THEORY OF DEVELOPMENT" that was undermining religious belief and contributing to the contemporary political and social unrest. Such views were not only suspect but downright dangerous in a context of home-grown Chartist agitation and Europe-wide revolution. The ambitious Owen reacted to these "hard epithets" (which were subsequently reinforced by Adam Sedgwick, the most vituperative critic of *Vestiges* and all such godless schemes of development) by forcing his evolutionism underground. He emerged openly as an evolutionist only when the more liberal and prosperous secular climate of the 1860s and the appearance of *The Origin of Species* made less hazardous the voicing of his (by then socially and intellectually outmoded) views (Richards 1987, 161–7).

Owen's major difficulty lay in the closeness of his views to those of the heretical and all too widely read *Vestiges*. It was the *Vestiges*, first published anonymously in 1844 and reissued in at least twelve subsequent English editions, that brought Geoffroy's teratological theory of evolution before the middle-class British public to an unprecedented extent. Its author, Robert Chambers, the Edinburgh publisher and essayist, superimposed Geoffroy's teratological mechanism onto a divergent von Baerian embryological model of evolutionary development. In Chambers's version, the length of the gestation period was the critical factor, with premature birth resulting in an arrest of development and monstrosity, while prolonged gestation led to an advance of development and transmutation of species. Owen dissented from the transmutationist and unilinear implications of Chambers's evolutionary mechanism, but he approved the general divergent embryological model of evolutionary development, and he endorsed Chambers's invocation of a "higher generative law," or preordained deviation from the normal course of development, as the ultimate teleomechanism of evolutionary change.[21]

When Owen finally made his own evolutionary views public in the 1860s, his "derivative hypothesis" incorporated the same divergent embryological model and the teleomechanism of a higher generative law. But Owen's theory stressed the role of preordained *premature* or "monstrous" birth, rather than prolonged foetation, for the origin of new species. This was a significant distinction that may be traced back to 1837: a notebook entry by Charles Darwin records that Owen had suggested to Darwin that the "production of monsters (which Hunter says owe their origin to very early stage and which follow certain laws according to species), present an analogy to production of species."[22] At that time Owen was engaged in cataloguing the Hunterian Collection and manuscripts in the Royal College of Surgeons, and his edited collection of John Hunter's papers on animal

---

21 Richards 1987. On the influence of *Vestiges* see Secord 1989. Like its forerunner, the law of parallelism, Karl Ernst von Baer's law of development (1828) assumed a correlation between ontogeny and taxonomy, but von Baer conceived both these latter in terms of divergent differentiation. See Russell [1916] 1972.
22 Charles Darwin, Notebook B, MS p. 161, in Darwin 1987, 210. Hereafter this work will be cited by the relevant notebook; corresponding pages in Barrett *et als.'* edition will appear in parentheses.

physiology, published in the same year, offers some important clues to Owen's early views on saltatory evolution.

As I have reconstructed matters in "A Question of Property Rights: Richard Owen's Evolutionism Reassessed" (Richards 1987), Owen seized upon Hunter's attribution of the production of monsters to the "condition of the original germ," and he counterposed this explanation and Hunter's authority to the politically suspect Geoffroy's exogenetic or environmental explanation of the production of monstrosities. By appropriating Hunter's endogenous explanation of teratological change and the evidence Hunter adduced in its support, Owen was able to oppose Geoffroy's materialistic emphasis on external causation and to validate his own teleomechanism of divinely preprogrammed evolutionary change in the embryo. According to Hunter, all foetal malformations are inherent in the germ, so all saltatory change must be preordained and not referable to any chance external causation (such as changing environmental conditions), as Geoffroy had alleged. The chief empirical evidence for Hunter's endogenous explanation of teratological change was put forward in what Owen described as "one of the most remarkable laws of aberrant formations" and reproduced as follows: "I should imagine that monsters were formed monsters from their very first formation, for this reason, that all supernumerary parts are joined to their similar parts, as a head to a head, etc. etc." I suggested that Hunter's law, embodied in the many instances of such "double monsters" in the Hunterian Collection, gave Owen the necessary empirical basis for confronting Geoffroy's teratological experiments and their exogenetic or materialistic interpretation. As well, Hunter's emphasis on the "very early stage" of the origin of foetal malformations (as Darwin reported) enabled Owen to dissociate them from the embryological law of parallelism and Geoffroy's associated lineal transmutationism and to bring such monstrous deviations into line with Karl Ernst von Baer's law and divergent evolutionary development.[23]

I want now to take this suggestion further and locate Owen's early evolutionary speculations in a particular double monster, the celebrated craniopagus skull of the Bengali boy. My evidence is circumstantial, for the characteristically cautious and secretive Owen wrote nothing directly on this significant monster. But we do have an indication of Owen's interest in this particular monstrosity from the published account of Carl Gustav Carus, the German comparative anatomist, craniologist, and exponent of *Naturphilosophie*.

In 1844, the same year in which the *Vestiges* burst upon the scene, and while Friedrich Engels was documenting the appalling conditions of the English working class, Carus made a splendid progress through England and Scotland in the retinue of Frederick Augustus, King of Saxony. Carus, who was acting as physician-in-ordinary to the Saxon king, met not only most of the British aristocracy, but also most of the leading British natural philosophers and anatomists,

---

23 Hunter 1837, xxvi. Owen discussed Hunter's "important" laws on monstrosities in much the same terms in his Hunterian Lectures of 1837: Lecture 4, 9 May 1837, in Sloan 1992, 185. See Richards 1987, 148–50.

including Owen, and he wrote a valuable ethnomethodological account of the condition and primitive beliefs of the strange tribe of British anatomists at this time. Carus's own sympathies lay with the plight of the British factory workers, especially the child workers, and their "dreadful" and "inhuman" conditions of employment. Although he enjoyed his "genuine English dinner" at Cambridge and the postprandial session where he passed the port to William Whewell and various other establishment luminaries, he made some scathing observations on the general state of British natural philosophy, where "every truth is decidedly repulsed, which is calculated to promote such a free spirit of inquiry or mental development, as might in the smallest degree interfere with or touch upon, any traditional, political, or orthodox ecclesiastical dogma."[24]

There were, however, some few British anatomists who were receptive to the "more philosophical mode of thinking and investigation" of the "*Natur-Philosophie* of the Germans." "Owen pleases me thoroughly," Carus wrote of their first meeting, where Owen introduced him to the riches and rarities of the Hunterian Collection: "a sensible, able man – deeply versed in what is old, and ready for the reception of what is new." They met together on several occasions, on one of which Carus measured Owen's head for his craniological tables and found that Owen's cranium measured up to the standards of "a truly distinguished scholar." On another occasion, in return for Carus's account of his system of "cranioscopy," Owen offered his own views of some "remarkable formations of skulls in his collection": "The most remarkable was a monstrous formation from India, in which another skull was joined to the head of a child in such a manner, that the two crowns were united. . . . The bony parts of the two united skulls were in Owen's hands, and we considered attentively this extraordinary malformation."[25]

My second piece of circumstantial evidence for the significance of the craniopagus skull to Owen's evolutionary views I have inferred from Owen's interestingly indiscreet letter of 1848 to the rationalist publisher John Chapman. In this letter Owen boasted to Chapman that he knew of at least "six possible secondary causes of species" and that "transmutation of species in the ascending course" was the "least probable of the six":

> When I remarked to the author of "Vestiges," the last time he visited the
> museum, how servilely the old idea had been followed by . . . Lamarck,
> and the author of "Vestiges" – viz. of "progressive development" – and
> that there were five more likely ways of introducing a new species, he

24 Carus 1846, 35. Carus made a point of visiting the anatomist William Lawrence, who had "allowed himself to be frightened" by the clergy and had given up his earlier work on physiology and psychology because of its supposed materialist implications (ibid., 88). See also Neuburger 1953.
25 Carus 1846, 60–2, 93–4. Carus's work was well known to Owen. His textbook on comparative anatomy was used as a source by Joseph Henry Green, Owen's mentor in the Royal College of Surgeons, and Owen may have derived his concept of the vertebrate archetype from Carus's theoretical work on the same. See Sloan 1992, 38n.

asked suddenly and eagerly, "What are they?" I declined to give him the information, but shortly after brought prominently under his notice the facts that might have suggested one, at least, of the more likely ways. He saw nothing of their bearing, and I shall refrain from publishing my ideas on this matter till I get more evidence.

(Rev. R. Owen 1894, 1: 309–10)

Given the central role that "anomalous monstrous births" played in Owen's early and mature speculations on the origin of species, and the uniquely available collection of monstrous productions in the Hunterian Museum, it is more than likely that these were the relevant "facts" that Owen tantalizingly placed under the very nose of the presumed author of the *Vestiges*. My inference that the celebrated craniopagus skull figured prominently on this occasion is strengthened by the probable identity of Owen's unnamed visitor: this was most likely the phrenologist George Combe, whom at this stage Owen regarded as the *Vestiges* author.[26] As a leading phrenologist, Combe may be safely presumed to have had a special interest in the museum's collection of crania. What more inevitable than that Owen should gratify this interest by bringing the symbolic craniopagus skull "prominently under his notice"? And, given the convolutions of Owen's transcendental-inspired theorizing, what more inevitable than that Combe (as Owen clearly expected) would make nothing of this complex clue?

The craniopagus skull of the Bengali boy was one of the most famous anatomical curiosities of Hunter's museum. It had been presented to Hunter by Everard Home and was the subject of two separate papers by Home in the *Philosophical Transactions* for 1790 and 1799. Home had obtained the monstrous skull through the offices of the East India Company; he gave a detailed account of the child's appearance while alive, with accompanying drawings (Figure 4.3), and, subsequently, of the crude postmortem examination that had been carried out by the agent of the East India Company who prepared the skull (Figure 4.2). Home's papers and illustrations were substantially reproduced by Owen in the 1831 volume of the catalogue of the contents of the Hunterian Museum that dealt with the "preparations of monsters and malformed parts."[27]

The celebrated craniopagus skull was the exemplary, if not the explicit, instance of Hunter's "most remarkable" teratological law that "all supernumerary parts are joined to their similar parts, as a head to a head, etc." Furthermore, the endogenous

---

26 See also Owen's second letter to Chapman, Rev. R. Owen 1894, 1: 310–11. Desmond has identified Owen's correspondent as Chapman and his visitor as Combe: Desmond 1982, 29, 210–11. He has suggested that these significant specimens bore on the phenomena of alternation of generations (ibid., 35). This possibility cannot be discounted; see note 30.

27 Home 1790, 1799. The history of Home's craniopagus skull is discussed in Bondeson 1988. I am indebted to Elizabeth Allen, curator of the Hunterian Museum, for this reference, and to Jan Ammerstadt for his translation of Bondeson's paper. Owen reproduced much of Home's work in Owen 1831, 68–74.

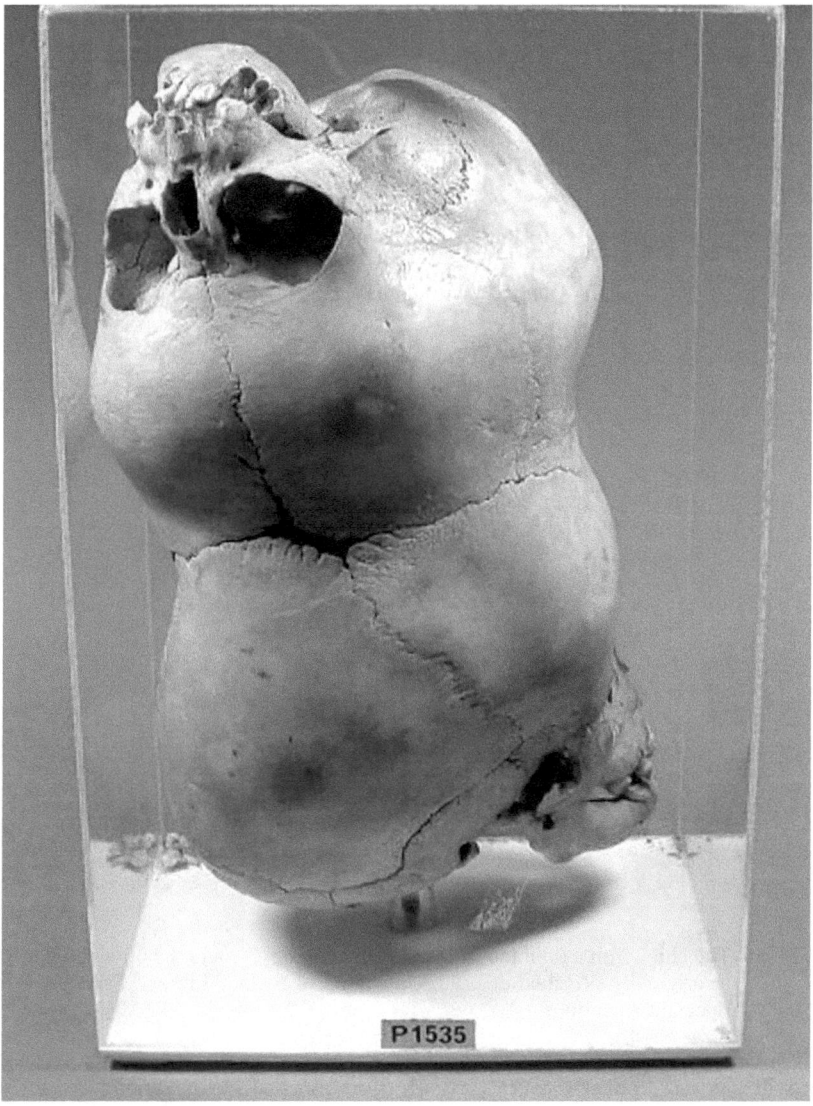

Figure 4.2 The craniopagus skull in the Hunterian Museum. Reproduced by permission of the Hunterian Museum of the Royal College of Surgeons.

origin of this exemplar of double monsters was certified via Home's account, reproduced by Owen in the Hunterian *Catalogue*. As Home stated, the mother of this unfortunate boy had declared to her East India Company interrogator "that no circumstance whatever of an uncommon nature had occurred: she had no fright,

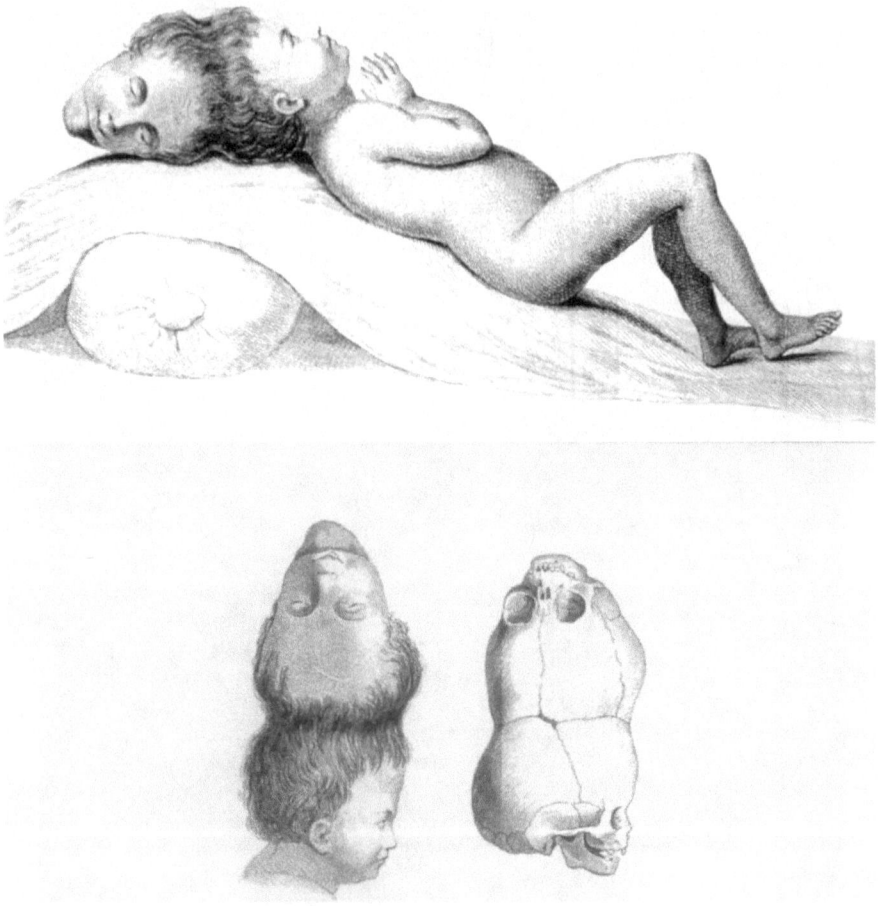

Figure 4.3 The child with a double head – the "Bengali boy" – and the craniopagus skull, as originally sketched in 1790. University of Sydney Library, Rare Books and Special Collections.

met with no accident, and went through the period of her pregnancy exactly in the same way as she had done with her other children."[28] This contradicted Geoffroy's assertion of a necessary correlation between an accident, disturbing what was until then a normal pregnancy, and the production of a monstrous offspring.

This leads me to the next (and this time quite conjectural) reason for the significance of this particular craniopagus skull to Owen. As we saw, Geoffroy too had

28 Owen 1831, 73. The boy was exhibited by his parents and died around the age of four years from the bite of a cobra.

been especially interested in the occurrence of double monsters. He had argued that the union of such double malformations always takes place between homologous parts and had proposed the law of attraction of similar parts or *soi pour soi*, which teratological law he had generalized into the materialistic law governing the formation of all organic and inorganic phenomena. The famous craniopagus skull posed a particular problem for Geoffroy's law of *soi pour soi*, for, as the accompanying figures show, the two heads are turned in "almost opposite directions, so that the left frontal bones of one articulate by suture with the right parietals of the other, and vice versa."[29] In other words, the heads are *not* joined by their homologous parts. However, they still satisfy Hunter's general rule that "all supernumerary parts are joined to their similar parts, as a head to a head." More to the point, the articulation of the craniopagus skull is consistent with Owen's theory of the vertebrate archetype and of the laws governing its development, which he was, at the time of his discussion of this monstrosity with Carus, in the process of formulating.

Owen's major theoretical work, *On the Archetype and Homologies of the Vertebrate Skeleton*, was published in 1848. This work, which owed a great deal to Oken's *Naturphilosophie* and Geoffroy's transcendental anatomy, contained Owen's vertebral theory of the skull and his adaptation of the Romantic principle of polarity to his synthesis of Geoffroyan morphology and Cuvierian teleology. Like Knox, Owen supposed that the development of all life was controlled by two antagonistic forces. In Owen's interpretation, the "specific organizing principle" is the teleological force, which shapes the living thing to its functions. It is opposed by the "general polarizing principle," which brings about the repetition of parts (or what Owen called "irrelative repetition") and similarity of forms – that is, all the signs of unity of organization and of the archetype. The extent to which the teleological principle overcomes the general polarizing force is an index of the grade of the species. These antagonistic forces not only control individual development or ontogeny, but also the development of life on earth (Owen 1848, 171–2, 1860, 506–7). As I have indicated, Owen was to hint at this process of evolutionary development in his popular exposition of his archetype theory, *On the Nature of Limbs*, in the following year.

In his mature theory of saltatory descent, Owen related the origin of a new species to the action of the specific teleological principle and illustrated this by reference to the fossil series he had established in 1851 – the development of the modem single-toed horse from the three-toed Paleotherium via the Hipparion, with its greatly reduced inner and outer toes. Owen attributed the three toes of the Paleotherium to the action of the general polarizing force, which brings about repetition of parts. The development of extra toes is suppressed or arrested through the agency of the teleological principle, which thus brings about the birth of the

---

29 See *Descriptive Catalogue of the Pathological Series in the Hunterian Museum of the Royal College of Surgeons of England*, Pt. 2 (Edinburgh/London: Livingstone, 1972), 234.

Hipparion and ultimately, through further arrests of development, the single-toed modern horse. Owen's "very significant" evidence for this was the occasional birth of a modern horse with the "supplementary ancestral hoofs." "In relation to actual horses," he wrote, "such specimens figure as 'monstra per excessum'; but, in relation to [the Hipparion] they would be normal, and those of the present day, would exemplify 'monstra per defectum.'" Evolutionary change thus occurs by preordained monstrous deviations or arrests of development, which Owen related to the premature birth of a *monstra per defectum*.[30]

All this goes to show that Owen, in his 1844 discussion with his sympathetic auditor, the *Naturphilosoph* Carus, may well have explained the craniopagus skull in the terms of his emerging archetypal theory and related laws of development – as an instance of a *monstra per excessum*, where the action of the specific teleological principle had been overcome by the transcendental "general polarizing force" that had produced this excessive growth by repetition of parts (or "vertebrae," according to Owen's vertebral theory of the skull) and thus this craniopagus or doubleheaded monster. Such "supernumerary" parts would be united, as Hunter had asserted, but there was no necessary union between strictly homologous parts, as Geoffroy's materialistic law of *soi pour soi* supposed.

There was therefore no need, as Owen wrote around this time (1846), to resort to "quite gratuitous systems of progressive transmutation and self-creative forces," as had Geoffroy and his radical materialist British followers. Rhetoric aside, we can see how well Owen's own teratological evolutionary theory might have served conservative institutional and social interests. I have pointed out elsewhere the ideological potential of such a theory. Owen's teleomechanism excluded all external causation, and organic development could proceed only according to a preconceived divine plan. All organic change was therefore inherently lawful and conservative – nothing new or radical could arise (Owen 1846, 147; Richards 1987, 157).

I have also argued that had it not been for the dangerous proximity of Owen's views to those expressed in that continuing source of anathema, the *Vestiges*, Owen might have been less cautious in voicing his related opinions and his established conservative audience might well have been more receptive to them. But Owen's "conserving reform" made him vulnerable to conservative attack. He could not associate himself with what they continued to perceive as an ideologically objectionable and politically dangerous theory of development without alienating his conservative backers. His public lambasting in the *Manchester Spectator* was a valuable political lesson for Owen, and for the next decade he retreated before it into a circumspect silence and a policy of self-censorship. He

---

30 Owen 1868, 3: 794–7. This emphasis on premature birth, I have suggested, is consistent with the relation Owen drew between such monstrous births and saltatory mutation in vertebrates and the phenomena of alternating generations in invertebrates. These latter phenomena had also deeply interested Owen in the 1840s, as his 1849 monograph *On Parthenogenesis* testifies. See Richards 1987, 145–50.

was a closet evolutionist who could find no audience for his views – and whose anatomical eminence indeed depended upon their suppression. I suggest that it was largely for this reason that the ultra-cautious Owen did not permit himself the intellectual luxury of theoretically developing the teratological argument for his "derivative hypothesis" – and why the craniopagus skull, which embodied so many of his earlier evolutionary conceptions, was never publicly deployed in their promotion. The famous craniopagus skull was (quite literally) a closet symbol, the well-known property of the ultraconservative Royal College of Surgeons, and Owen lacked the autonomy to subvert it to his own purposes. By the time Owen tried flounderingly in the 1860s to pull his earlier teratological views together, he had a more compelling model to hand in the form of the horse with the rudimentary supplementary toes – a model he could relate to his own intellectual property, the beautiful fossil series of horses that he had established in the 1850s. But by then it was already too late. Owen's fossil series of horses was in the process of expropriation to the Darwinian creed, shortly to be paraded by Huxley as the "demonstrative evidence" for evolution by natural selection.[31]

Owen, Knox, and the "Aztecque" children

In order to appreciate to the full the necessity of viewing the teratology of Knox and Owen in interrelated institutional, political, and ideological terms, it is instructive to compare their divergent interpretations of the so-called "Aztecque" children. This illustration is all the more cogent because of the overt similarities between the theories of Knox and Owen. As I have shown, both anatomists looked to teratology for the aetiology of the production of new species and viewed monsters or "deformations" as incipient species. Both adapted Geoffroy's teratological speculations to theories of nonlinear saltatory descent. Both anatomists dissented from Geoffroy's environmental explanation of teratological and evolutionary change and instead adopted endogenous explanations of these interrelated phenomena. Both incorporated versions of the transcendental principle of polarity or antagonistic forces into their theories in order to explain both individual development (or ontogeny) and the development of life (or phylogeny). But where Knox insisted on the materiality and non-teleological character of these endogenous developmental forces, Owen was just as insistent on their non-material and teleological origins.[32] Both sought to substantiate their views by reference to

---

31 Desmond 1982, 165–9. Another reason, which I have discussed elsewhere, for Owen's failure to develop his teratological argument was his major morphological preoccupation during this period with establishing the archetypal phenomena. Speculations on evolutionary processes were subsidiary to this dominating concern. See Richards 1987, 150–2.

32 These fundamental ideological differences are well brought out in Knox's critical response to the *Vestiges* (to which, as I have argued, Owen responded sympathetically). The *Vestiges,* stated Knox, "jumbled up the theory of human progress with the theory of development; its critics, the church and colleges, compelled its anonymous author to seek refuge in the doctrine of final cause;

discrete anomalous anatomical structures or monstrosities, Knox to the tiger arm and Owen to double monsters, most likely the craniopagus skull. I have argued that their divergent constructions and deployments of these monstrosities were not entirely based on biological phenomena but to some extent were motivated by and embodied their divergent political ideologies and social objectives. Their divergent reconstructions of the same set of "monsters," the "Aztecque" children, represent a critical instance supporting this contention.

These children, a boy and a girl, were exhibited in London by an enterprising showman during the 1850s as living representatives of the long-lost "Aztecque" race. They supposedly had been kidnapped, in stirring and dangerous circumstances, from an ancient Aztec city in Central America where they were the last remnants of the priestly Aztec caste, held in veneration by the local tribespeople. The children were abnormal in a number of ways, but their chief peculiarity lay in their compressed crania and in their facial features, which were held to be the same as those depicted in the ancient Aztec sculptures and carvings of Central America. Their presence in London excited a good deal of anatomical and ethnological interest, and both Knox and Owen examined the children and reported upon them.

The very different circumstances of their examinations and reports are themselves indicative of the different politico-institutional positions of the two authors and of the different audiences they were addressing. Owen carried out his private examination of the children at the invitation and under the auspices of the conservative and religious-oriented Ethnological Society of London, and he reported his findings and conclusions to a special meeting of the society in 1853; his report subsequently featured in the society's *Journal*. Knox (whose claims to ethnological expertise were very much better than Owen's) made his observations of the children without benefit of institutional backing. Knox's account of the "Aztecque" children was published in 1855 in the chief forum of the London medical reformers, the *Lancet*.[33]

Owen's report was a sober account of the dentition and other anatomical and physical features of the children. For Owen, the "chief and most striking characteristic of both children" was "due to the abnormal arrest of development of the brain and brain-case, which gives them the character of hemi-cephalous monsters." In the more leisurely and privileged circumstances of his examination of the children, he made careful cranial measurements in support of this contention.

---

a doctrine which the whole scope of the work repudiated. The doctrines of Geoffroy were in this work misstated, to serve a purpose": Knox 1850, 27.

33 See Owen 1853; Knox 1855c. Knox, indeed, may not have examined the children at all. He gave very little anatomical detail and seems to have drawn on Cull's and Owen's already-published accounts. See Richards 1989, 410–12. When the marginal and "savage radical" Knox first attempted to join the Ethnological Society, in 1855, he was blackballed by the dominant Quaker clique; he did not manage to gain entry until 1860, after the structure and orientation of the society had undergone some significant changes.

Owen's conclusion was that these "so-called Aztecs" were "instances of exceptional arrest of development, not representatives of any peculiar human race." He included in his published report an engraving of the skull of the "idiot" preserved in the museum of St. Bartholomew's Hospital, describing it as the "skull which offers the nearest approach to the peculiar form exhibited by the so-called Aztecs," and an engraving of the profile of the boy, Maximo, for comparative purposes (Figure 4.4). Owen also carefully dissociated the children from any comparison with the "brute," that is, the anthropoid apes, arguing that in "all the essential characters" they were "strictly human."[34]

Knox agreed about the essential humanity of the children and the anatomical and developmental gap separating them from the apes. Nor was he taken in by the myth of their fabulous Aztec origins. He wasted little time on anatomical details; his account introduced the children in the context of a general discussion of his law of generic descent and its application to race. Knox made the point that his concept of the "generic embryo" allowed for the sudden reappearance of previously extinct species or races if the appropriate deformations and the environmental conditions that could sustain them recurred. He related the physiognomy of the children both to the Central American carvings and to a "form of idiotic head, which occasionally appears in all the races of men." In short, for Knox, the "Aztecque" children represented a "race now extinct *as a race*, reproduced in these children by the law of 'interrupted descent.'" In support of this, Knox drew on the account of the mixed racial composition of the inhabitants of Central and South America "as described by that talented and acute observer, Mr. Darwin." The "blood of the ancient races of America, now extinct," would still be present in such a racial admixture, and it could therefore reappear from time to time, just as "Negro blood, once introduced into a family, will reappear in the descendants, after some hundred years, without any new infusion." Finally, Knox made it clear that this law of "interrupted descent" was to be assimilated to the more fundamental physical law of "generic descent":

> The reappearance of the species after a lapse of time is an event in no shape wonderful or miraculous, such being provided for in Nature's grand scheme, the varied races of men being but the product of physical laws, acting on an organization, the young – generically perfect, pliable, adaptive – above all, including within it all the forms which the natural family is destined to assume when developed and specialized in time and space.
>
> (Knox 1855c, 359)

---

34 Owen's clinching evidence for rejecting the Aztec origins of the children was an extract from the *Philadelphia Bulletin* for 13 July 1852, which contained a deposition from a "Mexican of Spanish origin," born in San Salvador, to the effect that he was the father of the two children in question; see Owen 1853. Knox also accepted this account of the children's origin.

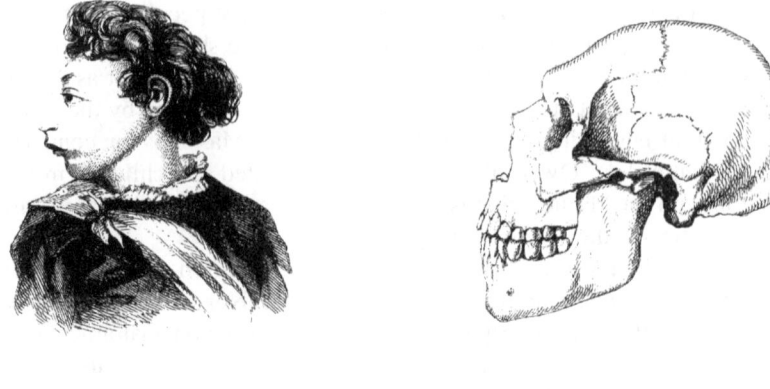

Maximo.
so-called Aztec boy.

Skull of an Idiot.
preserved in S.t Bartholomew's Hospital.

Figure 4.4 The "Azteque" boy, Maximo, and the skull of an idiot. *Journal of the Ethnological Society of London* (1856). University of Sydney Library.

The "Aztecque" children were not the harbingers of the reestablishment of the Aztec race in Central America; the cataclysmic geographical changes necessary for the reestablishment of the race had not taken place. But their "monstrous" reappearance by the law of "interrupted descent" was evidence of the higher-level law of generic descent and of the non-teleological and physical – that is, materialistic – character of both laws.

Knox's interpretation of the "Aztecque" children was entirely consistent with his radical materialist version of the teratological theory of saltatory descent. Owen's opposing view of these same children, his refusal to relate these "instances of exceptional arrest of development" to any extinct human race, was equally consistent with his ideologically conservative teleological version of saltatory descent. Where Knox's materialistic interpretation of generic descent logically implied the reappearance of extinct species should the appropriate conditions recur, Owen's teleological conception of the process required his rejection of such "gratuitous . . . selfcreating forces." Once a species had become extinct, it could not reappear. This was a possibility Owen had explicitly excluded as early as 1837, when he was first speculating on the monstrous origin of species. "The species must disappear with the extermination of the reproductive individuals," he then wrote, "for the genus has no power to reproduce the species." I suggest that Owen's extension of this teleological veto to the human races was the outcome of his earlier ideological objection to any materialization of the endogenous forces controlling both the ontogenetic and phylogenetic development of species.[35] Had he so wished, Owen might have speculated openly (with proper respect for biblical authority) about the Aztec or other racial origin or reappearance of these children, for monogenism (the derivation of the human races from a single common stock) was the fundamental premise of the dominant religious ethnological model of the period.[36] That he did not choose to do so is indicative of the ideological and theoretical constraints within which Owen structured his conceptions of the "Aztecque" children and other such monstrous productions, and of their continuing significance for his conservative teratological theory of saltatory descent.

The meaning of the monster

I think we gain a better understanding of evolutionary theorizing in the period under consideration (from the early 1820s to the late 1860s) if we view it in terms of a sociology of monsters. John Law has argued the necessity of retaining the slippery concept of power for such a sociology. Following Michel Foucault, Law stresses that power is more than a negative instance whose function is repression or domination. It is a productive network that runs through the social body. It produces things – pleasure, knowledge, discourse. Both "power over" and "power to" are crucial to the analysis of the social relations of science and technology. "Power in all its forms . . . is about the way in which objects are constituted and linked together." Susan Leigh Star has urged the need to describe the power in the positions of actors and actants, especially excluded or invisible ones, as *both* insiders and outsiders. Any analysis of power with respect to science and technology must examine the multiple memberships, marginalities, and exclusions, the complex relationships of actors with standardized conventions within and across communities of shared social practices or worlds. Metaphors are bridges between those different worlds. "Power is about *whose* metaphor brings worlds together, and holds them there. (Law 1991b, 185; Starr 1991, 52; Foucault 1980, 119). The metaphor of the monster – the outcast, the stranger, the marginal being who lives

---

35 R. Owen, first draft of May 1837 lecture, "M.S. of Lectures," Royal College of Surgeons, MS 67.b.12.E, fol. 20; see Lecture 5, 11 May 1837, in Sloan 1992, 222. I am indebted to Adrian Desmond for this reference. Unlike Knox, Owen believed the human races to constitute a single species. But Owen also flirted with the polygenist and racist Anthropological Society of London. See his references to the Aztec children and to the Neanderthal skull at the meeting of the Anthropological Society for 7 April 1863: *Anthropological Review*, 1863, 1: 187–91. Compare these views with those of Knox published posthumously in the same volume: Knox 1863; see also Richards 1989, Sect. II.

36 Thus Richard Cull, the secretary of the Ethnological Society, who preceded Owen's report with a historical account of the Aztec race, was open to this possibility in the case of the "Aztecque" children. Cull merely dissented from the alleged Aztec origin of the children's physiognomy, arguing for the Toltec origin of the Central American sculptures in question and relating the children's facial features to these. See Owen 1853.

in multiple worlds without delegation – is a particularly powerful one for making sense of the glue that holds bodies, entities, texts, and other material and social arrangements together.

To the emphasis on power for a sociology of monsters, I would add the indispensability of the even more slippery concept of ideology. Again, ideology is not to be understood simply in its classical Marxist sense as entailing the suppression of contradiction, or as false consciousness or "bad" science as opposed to truth or "good" science; I intend it, rather, more in the critical sense associated with the writings of Foucault on power, as relational and productive. Understood in this way, ideology is indissolubly linked with power in the sense elaborated by Law and Star. Ideologies, shared or contradictory systems of belief, locate actors and actants in the continuum of power that makes or breaks scientific knowledge in the different communities of practice within the shifting social order of the larger culture or society.[37]

A sociology of monsters, to begin with, permits us to shift focus from the great man and great idea to the "other" evolutionists of the nineteenth century, that is, to outsiders like Knox and Owen who failed to establish a professional scientific or wider audience for their evolutionary views. At the same time, it offers us a new opportunity for understanding and critiquing what we have come to take for granted: the privileged epistemological status of Darwinian evolutionary theory. A sociology of monsters also allows us to focus on the actants as well as the actors. In this case the actants are the historical monsters, the anatomical objects reconstituted and deployed for anatomical and larger political ends by Knox and Owen. These historical monsters thus may be understood at one and the same time both as anatomical objects and as the embodiments of different strategies of power pursued by the two anatomists against the dominant teleologists and special creationists who held the socially recognized authority to speak and act in anatomical matters. The tiger arm, the double monster of the craniopagus skull, and the "Aztecque" children were, in a very real sense, political anatomies.[38]

Viewed in these terms, Knox and Owen were themselves monsters, outsiders who partially inhabited a number of networks or worlds of power relations that simultaneously enabled and repressed or marginalized their particular teratological constructions and evolutionary ideas. Knox was more truly monstrous, an institutional and social outcast, a "kind of unviable hybrid . . . a radical and a

37 As Robert Young argued more than twenty years ago, ideology constitutes an "inescapable level of discourse" in the production, acceptance, and application of evolutionary biology: Young 1971. For a review of traditional approaches to ideology see Larrain 1979.

38 Cf. Armstrong's use of the term, following Foucault: Armstrong 1983, 2. Unlike Armstrong and Foucault, I am not concerned with abstract dealings in what Foucault has termed "conditions of possibility," or at what point a "thought became thinkable," but rather with concrete historical events wherein anatomical objects were politically reconstituted and deployed for particular political or ideological purposes. Although he does not use the specific term *political anatomy*, my approach is closer to that of Steven Shapin in his analysis of the role of anatomy in the Edinburgh phrenology disputes: Shapin 1979; see also Barnes 1977, 30.

racist."[39] His institutional autonomy enabled him to articulate his full-blown evolutionary schema, but his idiosyncratic ideology was out of step with the London-
based anatomists whose materialist and naturalistic platform was otherwise most
compatible with his own. Given their strong ideological and political commitment
to social and professional reform, Knox's attempt to replace their reforming and
improving Geoffroyan transmutationism with a theory that rigidly excluded environmentally induced change and emphasized radical non-directional saltation was
unlikely to recruit support from this particular network of power relations. The
social power available to Knox lay primarily outside the field of anatomy, in the
popular forum of his lectures on race as the key to scientific political legislation
and social procedure. But, here, the details of his teratological theory of evolution
were lost to view, as his radical "moral anatomy" was retooled to fit the growing
racism that accompanied the economic expansion of Britain and then subsumed
into the new evolutionary anthropology of the 1870s. Knox's tiger arm became
grist for the Darwinian mill, just one more piece of evidence in support of the
evolution of humans by natural and sexual selection (Richards 1989, 406–35;
Darwin 1871, 1: 28).

   Owen initially played both insider and outsider, depending on his audience, with
a certain degree of success. He could speculate privately with Darwin and other
open-minded individuals about the origin of species, while presenting himself in
more politically correct conservative persona before the distinguished audience
of his public Hunterian Lectures. But his attempt to gear his early teratological
speculations to the conservative social needs of his established anatomical audience and his institutional power base in the Royal College of Surgeons backfired
when his tentatively expressed evolutionary views were suppressed in the name
of those same conservative social needs in the face of the perceived social and
political threat posed by the monstrous *Vestiges* (literally abominated as a "filthy
abortion" and a "deformed progeny of unnatural conclusions" by Owen's self-
elected watchdog, Sedgwick).[40] Owen's retention of the sources of social power
available to him depended on his acquiescence to their suppression of his contentious evolutionism. The complexities of his simultaneous self-promotion as rock-
solid British empiricist and innovative transcendentalist were difficult enough to
negotiate in the late 1840s and early 1850s. He could not afford to commit professional suicide by openly declaring himself a deviant, dangerous, and putatively
godless evolutionist. For a while Owen managed to pack this monstrous image
back in the closet along with the craniopagus skull. And when he finally attempted
to rearticulate his teratological views in the more liberal and secular context of
mid-Victorian capitalism, he was forced to confront the competing anatomical
strategies of the ubiquitous and aggressively ambitious Thomas Henry Huxley.

---

39 I am indebted to an anonymous referee for this observation.
40 See Adam Sedgwick to Charles Lyell, 1845, in Clark and Hughes 1890, 2: 83; Desmond 1989,
    178–9.

In the end, their "fierce fight to construct reality" resulted in a resounding victory for Huxley and the assumption to cognitive authority in the field of anatomy by the more powerful Darwinians. Huxley's demolition of Owen's morphology and, with it, his anatomical reputation was a double ratification of Darwinian interests. In effect, Huxley played Frankenstein to Owen's ultimate monsterdom. Huxley, the self proclaimed "plain, prosaic inquirer into objective truth," not only promoted the social and professional credibility of the "young guard" Darwinians by contrasting their much-vaunted ideologically neutral and socially efficacious "natural knowledge" to Owen's ideologically laden transcendental "metaphorical mystifications" and "osteological extravaganzas"; at the same time, he also laid the foundations of the partisan positivist historiographic tradition that has only recently begun to be questioned and revised.[41]

In the "morality play" that constitutes our received interpretation of the Darwinian Revolution, Owen was set on stage to represent the forces of darkness, the embodiment of the other to the enlightening Darwinians, "the evil genius of the anatomy theatre, scheming always to pervert the course of true knowledge to reactionary religious ends of himself and his Oxbridge chums" (Ruse 1993b, 384–5). Owen was well and truly monstered, the evolutionary outcast who was not allowed in out of the cold.

The domestication of the monster

Historians no longer view the Darwinian Revolution in such black-and-white terms. In recent years scholars in and around the Darwin Industry have blurred the once clear-cut and inviolable boundaries between Darwinism and religion, between Darwinism and transcendentalism, between Darwin, Darwinism, and Social Darwinism, and between the content and context of Darwin's evolutionary theorizing.[42] In this process of revision and contextualization, the sociologically informed study of rejected or unsuccessful evolutionary knowledge serves as a valuable counterpoint to mainstream Darwin historiography.[43] The challenge is to extend such analyses symmetrically to accepted or successful evolutionary knowledge. I can do no more here than sketch out the explanatory possibilities of such an approach to Darwin's own teratological constructions in terms of a sociology of monsters that extends and extrapolates from my account of the rejected knowledge of the evolutionary losers, Knox and Owen.

41 See Huxley 1894, esp. p. 315. On Huxley's stratagem of divorcing Darwinism from metaphysics and ideology see Moore 1991. The phrase "fierce fight to construct reality" comes from Latour and Woolgar 1979, 243.

42 See Lenoir 1987; Kohn 1985; Moore 1989; Desmond and Moore 1991. Robert M. Young has been the major long-term exponent of the contextualist program in Darwin studies: Young 1971, 1985a, 1985b. But see Ingmar Bohlin's sympathetic but cogent critique of the limitations of Young's broad-brush historiographic approach: Bohlin 1991.

43 See, for instance, Adrian Desmond's illuminating reinterpretation of Darwin's much speculated about twenty-year delay in publishing his theory of natural selection: Desmond 1989, 398–414.

There are many references to monsters and anatomical abnormalities in Darwin's early transmutation notebooks. Most of these postdate Owen's comment to Darwin on the evolutionary significance of teratogeny. Darwin read Geoffroy St Hilaire's and, at Owen's prompting and with more care, Hunter's views on teratogeny as transcribed and edited by Owen. Darwin was, as M. J. S. Hodge first argued and Robert Richards has recently elaborated upon, a "lifelong generation theorist" (Hodge 1985; R. Richards 1992). From the opening of his first transmutation notebook (Notebook B) around mid-July 1837, he structured his evolutionary speculations around the complex phenomena of generation or reproduction, especially sexual reproduction. The unifying theme that ran through this early theorizing and impressed its mark on everything thereafter was Darwin's tendency to try to understand species change on the model of the development or evolution of individual organisms. He brought all the evidence and information he could find to bear on this essentially transcendental analogy, including instances and interpretations of monstrosity. But from a very early stage, even before he had read and assimilated Geoffroy's views and what he called "Hunter's law," Darwin was unpersuaded by Owen's suggestion that species change was akin to monstrous change. This was for a number of interconnected reasons, intellectual and social, which here can be indicated only sketchily and schematically.

First and foremost, for Darwin, generation was inherently conservative. Organisms generate or reproduce organisms of the same kind. They replicate themselves. Each individual generation is a condensed repetition of the sequence of changes undergone by its forebears over eons of time. However, sexual generation, which involves the propagation of one individual from two others, requires maturation and crossing. These processes lend themselves to adaptive variation in response to changing conditions of existence, and the resulting new characters may be passed on to the next generation. But because of the conservativism of generation and individual development, the innate tendency for the developing organism to replicate the embedded sequence of ancestral changes, in any cross the parent whose characters are older or more fixed in the breed will have a greater influence on the offspring than the parent whose characters are those of a younger breed. "What has long been in blood will remain in blood. – converse, what has not been, will not remain." And, "an animal is able to transmit only those peculiarities, to its offspring, which have been *gained slowly*" (Notebook D: 13, 16; Darwin 1987, 335–6). This was "Yarrell's law," which Darwin derived from the animal breeder William Yarrell. It guaranteed the stability of hereditary change and suited Darwin's conception of transformism. It meant that any major adaptive change could be acquired only gradually, through the slow accumulation of many small steps over many successive generations. This interpretation of Yarrell's law was reinforced by consultations with Yarrell and other influential breeders, which, Darwin thought, confirmed a significant distinction between the two ways in which domestic breeds were formed. One way was "natural" – through adaptation to local conditions, through migration or isolation, or through transport to another country; the other was through "picking," whereby breeders produced and maintained "artificial varieties" or even "great monstrosities" (Notebook C: 4, 106e; D: 20; Darwin 1987, 240, 271, 337; Hodge 1985, 218–21; Hodge and Kohn 1985, 189).

Following this thinking, Darwin consistently distinguished between "monstrosities" and what he called "necessary adaptations." Monstrosities were sudden, chance peculiarities that were not normally hereditary and could be maintained only by picking or artificial selection; adaptations were variations consistently elicited by the external conditions of which they were a necessary effect. Thus a puppy born with an exceptionally thick coat was a "monstrosity" in a warm country; but if a puppy were brought into a cold climate and there acquired a thick coat, then it was an "adaptation." "Dreadful monsters" were abortive, just like mules. According to the theory of special creation, new organisms, like monsters, must appear "at once," but according to the "ordinary laws" of inheritance such sudden monstrous introductions could not have offspring. "The creator would contradict his own law" (Notebook C: 66, 85; D: 14, 19; Darwin 1987, 259, 265, 335, 336).

Rather than sudden creation by divine fiat, Darwin reasoned, there must be some naturalistic "law, that whatever organization an animal has, it tends to multiply & IMPROVE on it . . . if generation is condensation of changes, then animal must tend to improve." A few days after this entry, and many months after his conversation with Owen, on 1 September 1838, Darwin finally came to grips with Hunter's law that monstrosities were "kind of determined by *age* of foetus." He returned to Hunter's views on monstrosity again and again over the next few critical weeks as he worried away at the connection between the generation of individuals and the generation of species. And, consistent with his prior reasoning, he decisively rejected Owen's idea that the production of monsters was analogous to the production of species. "The law of monstrosity," he wrote on 7 September, "not prospective, but retrospective as showing what organs are little fixed – (Hunter's law of monstrosity with regard to age of foetus. distinct consideration)" (Notebook D: 49, 57, 66–7; Darwin 1987, 347, 351–2, 355). What this meant was that Hunter's law that monstrous variations were produced at a very early stage of development confirmed Darwin's view that such variations were unlikely to be propagated. Monstrous change could neither lead to nor explain species change. It was not "prospective." He had argued previously that any "real adaptation" or heritable modification must be fitted to the whole life of the organism, not merely to its foetal phase. Such adaptations occurred generally only at the end of development, and most likely in the mature organism, as a stage additional to the fixed or embedded "condensation" of ancestral stages preserved in embryonic development. A monstrosity was a "retrospective" or abortive variation produced in utero, an "adaptation to unhealthy state of womb," as Darwin noted further on 14 September (Notebook C: 65, 83; D: 107; Darwin 1987, 264–5, 365).

Within another three weeks, Darwin was writing confidently: "No structure will last, without it is adaptation to *whole* life of animal, & not if it be solely to womb, as in monster, or solely to childhood, or solely to manhood, – it will decrease & be driven outwards in the grand crush of population." Darwin's reading of Thomas

Malthus's providential reproductive law had given him the explanation he was seeking for the aetiology of both individual and species generation, even to the minute changes that Darwin insisted upon. Malthusian superfecundity was the irresistible force that powered the adaptation of structure to changing conditions. It was a force "like a hundred thousand wedges," whose "final cause" was "to sort out proper structure & adapt it to change. – to do that for form, which Malthus shows is the final effect, (by means however of volition) of this populousness, on the energy of Man" (Notebook E: 9e; D: 135; Darwin 1987, 375–6, 399; Hodge and Kohn 1985, 194–7).

By the end of 1838 Darwin had formulated the "beautiful" analogy between nature's and man's selective breeding, and he then moved away from the idea of "necessary" adaptations to the view that variation in a state of nature, as under domestication, was "accidental" or occurred by chance. This gave him an even stronger reason for assuming that adaptive species change, having only rare chance variations to work on, was very slow and gradual (Notebook E: 71; Darwin 1987, 416; Hodge and Kohn 1985, 199, 201). The "very essence" of his theory was that "little change is produced." A monstrosity, by definition, could play no role in species formation but was relegated to the domestic domain, there to be dismissed by Darwin as "a mere monstrosity propagated by art." Darwin's mature theory of evolution of 1859 retained this crucial distinction. Domestic races "often have a somewhat monstrous character," while monstrosities in nature are "either injurious to or not useful to the species, and not generally propagated" (Notebook E: 151–2; D: 107; Darwin 1987, 441–2, 426, [1859] 1970, 78, 101).

Two final points will conclude this quick tour through Darwin's early evolutionary theorizing. The first is that he retained the teleological emphasis on perfect adaptation. Variations are produced by chance, but only those that have a "better chance" of surviving are propagated. By the continuous "wedging" action of the Malthusian struggle for existence, species become as perfectly adapted as any Paleyan-inspired natural theologian might insist. Darwin's theorizing did not eject teleology from comparative anatomy but, rather, reconciled the "traditional choice between chance and design." Second, Darwin's insistence on explaining major structural change through the gradual accumulation of minute changes was an explicit rejection of radical or revolutionary change in the political as well as the biological senses. When he came to integrate his conservative Yarrellian generational gradualism with his Lyellian geological gradualism, Darwin himself noted that his extension of the rule of constantly acting Malthusian population law from the human to all species "baffles the idea of revolution" in nature as in "government" and "institutions" (Notebook E: 3–6; Darwin 1987, 397–8; Hodge and Kohn 1985, 202–4; Hodge 1985, 226).

Darwin's domestication of the monster was one critical element in his long-drawn out negotiation of the highly political problem of packaging and presenting a theory associated with atheism and revolution in such a way as to make it scientifically and socially respectable. Unlike Owen, the wealthy Darwin was not financially tied to any scientific institution or patronage network, but he held serious scientific ambitions, and he wished to protect his family and his social standing. Openly to declare himself a transmutationist was tantamount to acting as his own Frankenstein: "it is like confessing a murder." During the twenty years between the inception of his theory and its publication, while he steadily accrued insider scientific status, the eminently respectable "squarson" of Down House sought out and carefully cultivated a more liberal audience for his deviant evolutionary views, especially amongst the new breed of professional scientists such as Hooker and Huxley.[44] With their energetic and increasingly powerful support, Darwin was more successful than Owen in juggling his overlapping social worlds and networks of power, capitalizing on the opportunities they presented and avoiding the dangers. He moved from nerve-wracked incipient monsterdom to safe establishment figure – a status symbolized by his burial in Westminster Abbey with the full panoply of church and state ritual. Ideological considerations may not have been the dominant determinants in this process, but they cannot be dissociated from it.[45] Theories of radical saltation or of divinely induced saltatory change were ideologically outmatched by the powerful social appeal of an emerging Darwinism that made unobstructed competition the guarantee of continuous social progress without revolutionary or radical change.[46]

Through the processes of consolidation and closure of the anatomical debate between teleologists and transcendentalists by the triumphant Darwinians, the field of anatomy was reconstituted. To state it succinctly, though at the risk of oversimplification: the new evolutionary anatomy retained the earlier emphasis on functional adaptation, while the transcendental focus on form and archetypes was reformulated as the search for common ancestors, primarily through the medium of embryology. Teratology and the anatomical objects of its study, deformations and monstrosities, lost the central explanatory role accorded them by the transcendentalist saltationists. They were neutralized and relegated to the category of the less interesting, to the domestic, the artificial, the unnatural. In sociological terms, biological monsters were marginalized or disempowered in the process of stabilization of a new set of power relations within which it was advantageous

44 Charles Darwin to Joseph Hooker, 11 January 1844, in Burkhardt and Smith 1987, 2. See Moore 1985.
45 Ingmar Bohlin questions a reading of Darwin's work and the related "delay problem" that privileges sociological or ideological determinants over others. There were, as he points out, a number of important developments in Darwin's evolutionary thinking between the early 1840s and 1859, notably the emergence of the principle of divergence. But these intellectual developments, while significant in their own right, do not invalidate an ideological component in the social and scientific success of Darwin's theory, nor in the explanation of the delay problem as initially developed by Desmond (1989) and in Desmond and Moore (1991). Darwin's subsequent intellectual development also may be consistently explained by invoking social as well as internalist factors. See Bohlin 1994, 1991.
46 James Moore has recently argued the need for historians to address the misleading separation of Darwinism from Social Darwinism, arguing that "Darwinism was social from the start" and that "Social Darwinism" is the "artefact of a professional discourse that increasingly pretended to divorce science from ideology": Moore 1986, 39. Silvan S. Schweber stresses the compatibility of Darwin's gradualism with contemporary British political economic thought in Schweber 1985, 35–69.

to conceive biological and social evolution as generally slow, steady, gradual, and continuous. Where Owen had failed, the Darwinians succeeded in taming the unruly transcendental monster and its radical social implications.

This interpretation may help to explain the continuing marginalization of those competing teratological strategies that have attempted to redeploy hopeful monsters in support of new theories of saltatory evolutionary change. It also problematizes the partiality of those analyses that eschew sociological explanation or apply it asymmetrically to "losers" or to those promoting deviant scientific theories. The contemporary hegemonic place of the Darwinian theory of evolution in the field of anatomy (or paleontology or genetics, for that matter) may not reflect the incontestable "reality" and ideological neutrality of the Darwinian or, indeed, neo-Darwinian anatomical constructions. Rather, as my analysis suggests, they, as much as any contemporary, ideologically "tainted" hopeful monster, also warrant full-blooded sociological analysis and reinterpretation as political anatomies.[47]

Acknowledgements

Earlier versions of this paper were read at the conference on Ideology and the Life Sciences, Harvard University, 1989; at the summer meeting of the British Society for the History of Science, Edinburgh, 1989; and at a series of seminars convened by the Wellcome Units for the History of Medicine at London University College and the Universities of Manchester and Cambridge in 1990. It benefited from comments and discussions on these occasions. I am also indebted to the comments of three anonymous referees. I should also like to thank the libraries of the Royal College of Surgeons of England, the British Library, and the British Museum (Natural History) for permission to study manuscript material. Elizabeth Allen, curator of the Hunterian Museum, Royal College of Surgeons, provided several useful sources and information on the craniopagus skull in the Hunterian Collection. This paper was supported in part by a grant from the Wellcome Trust for the History of Medicine.

Bibliography

Appel, Toby. 1987. *The Cuvier-Geoffroy Debate: French Biology in the Decades before Darwin*. Oxford: Oxford University Press.
Armstrong, David. 1983. *Political Anatomy of the Body: Medical Knowledge in Britain in the Twentieth Century*. Cambridge: Cambridge University Press.
Barnes, Barry. 1977. *Interests and the Growth of Knowledge*. London and Boston: Routledge and Kegan Paul.

---

47 Ruse 1993a, 144. Ideology, understood in the way I have indicated in the text, does not assume the possibility of an ideology-free biology or science (see note 37). By contrast, this possibility is implicit in Ruse's reference to an ideological "taint" in Gould's evolutionary theorizing.

Biddiss, Michael D. 1976. "The Politics of Anatomy: Dr. Robert Knox and Victorian Racism." *Proceedings of the Royal Society of Medicine* 69: 245–50.

Bohlin, Ingmar. 1991. "Robert M. Young and Darwin Historiography." *Social Studies of Science* 21: 597–648.

———. 1994. "Review of Adrian Desmond and James Moore, *Darwin*." *Science as Culture* 4: 425–39.

Bondeson, J. 1988. "Everard Homes fall av en craniopagus parasiticus: En teratologihistorisk studie." *Draco pro Medica* 2: 22–7.

Bowler, Peter J. 1976. *Fossils and Progress: Paleontology and the Idea of Progressive Evolution in the Nineteenth Century*. New York: Science History Publications.

Burkhardt, Frederick, and Sydney Smith, eds. 1987. *The Correspondence of Charles Darwin, Vol. 3: 1844–1846*. Cambridge: Cambridge University Press.

Carus, Carl G. 1846. *The King of Saxony's Journey through England and Scotland in the Year 1844*. London: Chapman & Hall.

Clark, J. W., and T. McKenny Hughes. 1890. *Life and Letters of Sedgwick*. Cambridge: Cambridge University Press.

Corsi, Pietro. 1988. *The Age of Lamarck: Evolutionary Theories in France, 1790–1830*. Berkeley: University of California Press.

Curtin, Phillip D. 1965. *The Image of Africa: British Ideas and Action, 1780–1850*. London: Macmillan.

Darwin, Charles. [1859] 1970. *On the Origin of Species by Means of Natural Selection, or the Preservation of Favoured Races in the Struggle for Life*. Edited by J. W. Burrow, reprint of 1st edn. Harmondsworth: Penguin.

———. 1871. *The Descent of Man, and Selection in Relation to Sex*. 2 vols. London: John Murray.

———. 1987. *Charles Darwin's Notebooks, 1836–1844*. Edited by Paul Barrett, Peter Gautrey, Sandra Herbert, David Kohn, and Sydney Smith. Ithaca, NY: Cornell University Press.

Desmond, Adrian. 1982. *Archetypes and Ancestors*. London: Blond & Briggs.

———. 1985. "Richard Owen's Reaction to Transmutation in the 1830's." *British Journal of the History of Science* 18: 25–50.

———. 1987. "Artisan Resistance and Evolution in Britain, 1819–1848." *Osiris*, 2nd Ser. 3: 77–110.

———. 1989. *The Politics of Evolution: Morphology, Medicine, and Reform in Radical London*. Chicago and London: University of Chicago Press.

Desmond, Adrian, and James Moore. 1991. *Darwin*. London: Michael Joseph.

Foucault, Michel. 1980. "Truth and Power." In *Power/Knowledge: Selected Interviews and Other Writings, 1972–1977*, edited by Colin Gordon, 107–33. Hassocks: Harvester.

Gedda, Luigi. 1961. *Twins in History and Science*. Springfield, IL: Thomas.

Geoffroy St Hilaire, Étienne. 1827. "Monstre." *Dictionnaire classique d'histoire naturelle* 11: 108–51.

———. 1828. "Mémoire où l'on se propose de rechercher dan quels rapports de structure organique et de parenté sont entre eux les animaux des ages historiques, et vivant actuellement, et les espèces antediluviennes et perdues." *Mémoires du Muséum d'Histoire Naturelle* XVII: 209–29.

———. 1833. "Le degré d'influence du monde ambiant pour modifier Jes formes animales; question interessant l'origine des espèces téléosauriennes et successivement celle des animaux de l'époque actuelle." *Mémoires de l'Académie des Sciences* XII: 63–92.

Goldschmidt, Richard. 1933. "Some Aspects of Evolution." *Science* 78: 539–47.

———. 1940. *The Material Basis of Evolution*. New Haven, CT: Yale University Press.

Gould, Stephen Jay. 1976. *Ontogeny and Phylogeny*. Cambridge, MA: Harvard University Press.

———. 1980. "Is a New and General Theory of Evolution Emerging?" *Paleobiology* 6: 119–30.

———. 1982. "The Meaning of Punctuated Equilibrium and Its Role in Validating a Hierarchical Approach to Macroevolution." In *Perspectives on Evolution*, edited by R. Milkman, 83–104. Sunderland, MA: Sinauer.

———. 1983. "Return of the Hopeful Monster." In *The Panda's Thumb*, 155–61. Harmondsworth: Penguin.

Haraway, Donna. 1987. "A Manifesto for Cyborgs: Science, Technology, and Socialist Feminism in the 1980s." *Australian Feminist Studies* 4: 1–42.

———. 1991. *Simians, Cyborgs, and Women: The Reinvention of Nature*. London: Free Association Books.

———. 1992. "The Promises of Monsters: A Regenerative Politics for Inappropriate/d Others." In *Cultural Studies*, edited by Lawrence Grossberg, Cary Nelson, and Paula Treichler, 295–337. New York and London: Routledge.

Hobsbawm, Eric J. 1975. *The Age of Capital, 1848–1875*. London: Weidenfeld and Nicolson.

Hodge, M. J. S. 1985. "Darwin as a Lifelong Generation Theorist." In *The Darwinian Heritage*, edited by David Kohn, 207–43. Princeton, NJ: Princeton University Press.

Hodge, M. J. S., and David Kohn. 1985. "The Immediate Origins of Natural Selection." In *The Darwinian Heritage*, edited by David Kohn, 185–206. Princeton, NJ: Princeton University Press.

Home, Everard. 1790. "An Account of a Child with a Double Head." *Philosophical Transactions of the Royal Society of London* 80: 296–305.

———. 1799. "Some Additions to a Paper, Read in 1790, on the Subject of a Child with a Double Head." *Philosophical Transactions of the Royal Society of London* 89: 28–30.

Huet, Marie-Helene. 1993. *Monstrous Imagination*. Cambridge, MA: Harvard University Press.

Hunter, John. 1837. *Observations on Certain Parts of the Animal Oeconomy, with notes by Richard Owen*. London: Longman.

Huxley, Thomas H. 1894. "Owen's Position in the History of Anatomical Science." In *Life of Richard Owen*, edited by Rev. R. Owen, vol. 2, 273–332, 2 vols. London: John Murray.

Kohn, David, ed. 1985. *The Darwinian Heritage*. Princeton, NJ: Princeton University Press.

Knox, Robert. 1841. "On the Occasional Presence of a Supra-condyloid Process in the Human Humerus." *Edinburgh Medical and Surgical Journal* 56: 125–8.

———. 1843. "Contributions to Anatomy and Physiology." *London Medical Gazette* 32: 529–32.

———. 1850. *The Races of Men: A Fragment*. London: Henry Renshaw.

———. 1852. *Great Artists and Great Anatomists*. London: John van Voorst.

———. 1855a. "Contributions to the Philosophy of Zoology." *Zoologist* 13: 4837–42.

———. 1855b. "Contributions to the Philosophy of Zoology, with Special Reference to the Natural History of Man." *Lancet* 2: 24–6, 45–6, 68–71, 162–4, 186–8, 216–18.

———. 1855c. "Some Remarks on the Aztecque and Bosjieman Children, Now Being Exhibited in London, and on the Races to Which They Are Presumed to Belong." *Lancet* 1: 357–60.

———. 1863. "On the Deformations of the Human Cranium, Supposed to Be Produced by Mechanical Means." *Anthropological Review* 1: 271–3.

Kranzler, Laura. 1988/89. "Frankenstein and the Technological Future." *Foundation* 44: 42–9.

Larrain, John. 1979. *The Concept of Ideology.* London: Hutchinson.

Latour, Bruno, and Steve Woolgar. 1979. *Laboratory Life: The Social Construction of Scientific Facts.* Beverly Hills, CA: Sage.

Law, John. 1991a. "Introduction: Monsters, Machines, and Sociotechnical Relations." In *Sociology of Monsters: Essays on Power, Technology, and Domination*, edited by John Law, 1–23. London and New York: Routledge.

———. 1991b. "Power, Discretion, and Strategy." In *Sociology of Monsters: Essays on Power, Technology, and Domination*, edited by John Law, 165–91. London and New York: Routledge.

Lenoir, Timothy. 1987. "Essay Review: The Darwin Industry." *Journal of the History of Biology* 20: 115–30.

Lonsdale, Henry. 1870. *A Sketch of the Life and Writings of Robert Knox the Anatomist.* London: Macmillan.

Moore, James R. 1985. "Darwin of Down: The Evolutionist as Squarson-Naturalist." In *The Darwinian Heritage*, edited by David Kohn, 435–81. Princeton, NJ: Princeton University Press.

———. 1986. "Socializing Darwinism: Historiography and the Fortunes of a Phrase." In *Science as Politics*, edited by Les Levidow, 38–80. London: Free Association Books.

———, ed. 1989. *History, Humanity, and Evolution.* Cambridge: Cambridge University Press.

———. 1991. "Deconstructing Darwinism: The Politics of Evolution in the 1860s." *Journal of the History of Biology* 24: 353–408.

Mosley, Nicolas. 1991. *Hopeful Monsters.* London: Minerva.

Neuburger, M. 1953. "C. G. Carus on the State of Medicine in Britain in 1844." In *Science, Medicine, and History*, edited by E. Ashworth Underwood, vol. 2, 263–73. London: Oxford University Press.

Ospovat, Dov. 1978. "Perfect Adaptation and Teleological Explanation: Approaches to the Problem of the History of Life in the Mid-Nineteenth Century." *Studies in the History of Biology* 2: 33–65.

Owen, Rev. R., ed. 1894. *The Life of Richard Owen.* 2 vols. London: John Murray.

Owen, Richard. 1831. *Catalogue of the Contents of the Museum of the Royal College of Surgeons in London, Pts. 5 and 6: Comprehending the Preparations of Monsters and Malformed Parts, in Spirit, and in a Dried State.* London: Richard Taylor.

———. 1846. *Lectures on the Comparative Anatomy and Physiology of the Vertebrate Animals.* London: Longman, Brown, Green and Longman.

———. 1848. *On the Archetype and Homologies of the Vertebrate Skeleton.* London: John Van Voorst.

———. 1849. *On the Nature of Limbs.* London: John Van Voorst.

———. [1853] 1856. "A Brief Notice of the Aztec Race, Compiled by Richard Cull, Hon. Sec., Followed by a Description of the So-called Aztec Children Exhibited on the Occasion by Professor Richard Owen, FRS, read at a special meeting, 6 July 1853." *Journal of the Ethnological Society of London* 4: 120–37.

———. 1860. "Darwin on the Origin of Species." *Edinburgh Review* 111: 487–532.

————. 1868. *On the Anatomy of Vertebrates*. 3 vols. London: Longmans, Green.

Park, Katharine, and Lorraine Daston. 1981. "Unnatural Conceptions: The Study of Monsters in Sixteenth- and Seventeenth-Century France and England." *Past and Present* 92: 20–54.

Rae, Isobel. 1964. *Knox the Anatomist*. Edinburgh: Oliver & Boyd.

Rehbock, Philip R. 1983. *The Philosophical Naturalists: Themes in Early Nineteenth-Century British Biology*. Madison, WI and London: The University of Wisconsin Press.

Richards, Evelleen. 1987. "A Question of Property Rights: Richard Owen's Evolutionism Reassessed." *British Journal for the History of Science* 20: 129–71.

————. 1989. "The 'Moral Anatomy' of Robert Knox: The Interplay between Biological and Social Thought in Victorian Scientific Naturalism." *Journal of the History of Biology* 22: 373–436.

————. 1990. "'Metaphorical Mystifications': The Romantic Gestation of Nature in British Biology." In *Romanticism and the Sciences*, edited by Andrew Cunningham and Nicholas Jardine, 130–43. Cambridge: Cambridge University Press.

Richards, Robert J. 1992. *The Meaning of Evolution: The Morphological Construction and Ideological Reconstruction of Darwin's Theory*. Chicago: University of Chicago Press.

Richardson, Ruth. 1987. *Death, Dissection, and the Destitute*. London and New York: Routledge and Kegan Paul.

Royle, Edward, and James Walvin. 1982. *English Radicals and Reformers, 1760–1848*. Brighton: Harvester.

Ruse, Michael. 1993a. "Is the Theory of Punctuated Equilibria a New Paradigm?" In *The Darwinian Paradigm: Essays on Its History, Philosophy, and Religious Implications*, 118–45. London and New York: Routledge.

————. 1993b. "Were Owen and Darwin *Naturphilosophen*?" *Annals of Science* 50: 383–8.

Russell, E. S. [1916] 1972. *Form and Function: A Contribution to the History of Animal Morphology*, reprint. Westmead: Gregg.

Schweber, Sylvan S. 1985. "The Wider British Context in Darwin's Theorizing." In *The Darwinian Heritage*, edited by David Kohn, 35–69. Princeton, NJ: Princeton University Press.

Secord, James. 1989. "Behind the Veil: Robert Chambers and *Vestiges*." In *History, Humanity, and Evolution*, edited by James R. Moore, 165–94. Cambridge: Cambridge University Press.

Shapin, Steven. 1979. "The Politics of Observation: Cerebral Anatomy and Social Interests in the Edinburgh Phrenology Disputes." *Sociological Review Monographs* 27: 139–78.

Shaw, Debbie. 1993. "In Her Own Image: The Constructed Female in Women's Science Fiction." *Science as Culture* 15: 263–81.

Sloan, Phillip R. 1990. "Deconstructing Evolution." *History of Science* 28: 419–29.

————. 1992. "Introductory Essay: On the Edge of Evolution." In *Richard Owen's Hunterian Lectures in Comparative Anatomy, May–June 1837*, edited by Phillip Sloan, 3–72. Chicago: University of Chicago Press.

Starr, Susan Leigh. 1991. "Power, Technology and the Phenomenology of Conventions: On Being Allergic to Onions." In *Sociology of Monsters: Essays on Power, Technology, and Domination*, edited by John Law, 26–56. London and New York: Routledge.

Stepan, Nancy. 1982. *The Idea of Race in Science: Great Britain 1800–1960*. London and Basingstoke: Macmillan.

Yeo, Richard. 1986. "The Principle of Plenitude and Natural Theology in Nineteenth-Century Britain." *British Journal for the History of Science* 19: 263–82.

Young, Robert M. 1971. "Evolutionary Biology and Ideology: Then and Now." *Science Studies* 1: 177–206.

———. 1985a. "Darwinism *is* Social." In *Darwinian Heritage*, edited by David Kohn, 609–38. Princeton, NJ: Princeton University Press.

———. 1985b. "The Historiographic and Ideological Contexts of the Nineteenth-Century Debate on Man's Place in Nature." In *Darwin's Metaphor: Nature's Place in Victorian Culture*, 164–247. Cambridge: Cambridge University Press.

Darwinian science, good wives, the "shrieking sisterhood", suffering animals, and radical birth control

5

DARWIN AND THE DESCENT
OF WOMAN

This is the Question

MARRY	NOT MARRY
Children – (if it please God) – constant companion, (friend in old age) who will feel interested in one, object to be beloved and played with – better than a dog anyhow – Home, and someone to take care of house – Charms of music and female chit-chat. These things good for one's health. Forced to visit and receive relations *but terrible loss of time* . . .	No children (no second life), no one to care for one in old age . . . Freedom to go where one liked – Choice of Society *and little of it.* Conversation of clever men at clubs . . . *Loss of time* – cannot read in the evening – fatness and idleness – anxiety and responsibility – less money for books etc . . .
Only picture to yourself a nice soft wife on a sofa with good fire, and books and music perhaps – compare this vision with the dingy reality of Grt Marlboro' St. Marry – Marry – Marry. Q.E.D.	Perhaps my wife won't like London; then the sentence is banishment and degradation [into] indolent idle fool –

– Charles Darwin, *Notes on the Question of Marriage, 1837–8*
(Barlow 1969, 232–33).

A growing number of social historians and sociologists of science have come to think of scientific knowledge as a "contingent cultural product, which cannot be separated from the social context in which it is produced", and they have begun to explore the possibility of there being direct "external" or what are generally regarded as "non-scientific" influences on the content of what scientists consider to be genuine knowledge. In their view, scientific assertions are "socially created and not directly given by the physical world as previously supposed". This is not to assert that science is merely a matter of convention – that the external world does not constrain scientific conclusions – but rather that scientific knowledge "offers an account of the physical world which is mediated through available cultural resources; and these resources are in no way definitive"

(Mulkay 1979, 79, 62, 60).[1] This view undercuts the special epistemological status generally accorded to scientific knowledge, whereby it is assumed to be value-free and politically and socially neutral. In this revised view of science, the basis of the traditional distinction between scientific and social thought is eliminated, and as a consequence, the customary contrast between "internal" intellectual and "external" social factors in the history of science loses its significance. It becomes possible to consider scientific knowledge as socially contingent and an understanding of the socially derived perspectives of the knowers and their purposes becomes essential to coherent historical explanation of scientific knowledge. This paper is an attempt to examine and explain Charles Darwin's conclusions on the biological and social evolution of women in the light of this revised view of scientific knowledge.

The Darwinian theory of evolution is the subject of a large and growing literature, but most historians have treated its content and its reception as independent of the social context in which it was conceived and accepted into the body of scientific knowledge. With few exceptions, Darwin is presented as the young naturalist of the "Beagle", subsequent pigeon breeder and barnacle dissector and, above all, detached and objective observer and theoretician – remote from the political concerns of his fellow Victorians who misappropriated his scientific concepts to rationalize *their* imperialism, laissez-faire economics and racism. The congruence of his writings, especially *The Descent of Man*, with the flourishing Social Darwinism of the late Victorian period, is either ignored or tortuously explained away and Darwin himself absolved of political and social intent and his theoretical constructs of ideological taint.[2]

The handful of Darwin studies like those of Young (1969, 1971a, 1971b, 1973) and Gale (1972) which does not conform to this historical orthodoxy but has been concerned to depict Darwin's evolutionary theory as embedded in an ideological context, has focussed on the concept of natural selection and the associated themes of struggle and adaptation. As far as I am aware, no similar "contextualist" or "naturalistic"[3] study has been made of Darwin's concept of sexual selection and his related conclusions on the biological and social evolution of women. In fact, these have received scant attention from more orthodox scholars, who have also focussed on natural selection. Michael Ghiselin is one of the few of the orthodox to have dealt in any detail with sexual selection, which he did in his work, *The Triumph of the Darwinian Method* (1972, 214–31). Ghiselin's analysis has the virtue of taking into account the whole corpus of Darwin's writings, including *The Descent of Man* and the early Notebooks, but is skewed

---

1 See also Barnes and Shapin 1979, 9–13; Macleod 1977, 189–95; Johnston 1976, 193–203.
2 For a perceptive analysis of historiographic representations of Darwin's relation to Social Darwinism, see Barnes and Shapin 1979, 125–42.
3 "Contextualism" is the term adopted by Young (1971a, 1971b) and Johnston (1976) to describe the sociocultural history of scientific knowledge they advocate, and is to be preferred to that of "naturalism" adopted by Barnes and Shapin (1979) for the same purpose.

by his determination to present Darwin as an unswerving scientific adherent of the hypothetico-deductive method and a good Popperian, like Ghiselin himself.[4] Thus social and political factors are systematically excluded from his account, and not surprisingly, sexual selection emerges as Darwin's "brilliant" value-free hypothesis, deductively consistent with his over-all evolutionary thesis (Ghiselin 1972, 214–231). Ghiselin manages the *tour de force* of an analysis of sexual selection and *The Descent of Man* without ever coming to grips with Darwin's extension of sexual selection to human biological and social evolution, which I shall show was the main thrust of *The Descent*. This deficiency however has been more than amply remedied in Ghiselin's subsequent work *The Economy of Nature and the Evolution of Sex* (1974) where he has turned his hand to applying Darwin's theory to society and reveals himself as the ultimate Social Darwinist, or, more correctly, defender and advocate of genetic capitalism.[5] Ghiselin introduces his book as a "cross between the *Kama Sutra* and the *Wealth of Nations*" and deals in such provocative chapter headings as "The Copulatory Imperative", "Seduction and Rape . . ." and "First Come, First Service . . .". As these headings indicate, the book is largely a vindication and extension of Darwin's "long-neglected" idea of sexual selection. For Ghiselin, if we are to understand why men and women behave as they do, we must treat them as the products of reproductive competition – of a prolonged and enduring sexual contest. This conclusion becomes inescapable, once we have accepted Darwin's theory. Even our moral sentiments subserve reproduction:

> [O]ne would predict that there should be certain kinds of sexual dimorphism in our ethical attitudes. Females know who are their offspring; hence it is expedient for them to play favourites. Males, in so far as they find it difficult to know who fathered whom, would perhaps benefit more from a general contribution to the welfare of their group. Loyalty should thus be a feminine virtue, justice a masculine one . . . Recent research has brought to

---

4 See J.C. Greene's critique of *The Triumph of the Darwinian Method*; Greene 1975, 243–73, 254–59.
5 To wit: "The evolution of society fits the Darwinian paradigm in its most individualistic form. Nothing in it cries out to be otherwise explained. The economy of nature is competitive from beginning to end. Understand that economy, and how it works, and the underlying reasons for social phenomena are manifest. They are the means by which one organism gains some advantage to the detriment of another. No hint of genuine charity ameliorates our vision of society, once sentimentalism has been laid aside. What passes for cooperation turns out to be a mixture of opportunism and exploitation. . . . Where it is in his own interest, every organism may reasonably be expected to aid his fellows. Where he has no alternative, he submits to the yoke of communal servitude. Yet given a full chance to act in his own interest, nothing but expedience will restrain him from brutalising, from maiming, from murdering – his brother, his mate, his parent, or his child. Scratch an 'altruist' and watch a 'hypocrite' bleed". (Ghiselin 1974, 247.) For a critique of this work see Sahlins 1977, 71–91.

light quite a number of differences between the sexes in moral attitudes, at least some of which seem to be inherited . . . (Ghiselin 1974, 256).

It has been left to feminist scholars who are concerned with disputing evolutionary arguments like Ghiselin's, to explore the social dimension of Darwin's writings on the biological and social evolution of women. They are unanimous in their categorization of them as catering to and supporting a prejudiced and discriminatory view of women's abilities and potential – one unsupported by evidence and based upon Victorian sexist ideology.[6] The small section of the appropriately named *Descent of Man*, where Darwin deduced the natural and innate inferiority of women from his theory of evolution by natural and sexual selection, is fast becoming notorious in feminist literature.

The most extensive feminist critique of Darwin has been undertaken by Ruth Hubbard, Professor of Biology at Harvard. Hubbard has been readily able to point to passages in Darwin's writings to support her charge of "blatant sexism" (Hubbard 1979, 16). She places late-Victorian scientific sexism and its contemporary re-emergence in ethology and sociobiology squarely at Darwin's door. Contemporary ethologists and sociobiologists she asserts, are conducting their arguments within the context of nineteenth-century anthropological and biological speculation. Nineteenth-century anthropology and biology were dominated by Darwin, whose *Origin of Species* and *Descent of Man* provided the theoretical framework within which anthropologists and biologists have ever since been able to endorse the social inequality of the sexes.

Where Ghiselin sees only clear-eyed scientific judgement and a vindication of his own values, Hubbard sees only cloudy male bias and confirmation of her own perspective of male domination and female exploitation. If Ghiselin refuses to concede any but intellectual and theoretical constraints on Darwin's constructs, Hubbard as systematically excludes them. She goes so far as to imply that Darwin's theory of sexual selection was generated as a male scientist's response to the perceived threat of nineteenth-century feminism.[7]

This paper goes beyond Hubbard's charge of sexism and anti-feminism by locating Darwin's theoretical constructs and Darwin himself in their larger social, intellectual and cultural framework. Without this framework the larger social, political and epistemological questions are never confronted and the issues dwindle to ones of personal bias. While I agree with Hubbard that Darwin's concept of sexual selection and his application of it to human evolution were contingent upon his socially derived perceptions of feminine characteristics and abilities, I argue in this paper that it is not only historically incorrect to impute an anti-feminist motive to Darwin, but unnecessary.

---

6 See Hubbard 1979, 7–35; see also the Introduction, p.xv. Darwin's views on the inferiority of women are also discussed by Sleeth Mosedale (1978) and by Alaya (1977). See also Crooke 1973.
7 This seems to be the gist of Hubbard's remarks (Hubbard 1979, 26 and 35n30).

It is historically incorrect, because Darwin's conclusions on the biological and social evolution of women were as much constrained by his commitment to a naturalistic or scientific explanation of human mental and moral characteristics as they were by his socially derived assumptions of the innate inferiority and domesticity of women, as I argue in Section 1. It is unnecessary, because in order to demonstrate that Darwin's reconstruction of human evolution was pervaded by Victorian sexist ideology, one has only to examine his lived experience as Victorian bourgeois husband and father, as I do in Section II of this paper, and relate it to his theoretical arguments. Generally, the domestic relations of Charles and Emma Darwin have been of interest to historians only in so far as Charles' deference to Emma's religious beliefs offers a ready-made explanation of the twenty year delay between the inception of his theory of evolution and its publication. However, I argue that his relations with Emma had a more fundamental and enduring effect on his theory of evolution than this. Just as contextualists have argued that Darwin's concepts of artificial and natural selection were not directly based on biological phenomena, but were in some degree taken over from the practical activities of the plant and animal breeders with whom he associated and whose commercial criteria and interests he absorbed,[8] so I argue that Darwin's experience of women and his practical activities of husband and father entered into his concept of sexual selection and his associated interpretations of human evolution. To this end I demonstrate in Section II that Darwin's domestic relations in no way called into question Victorian sexual stereotypes but entirely conformed with them.

In Section III I carry this analysis further and locate both the content of Darwin's theory of human evolution and his domestic relations in the larger context of Victorian society. Here, both feminism and Darwinism are related to the nineteenth-century naturalist movement, which was concerned with bringing the whole of nature and society under the sway of natural law and improving the social standing of science. In the process, naturalism was brought into opposition to the traditional authority and status of religion and into line with those of the newly-powerful bourgeoisie, whose interests it promoted and rationalized under the universality and inevitability of natural law. Darwin's *Origin of Species* and *Descent of Man* and the intense public debate they engendered in the mid-Victorian period, are viewed as central to this transition and were shaped and constrained by it. When the bourgeois social order began to perceive the growing feminist movement as a threat, late-Victorian Darwinism was brought into conflict with feminism and imposed naturalistic scientific limits to the claims by women for political and social equality, thus effectively undermining feminism, which subscribed to the same naturalistic ideology. Finally, Darwin's role in late-Victorian scientific opposition to feminism is assessed in the light of the above analysis.

8 Mulkay 1979, 100–108; Sandow 1938, 315–26. See also Young 1971b.

My analysis thus proceeds on three inter-related levels and is organized in conformity with this.

I. The Descent of Woman

> Even the preliminary knowledge, what the differences between the sexes now are, apart from all questions as to how they are made what they are, is still in the crudest and most incomplete state. Medical practitioners and physiologists have ascertained, to some extent, the differences in bodily constitution . . . Respecting the mental characteristics of women, their observations are of no more worth than those of common men. It is a subject on which nothing final can be known, so long as those who alone can really know it, women themselves, have given but little testimony, and that little, mostly suborned.
>
> – J. S. Mill (and Harriet Taylor Mill),
> *The Subjection of Women*, 1869[9]

In *The Descent of Man, and Selection in Relation to Sex* (1871), Darwin applied himself for the first time in his published writings to the highly contentious problem of human evolution. Twelve years earlier, in *The Origin of Species*, he had made only one brief allusion to the topic: "light will be thrown on the origin of man and his history". But where Darwin had hesitated, others had not, and by 1871 various "Darwinians" (including prominent naturalists, anthropologists, and social theorists) had published their views on "man's" origin and offered speculative reconstructions of "his" history. To some extent Darwin was pre-empted, but in several significant respects he was not.

He was, after all, the author of *The Origin* and a number of other respected scientific works, whose hard-earned reputation was acknowledged even by his critics, while his increasing number of converts might be expected to treat his long-awaited views on human evolution as authoritative. By the late 1860s, Darwin was under considerable pressure to reveal these views.[10]

Secondly, these views had matured over a very long period of time. More than thirty years earlier Darwin had begun to record his ideas and notes on transmutation, and from the first he was convinced that humanity was part of the evolutionary process. The questions he then posed on the evolution of human instinct, sexual differences, emotion, language, intelligence and sociability, and which were crucial to the formation of his theory of evolution, were suppressed while he very consciously drained his argument of references to human evolution for

---

9 Extract from Rossi 1970, 150. Rossi asserts the joint collaboration of Harriet Taylor and Mill in their essays on sex equality. *The Subjection of Women* was written by Mill after Taylor's death, but was based on their previous intellectual collaboration on the issue (ibid., 31–45).

10 See his comments to de Candolle in 1868, in F. Darwin 1888, 3: 100.

presentation to his scientific and lay audience. With the resolution of the post-*Origin* debates of the 1860s more or less in favour of evolution, and the dwindling of hard-core opposition to the theory, the time had come to reinsert men and women alongside pigeons, barnacles and orchids, and subject them to the same evolutionary processes. The Notebooks, especially those on "Man, Mind and Materialism" that Darwin began to keep in the late 1830s were the basis of *The Descent*.[11] They are a repository of observations and reflections on the continuity between human and other animals, and they document Darwin's growing conviction that only a materialist philosophy of nature can support the treatment of human development in a natural scientific manner. They were, in effect, a testing ground for the disputes of the 1860s, which revolved around just these issues. *The Descent* is the logical extension of these notebook constructions.

Darwin had a further impetus towards publication in the failure of two of those he had most counted on to promote his views on human evolution. In 1863, his long-standing patron Charles Lyell had burked the issue in his *Antiquity of Man*. Despite his private reassurances to Darwin that he was prepared to "go the whole orang", Lyell, when it came to the point, suggested that man was the result of a leap of nature separating him at one bound from the next highest species, the whole being "the material embodiment of a pre-concerted arrangement".[12] Darwin was bitterly disappointed by the equivocation of the extremely influential but conservative Lyell. However, the following year his hopes were raised by Alfred Russel Wallace, co-founder with Darwin of the theory of natural selection. In 1864, Wallace, at this stage strongly influenced by Herbert Spencer, published an article in the *Anthropological Review*, in which he argued the central role of natural selection in the intellectual and moral progress of humanity.[13] Darwin was greatly impressed by Wallace's paper and wrote his approbation, going so far as to offer him his own notes on "Man" and a few suggestions on the origin of the different races via sexual selection.[14] Whatever hopes Darwin may have entertained of Wallace in this respect were quickly dashed. Wallace not only rejected

11 C. Darwin, "M and N Notebooks and Old and Useless Notes", were originally published in H. E. Gruber, *Darwin on Man* (London, 1974); they and "Darwin's Notebooks on Transmutation of Species" (originally edited by Gavin de Beer and published in the *Bulletin of the British Museum (Natural History)* in 1960–61 and1967), were re-edited by Barret et al. and published collectively as *Charles Darwin's Notebooks* in 1987; references are to this later edition. See also: Schweber 1977; Manier 1977; Herbert 1974, 1977. An excellent analysis of the relation of *The Descent* to the earlier Notebooks and to contemporary social thought is contained in Jones 1978; see also Greene 1977. I am especially indebted to the work of Randall Albury for his examination of Darwin views on women and his clarification of Darwin's relationship to the contemporary writings of J. S. Mill in his paper, "The Descent of Man and the Subjection of Women: Science and Ideology in Darwin's Answer to Mill", a version of which was published as Albury 1975.

12 *The Correspondence of Charles Darwin* (Burkhardt et al. 1985-, hereinafter *CCD*), *CCD* 11: 230–31; Himmelfarb 1967, 259. For Darwin's response to Lyell, see *CCD* 11: 166, 172–73, 207.

13 Wallace 1864. For a discussion of Wallace's paper and Darwin's response to it, see Green 1977.

14 F. Darwin 1888, 3: 89–91; F. Darwin and Seward 1903, 2: 31–37 (*CCD* 12: 216–17, 220–22).

his ideas on the part played by sexual selection in human evolution, but within a remarkably short time retracted his belief in the all-sufficiency of natural selection in human physical, social, and mental development. By 1869, Wallace inspired by his growing socialist and spiritualist beliefs, was suggesting that a "higher intelligence" had guided the development of the human race and anticipated its needs.[15]

The recourse by two of his most prominent scientific supporters to supernatural explanations (however different) of human faculties and abilities undoubtedly reinforced Darwin's determination to demonstrate that there was "no necessity", as he wrote to Wallace, "for calling in an additional and proximate cause in regard to man".[16] For Darwin, the human races were the equivalent of the varieties of plants and animals which formed the materials of evolution in the organic world generally, and they were subject to the same main agencies of struggle for existence and the struggle for mates. Human evolution could be entirely explained in terms of natural evolutionary processes and the continuity between the complex human faculties and their animal ancestry established.

This leads us to Darwin's emphasis on the overriding importance of sexual selection in human evolution. In fact, the major theme of *The Descent*, as the full title indicates, was sexual selection, with the greater part of the work being devoted not to human evolution, but to an elaboration of the principles of sexual selection and its exhaustive application to the various members of the animal kingdom, humanity included. For in Darwin's view, sexual selection was primarily responsible for human racial and sexual differences, not just physical differences, but what he called differences in "the mental powers", that is, emotional, intellectual and moral differences.

Darwin had briefly discussed sexual selection in *The Origin*, and carefully distinguished it from natural selection:

> [Sexual selection] depends, not on a struggle for existence, but on a struggle between the males for possession of the females; the result is not death to the unsuccessful competitor, but few or no offspring. Sexual selection is, therefore, less rigorous than natural selection. Generally, the most vigorous males, those which are best fitted for their places in nature, will leave most progeny. But in many cases, victory will depend not on general vigour, but on having special weapons, confined to the male sex (Darwin [1859] 1968, 136).

Apart from male combat for possession of the females, Darwin recognized another aspect of sexual selection – female choice. This occurred especially among birds, where the males competed with one another in brilliance of plumage, song, etc., in their wooing of the female during courtship. Sexual selection could be invoked to

---

15 Wallace 1869. For a discussion of Wallace's socialism and spiritualism and their effects on his evolutionary arguments, see Smith 1972; Durant 1979.
16 F. Darwin 1888, 3: 116 (*CCD* 17: 175).

explain a great deal that otherwise seemed inexplicable in terms of natural selection, such as the bright plumage of many male birds that renders them more conspicuous to predators, or the disadvantageously long, curved horns of an antelope. Such structures did not confer any advantage in the struggle for existence, but they were advantageous in the struggle for mates and thus gave their possessors a better chance of reproducing themselves, of leaving more offspring than other less well-endowed males. As Darwin succinctly expressed it in *The Origin:*

> [W]hen the males and females of any animals have the same general habits of life, but differ in structure, colour, or ornament, such differences have been mainly caused by sexual selection; that is, individual males have had, in successive generations, some slight advantage over other males, in their weapons, means of defence, or charms; and have transmitted these advantages to their male offspring (Darwin [1859] 1968, 137–38).

Sexual selection was vital to Darwin's defence of natural selection against the established theory of special creation. Apart from its importance in explaining the persistence of seemingly disadvantageous or useless characteristics, it enhanced the action of natural selection by ensuring that the fittest males ("the most vigorous males, those which are best fitted for their places in nature") were reproduced. The accumulation of advantageous variation would therefore be all the most probable. Thus, although so little space was given to sexual selection in *The Origin*, it was of considerable importance to Darwin's theory of evolution.

At this stage, it should be noted that in Darwin's initial presentation of sexual selection, attention is focussed on the *males* who compete actively with one another for the females. Even in cases of female choice, males compete to display before the females "which standing by as spectators, at last choose the most attractive partner"; though of a "more peaceful character" it is still a contest and it is the males who play the active role, who "struggle", female choice being depicted as passive. In *The Origin* sexual selection is a process whereby males compete with other males by means of weapons or charms to reproduce themselves. The female role is merely on of submission to and transmission of these male characteristics. As a description of sex roles in reproduction, it is undeniably androcentric.[17]

When it came to human evolution, Darwin's androcentric bias became even more pronounced, with female choice, however passive, being all but swamped by male combat and male aesthetic preference in the shaping of racial and sexual differences. As Darwin first put it to Wallace in his letter of 1864:

> I suspect that a sort of sexual selection has been the most powerful means of changing the races of man. I can show that the different races have a widely different standard of beauty. Among savages the most powerful

---

17 Male-centred or sexist. See Hubbard 1979, 16.

men will have the pick of the women, and they will generally leave the most descendants.

A post-script intimated the Victorian class and cultural overtones of Darwin's perception of primitive human behaviour:

> P.S. Our aristocracy is handsomer (more hideous according to a Chinese or Negro) than the middle classes, from [having the] pick of the women . . .[18]

Wallace, the incipient socialist, dissented from both points of view by return of post, and touched off a long-standing dispute between the co-founders of natural selection on the efficacy of sexual selection in accounting for sexual and racial differentiation. Over the years the letters went back and forth: Wallace opting for the primacy of natural selection in the evolution of female protective colouration and other characteristics; Darwin continuing to focus on the evolution of male sexual differences through sexual selection, badgering naturalists and breeders for corroborative evidence and opinions. By the beginning of 1867, Darwin had accumulated so much material on sexual selection and was so convinced of its essential role in human evolution, that he decided to assemble his notes into an "essay on Man", to fulfil the overall task that *The Origin* had set. He wrote of his intention to Wallace in February 1867:

> The reason of my being so much interested just at present about sexual selection is, that I have almost resolved to publish a little essay on the origin of Mankind, and I still strongly think (though I failed to convince you, and this to me is the heaviest blow possible) that sexual selection has been the main agent in forming the races of man (F. Darwin 1888, 3: 95; *CCD* 15: 93).

The following month, Darwin again wrote to Wallace of his "essay on Man":

> [M]y sole reason for taking it up, is that I am pretty well convinced that sexual selection has played an important part in the formation of races, and sexual selection has always been a subject which has interested me much.[19]

Whatever their order of priority, it is clear that for Darwin human evolution and sexual selection had become inextricably linked together, and the structure of *The Descent* bears this out. It is divided into three sections. The first part deals with "The Descent or Origin of Man" and the main thrust of this section was to demonstrate that there was no fundamental difference between humanity and the higher animals – above

18 F. Darwin1888, 3: 91; F. Darwin and Seward 1903, 2: 33–34 (*CCD* 12: 216–17).
19 F. Darwin 1888, 3: 97 (*CCD* 15: 141]. For the Darwin/Wallace correspondence on selection see ibid., 89–100, and F. Darwin and Seward 1903, 55–97.

all, that the "difference in mind between man and the higher animals, great as it is, certainly is one of degree not of kind". Thus Darwin saw the seeds of intelligence and social organization in the higher animals, and from these rudimentary beginnings evolved the complex human intellectual and moral characteristics that his critics argued were unique and lay outside the scope of evolutionary explanation. To this end he insisted that mental and moral differences were heritable and that natural selection, aided by the inherited effects of mental and moral exercise,[20] had acted on them throughout history in the competition of individuals, tribes, nations and races:

> All that we know about savages . . . shew that from the remotest times successful tribes have supplanted other tribes. . . . At the present day civilised nations are everywhere supplanting barbarous nations, excepting where the climate opposes a deadly barrier; and they succeed mainly, though not exclusively, through their arts, which are the products of intellect. It is, therefore, highly probable that with mankind the intellectual faculties have been mainly and gradually perfected through natural selection; and this conclusion is sufficient for our purpose.[21]

Similarly, the 'social and moral faculties" such as sympathy, fidelity and courage "were no doubt acquired . . . through natural selection aided by inherited habit". Those who practised mutual aid would benefit and this would foster the habit of aiding one's fellows and strengthen feelings of sympathy and altruism. Such habits, followed during many generations, "probably tend to be inherited" (Darwin 1889, 130–31).

Darwin's insistence on the biological basis of intellectual and moral differences brought him into conflict with environmentalists like John Stuart Mill, who had argued in his *Utilitarianism* that the moral feelings are not innate but acquired. In a footnote, Darwin discussed his differences with Mill, but remained adamant:

> It is with hesitation that I venture to differ at all from so profound a thinker, but it can hardly be disputed that the social feelings are instinctive or innate in the lower animals; and why should they not be so in man? Mr Bain . . . and others believe that the moral sense is acquired by each individual during his lifetime. On the general theory of evolution it is at least extremely improbable. The ignoring of all transmitted mental

20 By this stage, for a number of reasons, but primarily because of the lack of a satisfactory theory of heredity, Darwin was allowing an increasingly greater role for mechanisms other than natural selection in the evolution of organisms. He employed use-inheritance generously throughout *The Descent* on the basis of his controversial theory of pangenesis, which allowed for the inheritance of acquired characteristics. See Vorzimmer 1970, especially Chs 5 and 6. This also possibly explains why Darwin came to rely so heavily on the mechanism of sexual selection in accounting for human evolution.

21 Darwin 1889 (2nd ed.), 128. All quotations from *The Descent* are taken from this edition. The relevant passages have been checked against the first edition (2 vols, London, 1871) for variations, and any such variations are indicated in the Notes.

qualities will, as it seems to me, be hereafter judged as a most serious blemish in the works of Mr. Mill.[22]

This emphasis on nature rather than nurture as the source of complex human behaviour, inevitably led Darwin into contradiction, which, as John C. Greene has pointed out, remained unresolved in *The Descent*.

> On the one hand, natural selection had operated to strengthen the social and sympathetic feelings among men. On the other, these feelings had acted to inhibit the operation of natural selection in civilised societies, thereby posing a threat to the continued progress of mankind. Here was the dilemma Darwin was to wrestle with in *The Descent of Man* without achieving a resolution (Greene 1977, 11).

The result was that while Darwin acknowledged the influence of purely social and cultural factors in social evolution, he was convinced that in the long run social progress could not occur through environmental improvements alone; a severe competitive struggle was necessary to prevent humanity from sinking into moral and intellectual degeneracy, and he urged a Malthusian prescription for social improvement in the General Summary of *The Descent*:

> [A]ll ought to refrain from marriage who cannot avoid abject poverty for their children; for poverty is not only a great evil but tends to its own increase by leading to recklessness in marriage. On the other hand, as Mr. Galton has remarked, if the prudent avoid marriage, whilst the reckless marry, the inferior members tend to supplant the better members of society. Man, like every other animal, has no doubt advanced to his present high condition through a struggle for existence consequent on his rapid multiplication; and if he is to advance still higher, it is to be feared that he must remain subject to a severe struggle, otherwise he would sink into indolence and the more gifted men would not be more successful in the battle of life than the less gifted. Hence our natural rate of increase, though leading to many and obvious evils, must not be greatly diminished by any means. There should be open competition for all men; and the most able should not be prevented by laws or customs from succeeding best and rearing the largest number of offspring.[23]

In Darwin's hands, natural section and the inheritance of acquired characteristics could therefore be invoked to explain a good deal more than mere genetic continuity with the lower animals. They explained and endorsed a number of

---

22 Darwin 1889, 98. The last sentence of this quotation does not appear in the first edition of *The Descent* (1871, I: 71). Its addition to the second edition (first published in 1874) suggests a hardening of Darwin's opposition to environmental explanations such as those offered by Mill.

23 Darwin 1889, 618. See also Greene's comments, Greene 1977.

assumptions which had assumed considerable social and political significance by 1871 – the superiority of the Anglo-Saxon (especially middle class Anglo-Saxons), the inevitable triumph of the more intellectual and the moral races over the lower and more degraded ones, the primitive evolutionary status of the "inferior" races and the continuing beneficent effects of competitive struggle in "civilized" societies. However there were limits to their explanatory power, particularly in the areas of racial and sexual differentiation, and these too were areas of major social and political concern in mid-Victorian England. Here sexual selection assumed a prominence which was to dominate *The Descent*.

Darwin initially introduced sexual selection in *the Descent* at the close of Part I, as an explanation of racial differences such as skin colour, hair, shape of skull, proportions of the body, etc., which he assumed to be of no evident benefit and not to correlate with climate and racial habits and customs. However, like natural selection, sexual selection took on a much wider role in human evolution. Darwin summed up its effects in the General Conclusion:

> He who admits the principle of sexual selection will be led to the remarkable conclusion that the nervous system not only regulates most of the existing functions of the body, but has indirectly influenced the progressive development of various bodily structures and of certain mental qualities. Courage, pugnacity, perseverance, strength and size of body, weapons of all kinds, musical organs, both vocal and instrumental, bright colours and ornamental appendages, have all been indirectly gained by the one sex or the other through the exertion of choice, the influence of love and jealousy, and the appreciation of the beautiful in sound, colour or form; and these powers of the mind manifestly depend on the development of the brain (Darwin 1889, 617).

Thus, apart from its primary function of explaining the persistence of seemingly non-beneficial human racial and sexual physical differences, sexual selection explained the utility of the aesthetic sense, and accounted for its high human development. It also accounted for the evolution of other uniquely human traits such as speech and music, for Darwin argued that these derived from the courtship behaviour of our "ape-like progenitors", females for instance, having acquired sweeter voices to attract the male; human speech having arisen from the probably effects of the long-continued use of the vocal organs of the male under the excitement of love, rage and jealousy. Sexual selection also of course accounted for the social inequality of the sexes, that aspect of its application with which this paper is most concerned and with which I shall deal in detail.

In all, there was a good deal riding on the efficacy of sexual selection in human evolution, and it becomes clear why Darwin devoted Parts II and III which comprise the major portion of *The Descent* to the demonstration of the general action of sexual selection throughout the animal kingdom and ultimately its extension to human evolution. *The Descent* does not comprise two books (one on human evolution and one on sexual selection) as has often been asserted, but is *one* book. Nor is its subject sex, as Ghiselin alleges (1972, 214). Its subject is human evolution. The extensive middle section on sexual selection is there as part of Darwin's overall strategy in arguing towards a natural scientific explanation of all aspects of human evolution; an explanation that extends from animal behaviour to human society and devolves on analogous courtship patterns of male combat and aesthetic preference in animals and humans.

Of course, as previously noted, Darwin conceded certain differences between animal and human courtship behaviour. In human evolution, aesthetic choice was exerted by the male, rather than the female as with the lower animals. The differing standards of beauty of the various races offered the explanation, via male aesthetic preference, of racial differentiation. "Monstrous" as it might seem that the "jet-blackness of the negro should have been gained through sexual selection", Darwin was convinced that it was so (Darwin 1889, 604). He was also certain that women's sweeter voices, absence of body hair, long tresses and greater beauty had all been acquired by male selection. The only physical trait he was inclined to attribute to female selection was that splendid Victorian emblem of virility, the beard (ibid., 566–606). As he explained it to Wallace in a passage redolent with Victorian values:

> A girl sees a handsome man, and without observing whether his nose or whiskers are the tenth of an inch longer or shorter than in some other man, admires his appearance and says she will marry him. So, I suppose, with the pea-hen; and the tail has been increased in length merely by, on the whole, presenting a more gorgeous appearance.[24]

Apart from this limited concession to feminine influence, Darwin held to the conviction that male selection predominated among humans. This role reversal caused him some bother, as he indicated to Wallace who was still insisting on the "greater, or rather, the more continuous, importance of the female (in the lower animals) for the race":

> Nothing would please me more than to find evidence of males selecting the more attractive females [among the lower animals]: I have for months been trying to persuade myself of this. There is the case of man in favour of this belief. . . . Perhaps I may get more evidence as I wade through my twenty years" mass of notes (F. Darwin and Seward 1903, 2: 76 [*CCD* 16, 2: 452]).

---

24 Darwin to Wallace, 1869, in F. Darwin and Seward 1903, 2: 63 (*CCD* 16, 1: 291–92). Darwin's practice of arguing by analogy from human to animal behaviour and his resultant anthropomorphism are here beautifully illustrated. The more so, because a year later in another letter to Wallace he reversed the analogy and circled back to human sexual selection: "It is an awful stretcher to believe that a peacock's tail was thus formed; but, believing it, I believe in the same principle somewhat modified applied to man", Darwin to Wallace, 1869, F. Darwin and Seward 1903, 2: 90 (This letter has now been redated as written in 1864, *CCD* 12: 248–49).

The problem was, as Darwin expressed it, that the male was the "searcher" who had "required and gained more eager passions than the female" – this made him ready to seize on any or many females without much regard to aesthetic prefer-ence (*CCD* 16, 2: 452; Darwin 1889, 221). How then had male humans become more discriminating? Without doubt they too were "searchers", more passion-ate and eager than women, in fact natural polygamists, as Darwin argued in *The Descent*. The answer, as given in *The Descent*, was that man had seized the power of selection from woman:

> Man is more powerful in body and mind than woman, and in the savage state he keeps her in a far more abject state of bondage, than does the male of any other animal; therefore it is not surprising that he should have gained the power of selection (Darwin 1889, 597).

This in turn, invited the question: How had man become "more powerful in body and mind than woman"? For it is not probable, as Darwin himself argued, that these differences had arisen through natural selection or through the inherited effects of men having worked harder for their subsistence than women: "for the women in all barbarous nations are compelled to work at least as hard as the men". The answer again lay in sexual selection, but in this case, through the alternative variant – male combat. Thus man's "greater size and strength . . . courage and pugnacity" had been acquired during the "long ages of man's savagery, by the success of the strongest and boldest men, both in the general struggle for life and in their contest for wives; a success which would have ensured their leaving a more numerous progeny than their less favoured brethren" (Darwin 1889, 563).

Here Darwin could invoke the analogy with animal courtship patterns with confidence. There is evidence of male combat or contest for wives among existing savages, "but even if we had no evidence on this head, we might feel almost sure, from the analogy of the higher Quadrumana, that the law of battle had prevailed with man during the early stages of his development" (Darwin 1889, 561–62).

As for the mental differences between the sexes, here Darwin was aware that he was venturing on a contentious issue. He had read *The Subjection of Women* where Harriet Taylor and John Stuart Mill had argued that such differences as could be ascertained were culturally conditioned, not innate.[25] But, consistent

---

25 Mill wrote: "I consider it presumption in any one to pretend to decide what women are or are not, can or cannot be, by natural constitution. They have always hitherto been kept, as far as regards spontaneous development, in so unnatural a state, that their nature cannot but have been greatly distorted and disguised; and no one can safely pronounce that if women's nature were left to choose its direction as freely as men's, and if no artificial bent were attempted to be given to it except that required by the conditions of human society, and given to both sexes alike, there would be any material difference, or perhaps any difference at all, in the character and capacities which would unfold themselves. I shall presently show, that even the least contestable differences which now exist, are such as may very well have been produced merely by circumstances, without any difference of natural capacity." *The Subjection of Women* (Rossi 1970, 190). Darwin referred to

MR. PUNCH'S DESIGNS AFTER NATURE.

GRAND BACK-HAIR SENSATION FOR THE COMING SEASON.

Figure 5.1 Punch's commentary on Darwin's theory of sexual selection in the just-published *Descent of Man*. A modish young woman displays the full plumage of a peacock for masculine appreciation, replicating the contradictions of Darwin's naturalization of female choice among birds, while normalizing male choice in humans. *Punch*, April 1, 1971. University of Sydney Library Collection.

with his earlier opposition to Mill on the heritability of the "moral faculties", Darwin insisted that the "differences in the mental powers of the two sexes" (and he emphasized considerable differences) were biologically based. Again he invoked the analogy with lower animals:

> I am aware that some writers doubt whether there is any such inherent difference; but this is at least probable from the analogy of the lower animals which present other secondary sexual characters. No-one disputes that the bull differs in disposition from the cow, the wild-boar from the sow, the stallion from the mare, and, as is well known to the keepers of menageries, the males of the larger apes from the females (Darwin 1889, 563).

On this basis Darwin proceeded to assert the instinctive maternal traits of the human female and the human male's innate aggressive and competitive characteristics. Woman's maternal instincts lead her to be generally more tender and altruistic than man whose "natural and unfortunate birthright" is to be competitive, ambitious and selfish. But above all man is more intelligent than woman:

> The chief distinction in the intellectual powers of the two sexes is shewn by man's attaining to a higher eminence in whatever he takes up, than can woman – whether requiring deep thought, reason, or imagination, or merely the use of the senses and hands (Darwin 1889, 564).

For Darwin, the intellectual differences between the sexes were entirely predictable on the basis of a consideration of the long-continued action of natural and sexual selection, reinforced by use-inheritance. Male intelligence would have been consistently sharpened through the struggle for possession of the females, through hunting and other male activities such as defence of the females and young. Intelligence thus acquired by males after sexual maturity would be inherited by male offspring at a corresponding period. Male pre-eminence has thus come about:

> ... partly through sexual selection, – that is, through the contest of rival males, and partly through natural selection, – that is, from success in the general struggle for life; and as in both cases, the struggle will have been during maturity, the characters gained will have been transmitted

---

The Subjection of Women in a footnote to this section (*The Descent of Man*, Darwin 1889, 564n): "J. Stuart Mill remarks (*The Subjection of Women*, 1869, p. 122), 'The things in which man most excels woman are those which require most plodding, and long hammering at single thoughts'. What is this but energy and perseverance?" Compare this with Darwin's description of his own "mental qualities" where he attributed his success as a "man of science" to, among other qualities, "unbounded patience in long reflecting over any subject – industry in observing and collecting facts". (*The Autobiography of Charles Darwin*, Barlow 1969, 139–145.) See also his letter to Francis Galton of 1870, where he stressed the importance of "zeal and hard work" in intellectual achievement (quoted in full, Note 57 below).

more fully to the male than to the female offspring. . . . Thus man has ultimately become superior to woman (Darwin 1889, 565).

Reference must here be made to Darwin's notion of inheritance, which he had made clear in the earlier section on sexual selection. In brief, the tendency was for "characters acquired by either sex late in life, to be transmitted to [offspring of] the same sex at the same age, and of early acquired characters to be transmitted to both sexes".[26] These rules, however, as Darwin acknowledged, did not always hold good. Indeed it was fortunate that they did not, and that in mammals late acquired characteristics were sometimes transmitted to both sexes "otherwise it is probable that man would have become as superior in mental endowment to woman, as the peacock is in ornamental plumage to the peahen". If they always held good, Darwin wrote, we could draw certain social conclusions from them "(but here I exceed my proper bounds)". Nevertheless, he proceeded to argue that the inherited effects of the early education of boys and girls would be transmitted equally to both sexes, so a similar early education would do nothing to equalize the current intellectual differences between the sexes which would be maintained by the inherited effects of their very different mature roles; nor, for the same reason, could these differences be attributed to the different early training of boys and girls. Rather, Darwin proposed:

In order that woman should reach the same standard as man, she ought, when nearly adult, to be trained to energy and perseverance, and to have her reason and imagination exercised to the highest point; and then she would probably transmit these qualities chiefly to her adult daughters (Darwin 1889, 565).

The difficulty was that in order for the general level of feminine intelligence to be raised, such educated women would need to produce more offspring over many

26 Darwin 1889, 565 and 227–39. These rules of inheritance were quite consistent with Darwin's belief that acquired characters could be inherited and that sex was determined by the relative contributions of the parents. Thus, if males required more intelligence for hunting and other male activities, this would extend their intelligence during their life and would be inherited by their offspring. If the male contributed a greater complement to the individual offspring, then it would be male and would be more likely to inherit the higher intelligence acquired by the father. Of course, if the child were a daughter, some of the characteristics of the father would be inherited, but these would be in a smaller proportion than in the case of a son. Over successive generations, slight increases in intelligence acquired by males would gradually accumulate and become proportionately greater in males than in females. Thus Darwin's views were compatible with his ideas on inheritance. The real issue is whether males are in fact superior in intelligence to females, and Darwin gave no factual support to this assumption, arguing entirely in social terms and citing the platitudes of his time. It is interesting to note that even if hunting man did require more intelligence for his male pursuits (which is dubious), on current theories of inheritance, any intelligence giving the hunting man an advantage would be inherited equally by his daughters and sons. I am indebted to my colleague Margaret Campbell for her clarification of this point.

generations than their less educated sisters. The implication was that this was unlikely. Meanwhile, although male combat was no longer in operation in civilized societies, male intelligence would be constantly enhanced by the severe competitive struggle males necessarily underwent in order to maintain themselves and their families, and "this will tend to keep up or even increase their mental powers, and, and a consequence, the present inequality between the sexes" (Darwin 1889, 565–66). The conclusion to be drawn from this was that the higher education of women could have no long-term impact on social evolution and was, biologically and socially, a waste of resources.

It is noteworthy that in support of his assertion of male intellectual superiority. Darwin did not deploy his favourite tactic of arguing by analogy from the lower animals. He argued solely in social terms of the lack of feminine eminence in the arts and sciences:

> If two lists were made of the most eminent men and women in poetry, painting, sculpture, music . . . history, science, and philosophy . . . the two lists would not bear comparison (Darwin 1889, 564).

Again, while he conceded that "with woman the powers of intuition, of rapid perception, and perhaps of imitation, are more strongly marked than in man", he dismissed these faculties as "characteristic of the lower races, and therefore of a past and lower state of civilisation" (Darwin 1889, 564).

In order to understand the sense of this statement by Darwin, it is necessary to turn to the theory of recapitulation. This theory, epitomized in the unqualified and misleading slogan "Ontogeny recapitulates Phylogeny" by the German morphologist and Darwinian Ernst Haeckel in 1866, became the cornerstone of late Victorian evolutionary theory. It functioned as the organizing principle for generations of work in comparative embryology, physiology, morphology and palaeontology. In its pervasive influence on nineteenth-century social theory, psychology and anthropology, it was outstripped only by natural selection itself (Gould 1977, 115–66). The idea that individual development is a recapitulation of ancestral stages was implicit in *The Origin* and Darwin himself had placed considerable emphasis on this embryological evidence of evolution. By the time *The Descent* appeared, the majority of Darwinians had uncritically adopted recapitulation and it figured prominently in Darwin's argument for the animal ancestry of humanity. More significantly, it underlay his conception of the development of human mental, social and ethical faculties. [27] For the study of human developmental stages was a method that allowed the reconstruction of human "ancestors" and

---

27 "In the next chapter I shall make some few remarks on the probable steps and means by which the several mental and moral faculties of man have been gradually evolved. That such evolution is at least possible, ought not to be denied, for we daily see these faculties developing in every infant; and we trace a perfect gradation from the mind of an utter idiot, lower than that of an animal low in the scale, to the mind of a Newton" (Darwin 1889, 127). For a discussion of the role

the ranking of races, depending on how closely their modern descendants could be correlated with the primitive forms revealed by the ontogeny of "higher" races.

The recapitulatory argument for ranking extended beyond race to sex. It was a standard claim of recapitulationists that woman's development was arrested at the level of the child and the negro:

> In the brain of the Negro the central gyri are like those in a foetus of seven months, the secondary are still less marked. By its rounded apex and less developed posterior lobe the Negro brain resembles than of our children, and by the protuberance of the parietal lobe, that of our females (Vogt 1864, 183).

This quotation is taken from the work of Carl Vogt, the German Darwinian and polygenist,[28] whose *Lectures on Man* was published in English translation in 1864 by the racist Anthropological Society of London. Darwin was impressed by Vogt's work and proud to number him among his advocates (Darwin 1889, 1). He cited Vogt's morphological arguments on racial and sexual differences and inequalities on several occasions in *The Descent*. He agreed with Vogt that the mature female, in the formation of her skull, is "intermediate between the child and the man" and that woman's anatomy generally, was more child-like or "primitive" than man's (Darwin 1889, 557). It was an extension of Vogt's woman-as-child-as-primitive argument that provided the sole scientific underpinning of Darwin's conclusions on the futility of higher education for women. In a footnote to his assertion that the present sexual inequalities could only by enhanced rather than diminished by social progress, Darwin wrote:

> An observation by Vogt bears on this subject: he says, "It is a remarkable circumstance, that the difference between the sexes, as regards the cranial cavity, increases with the development of the race, so that the male European excels much more than the female, than the negro the negress" (Darwin 1889, 566n).

Darwin cited further evidence from measurements of negro and German skulls in support of this contention, but scrupulously added Vogt's qualification that more

---

of recapitulatory theory in the development of Darwin's evolutionary theory, see Richards 1976; also Oppenheimer 1959.

28 The polygenists of the nineteenth century generally believed that the human races were aboriginally distinct, in opposition to the monogenists who advocated an original racial unity in terms consistent with the biblical account. Vogt managed to reconcile his racist polygenist belief with his Darwinism by arguing that the human races were actually different species whose separate lines of evolution might be traced back into the very remote past to a common ancestry, but whose current differences were so great as to be virtually unbridgeable. Although Darwin did not agree with the polygenist categorization of human races as distinct species, he seems to have agreed with Huxley that the Darwinian theory satisfactorily reconciled the monogenist emphasis on human unity with the polygenist insistence on the maximum of racial divergence consistent with an extremely remote common ancestry. See Stocking 1968, 1971; Darwin 1889, 176–78. See also Note 34 below.

observations were requisite before it could be accepted as generally true. Nevertheless, Vogt had been as ready as Darwin to found contemporary sexual inequalities on this admittedly inadequate evidence, and to proscribe any possibility of future sexual equality. Immediately after the above statement cited by Darwin, Vogt had written in his *Lectures on Man*:

> It has long been observed that, among peoples progressing in civilization, the men are in advance of the women; whilst amongst those which are retrograding, the contrary is the case. Just as, in respect of morals, woman is the conservator of old customs and usages, of traditions, legends, and religion; so in the material world she preserves primitive forms, which but slowly yield to the influences of civilization. We are justified in saying, that it is easier to overthrow a government by revolution, than alter the arrangements in the kitchen, though their absurdity be abundantly proved. In the same manner woman preserves, in the formation of the head, the earliest stage from which the race or tribe has been developed, or into which it has relapsed. Hence, then, is partly explained the fact, that the inequality of the sexes increases with the progress of civilization (Vogt 1864, 81–82).

There can be little doubt that Darwin shared Vogt's conclusion that sexual inequality was the hallmark of an advanced society, and his previous relegation of certain of woman's mental traits to a "past and lower state of civilization" may also be attributed to this source.

In all, the evidence Darwin marshalled in support of his argument for the innate and continuing inferiority of women through the combined action of natural and sexual selection was scanty and primarily socially derived. The familiar analogy with the animals was conspicuously lacking (where were those examples of greater male intelligence among the higher Quadrumana?) and such morphological evidence as could be cited was as yet unsubstantiated (and never to be) (Gould 1977). The whole was a triumph of ingenuity in response to theoretical necessity in the face of a dearth of hard evidence, fed by Victorian assumptions of the inevitability and rightness of the sexual division of labour: of woman's role as domestic moral preceptor and nurturer and man's role as free-ranging aggressive provider and jealous patriarch. Consistent with this, Darwin went to some lengths in *The Descent* to defend what he called the "natural and widely prevalent feeling of jealously, and the desire to each male to possess a female for himself" (Darwin 1889, 594). In the process he attacked the contemporary anthropological notion of primitive promiscuity and the even more unnatural "perversion" of polyandry, even though he admitted anthropological evidence of both practices among existing savages. Here he swept aside anthropology and reverted to the animal analogy:

> At a very early period, before man attained to his present rank in the scale, many of his conditions would be different from what now attains

amongst savages. Judging from the analogy of the lower animals he would then either live with a single female, or be a polygamist. The most powerful and able males would succeed best in obtaining attractive females.[29]

As the quotation indicates, Darwin was not so much promoting patriarchy as defending sexual selection which he could only envisage as operative in some system of male dominance where males held the power of selection and females were valued for their charms.

If Darwin was, in fact, "in the grip of the system he had constructed" (Jones 1978, 16), the relevancy of *The Descent* to predominant Victorian social and political concerns is none-the-less real and must be faced. It is not necessary to assume that Darwin's reconstruction of human evolution was primarily a political ploy, in order to argue that Darwin was deeply influenced by certain social and political assumptions, which coloured his ideas about nature and society and directed his attention to certain contentious areas. The derivative character of *The Descent* and Darwin's practice of sorting and sifting the information he collected into support for or opposition to his theory has been asserted by a number of scholars (Jones 1978; Greene 1977), and I shall return to this. For my immediate purposes, it is essential to see Darwin's work as part of a more general tendency of nineteenth-century thought to treat human mental and social development more scientifically or naturalistically. In this light, what might seem to be mere appropriation on Darwin's part, may be more correctly considered as reciprocal borrowings from a related trend. Thus Vogt's recapitulatory arguments for woman's inferiority can be found in embryo, so to speak, in Darwin's Notebook entry of 9 September 1838:

> It is worthy of observation that in insects where one of the sexes is little developed, it is always female which approaches in character to the larva, or less developed state. –
> The female & young of all birds resemble each other in plumage. – (That is where the female differs from the male?) children & women – "women recognized inferior intellectually".[30]

It is clear from this entry that Darwin had already arrived at the woman-as-child-as-primitive equation, and that in considering human sexual differences he assumed intellectual as well as physical juvenility, hence, inferiority in women. Vogt's basic premise was not new to Darwin, but Vogt had given it a limited

---

29 Darwin 1889, 594; see also 46–7, and 216.
30 Notebook D 76; Barrett et al. 1987, 358. At this stage, Darwin was still doubtful of the generality of this argument, as his question mark indicates. He wrote after the above: "Opposed to these facts are effects of castration on males and of age or castration in females". By the time of *The Descent*, he presented the effects of male castration as "striking" confirmation of his argument for the intellectual inferiority of women via sexual selection (Darwin 1889, 565).

empirical basis and an overt social content, which Darwin could hook on to the contemporary controversy on higher education for women. When he linked it with the concepts of sexual and natural selection (themselves heavily freighted with social and cultural values) he could prescribe as well as interpret and justify the existing social inequality of the sexes on this "naturalistic" basis.

Another Notebook entry made a few days after the above, will serve to illustrate Darwin's theoretically directed practice of arguing analogically from humans to animals:

> September 13th. The passion of the doe to the victorious stag, who rubs the skin of[f] horns to fight, is analogous to the love of women (as Mitchell remarks seen in savages) to brave men.[31]

Such analogy, as we have seen, was necessary to Darwin's argument that the higher human faculties had evolved from instinctive animal behaviour. He instituted and defended the practice in the Notebooks: "Arguing from man to animals is philosophical".[32] Although he was aware of some of the pitfalls that might attend such subjective description of behaviour ("I must be very cautious"),[33] it led directly to some of the more absurd aspects of *The Descent*, such as where Darwin pictured animal sexual behaviour in terms consistent with Victorian sexual morality – where female animals were depicted as coyly Victorian, with as little inclination for sexual encounters as their human counterparts were generally considered to have:

> The female, on the other hand, with the rarest exceptions, is less eager than the male. As the illustrious Hunter long ago observed, she generally "requires to be courted"; she is coy, and may often be seen endeavouring for a long time to escape from the male. Every observer of the habits of animals will be able to call to mind instances of this kind. It is shown by various facts, given hereafter, and by the result fairly attributable to sexual selection, that the female, though comparatively passive, generally exerts some choice and accepts one male in preference to others. Or she may accept, as appearances would sometimes lead us to believe, not the male which is the most attractive to her, but the one which is the least distasteful (Darwin 1889, 222).

It is such value-laden description that prompted Ruth Hubbard to comment:

> Make no mistake, wherever you look among animals, eagerly promiscuous males are pursuing females, who peer from behind languidly

---

31 Notebook D 99; Barrett et al., 1987, 362.
32 Notebook N 49; Barrett et al., 1987, 576.
33 Notebook N 10; Barrett et al., 1987, 565.

drooping eyelids to discern the strongest and handsomest. Does it not sound like the wishfulfillment dream of a proper Victorian gentleman? (Hubbard 1979, 18–19)

When such anthropomorphic description was analogically reapplied to human behaviour and social institutions, it inevitably provided naturalistic corroboration of Victorian values.

Further, Darwin's androcentric description of animal courtship practices, where the initiation of all activity was assigned to the male and females (although possessed of some rudimentary aesthetic sense which they exercised in the selection of male charms) remained passive "spectators" of male combat and display, paved the way for Darwin's analogical role reversal from animal female to human male aesthetic selection.

In *The Descent* the human male became more the analogue of the animal breeder, who exercises his caprice in varying the appearance of the breed:

> Each breeder has impressed . . . the character of his own mind – his own taste and judgment – on his animals. What reason, then, can be assigned why similar results should not follow from the long-continued selection of the most admired women by those men of each tribe who were able to rear the greatest number of children? (Darwin 1889, 596)

As the breeder selects and shapes his domestic productions, so man has moulded woman to his fancy. In illustration of this, Darwin credulously offered the unforgettable picture of the Hottentots (courtesy of Burton) who "are said to choose their wives by ranging them in a line, and by picking her out who projects farthest *a tergo*. Nothing can be more hateful to a negro than the opposite form" (Darwin 1889, 579).

In the earlier work of James Cowles Prichard (1813) there is historical precedent for the agency of male aesthetic preference in the shaping of human variety. Prichard also argued analogically from artificial selection and it is possible that Darwin was familiar with Prichard's argument.[34] However there is no reason to

---

34 Prichard 1973, 41–6. Prichard was no biological evolutionist but more a "diffusionist" who was concerned with the problem of explaining human variation in terms consistent with the biblical account that all humanity had descended from a single human family – presumably that of Noah. His concept of "sexual selection", while in some respects similar to Darwin's, was advanced in thoroughly teleological terms. Moreover, Prichard's views were modified in response to social pressure, so that in subsequent editions of *The Researches* he dropped his emphasis on sexual selection as a forming factor of race and developed his argument in terms of a correlation of climate and physical type. Whether or not Darwin was familiar with Prichard's earlier ideas is not clear. As far as can be ascertained, he read only the third and fourth editions of Prichard's Researches (F. Darwin and Seward 1903, I: 46. But see Greene 1977, 4). The notion of aesthetic preference as a factor in racial variation was not unique to Prichard. It was also suggested by Edward Blyth (1835), whose work was certainly familiar to Darwin (Eiseley 1979, 106). From the

suppose that Darwin could not have arrived at this conception of human variation independently of Prichard.[35] Darwin's dependency on the analogy of artificial
selection to illustrate, explain and endorse the action of natural selection is too
well known to require elaboration here (Ruse 1975; Young 1971a). It was inevitable that he would see in the notion of aesthetic choice an even closer analogy with
artificial selection. Darwin regarded humans as pre-eminently a domesticated species, and was fond of comparing civilization to the process of domestication (Darwin 1889, 172). This was consistent with his insistence on the biological basis of
mental and moral qualities. The domestication of animals is brought about not
through training, but by a process of selection and breeding for the required traits.
In his correspondence with Wallace on sexual selection, Darwin wrote: "I lay
great stress on what I know takes place under domestication".[36] So I agree with
Ghiselin that "the theoretical elaboration and verification of sexual selection drew
strongly upon the study of artificial selection and embryology" (Ghiselin 1972,
220). But I would go further than Ghiselin and argue that in the case of human
selection, Darwin identified the human male with the breeder – that he put into
men's hands the modifying and shaping power of the breeder, and that he did so
for the purely cultural reason that it was inconceivable to this proper Victorian that
human evolution could have been modified and shaped by female caprice or by
female sexuality and passion. Where Ghiselin sees only theoretical consistency in
Darwin's overall concept of sexual selection and defends Darwin from the charge
of anthropomorphism (Ghiselin 1972, 218), I concede the theoretical constraints,
but argue that the concept of sexual selection and Darwin's application of it to
human evolution is pervaded by Victorian sexist ideology. Where Ghiselin asserts
that *The Descent* "owes its success to the power of abstract reasoning that gave

---

context of Blyth's remarks, it seems he adopted the idea from Prichard or possibly from William
Lawrence's *Lectures on Physiology. Zoology and the Natural History of Man* (1819), with which
Darwin was also familiar. While on the subject of historical precedent, the initial stimulus for Darwin's interest in sexual selection (though not for the notion of aesthetic preference) undoubtedly
came from the *Zoonomia, or, the laws of Organic Life* (1791) of his grandfather Erasmus Darwin.
Erasmus wrote of the effect of male combat in ensuring the propagation of the "strongest and
active" males, and this is clearly the source of Darwin's contention that sexual selection via male
combat enhances the action of natural selection. See Ghiselin 1976, 127; also Gruber's remarks on
the "family Weltanschauung" shared by the two Darwins (Gruber 1974, 49–52).

35 The evidence of the Notebooks, sketchy though it is, supports the contention that Darwin charted his
own course to sexual selection with the help of a few nudges from his predecessors. The Notebooks
catalogue numerous observations on human and animal sexual behaviour and sexual differences,
and some speculation on the role of aesthetic factors in reproduction, but not the concept of sexual
selection which did not appear in Darwin's account of evolutionary processes until the "sketch"
of 1842, to be expanded in the "Essay" of 1844. Both these early accounts of sexual selection are
entirely androcentric in their description of animal sexual behaviour, and the discussion of sexual
selection in the "Essay" concludes with the analogy with artificial selection: "This natural struggle
among the males may be compared in effect, but in a less degree, to that produced by agriculturalists
who pay less attention to the careful selection of all the young animals which they breed and more to
the occasional use of a choice male" (F. Darwin and Seward 1903, 93, see also 10).

36 Darwin to Wallace 1868, in F. Darwin and Seward 1903, 2: 84 [*CCD* 16, 2: 762].

rise to it" (Ghiselin 1972, 230), I would argue that *The Descent* owed its success primarily to the fact that it had social and political sanction.

Clearly *The Descent* did much more than proffer a naturalistic or scientific explanation of human evolution as an intellectual *tour de force*. It proffered social interpretation, justification and prescription. The congruence of *The Descent* with dominant Victorian social and political assumptions arose partly from Darwin's persistent practice of arguing analogically from humans to animals which led to anthropomorphism and ultimately to circularity when such arguments were reapplied to human behaviour and social arrangements; partly from Darwin's need to seek out and consolidate alliances with a related intellectual tradition that had a more explicit social and political content as in the writings of Vogt and Spencer. Darwin borrowed widely from this tradition for *The Descent*, reinforced it, and thereby strengthened his own values which he had held from his earliest Notebook jottings.

I shall now turn to the consideration of how Darwin, as an individual, came to hold his beliefs on feminine abilities and differences and how these matched up with and fed into the general Victorian image of the female role. In the absence of any other historical evidence, and for the reasons outlined in the introduction, it is necessary to reconstruct, as far as possible, Darwin's relations with the woman with whom he lived on close and harmonious terms for forty-three years – his wife Emma.

II. Emma

> *The most favourable case which a man can generally have for studying the character of a woman, is that of his own wife: for the opportunities are greater, and the cases of complete sympathy not so unspeakably rare. And in fact, this is the source from which any knowledge worth having on the subject has, I believe, generally come. But most men have not had the opportunity of studying in this way more than a single case: accordingly one can, to an almost laughable degree, infer what a man's wife is like, from his opinions about women in general.*
>
> – J. S. Mill (and Harriet Taylor Mill), *The Subjection of Women* (Rossi 1970, 151)

Having duly weighed the pros and cons in favour of marriage, Charles Darwin soon found his "nice soft wife on a sofa" in his cousin Emma Wedgwood, although throughout their life together it was the semi-invalid Charles who occupied the sofa, not Emma. Emma hardly had the chance. As their daughter Henrietta recorded:

My mother had ten children and suffered much from ill-health and discomforts during those years. Many of her children were delicate and difficult to rear, and three died. My father was often seriously ill and always suffering, so that her life was full of care, anxiety, and hard work. But she was supported by her perfect union with him, and by the sense that she made every minute of every weary hour more bearable to him (Litchfield 1915, 2: 45).

Even against the "little woman behind the great man" stereotype, Emma stands out in her total submergence of self in the great man's well-being and his projects. Ever solicitous of Darwin and his numerous ailments through his forty years of invalid existence, utterly devoted to his interests (although she in no way shared them), she created and preserved the orderly, quiet, entirely domestic environment Darwin desperately craved for his work and health. Her days were planned out to suit him and the elaborate routine he devised to achieve the maximum of work with the least possible distress to his delicate constitution. Emma was ready to read aloud to him during his periods of rest on the sofa, to write his letters at his dictation, go for walks with him, and be constantly at hand to alleviate his daily discomforts. She helped proof *The Origin* and dutifully watched over his experiments. But she had little interest in science, only in the scientist. She was deeply religious and many of his opinions were painful to her, yet it was Emma whom Darwin entrusted to carry out the publication of the preliminary version of his "species Theory" in the event of his death. It proved unnecessary (he lived for another thirty-eight years), but there is no doubt that Emma would have loyally carried out his wishes.[37]

With the possible exception of her religious beliefs, there is no evidence whatever that Darwin was not more than content with Emma's circumscribed role of perfect nurse and loyal helpmate. Before their marriage, he defined her proper sphere: Emma was to "humanize" him, to teach him that there was greater happiness in life than "building theories and accumulating facts in silence and solitude" (Litchfield 1915, 2; 23). He had not expected intellectual companionship in marriage, and in fact discouraged it. While she was still his fiancée, he dissuaded Emma from reading Lyell's *Elements of Geology* which she had embarked upon under the impression that she should "get up a little knowledge" for him. In Darwin's experience, science was an exclusively male preserve, which women entered, if they entered at all, only as spectators – at the most as fashionable dabblers, not to be taken seriously. He did not expect or want women to converse intelligently about science, but rather to be tolerant of masculine preoccupation with it, like "poor Mrs Lyell" who sat by, a "monument of patience", while Darwin and Lyell talked "unsophisticated geology" for half an hour (Litchfield 1915, 2: 13, 24).

The one occasion we know of when Darwin set aside these conventional views of his "nice soft wife" was when he decided to disregard his father's advice and discuss his loss of religious faith with Emma soon after they married. The result

37 Litchfield 1915, passim; see also Himmelfarb 1967, 196–97.

was not happy. Emma was evidently seriously distressed by Darwin's religious doubts, so much so that she set down her concern in writing – a carefully phrased letter which Darwin preserved. She suggested that he had been unduly influenced by his brother Erasmus, that the scientific habit of "believing nothing until it is proved" ought not be extended to matters of faith, and expressed her belief in the value of prayer. The letter is at once an expression of diffidence at opposing her "feeling" to his "reasoning" and of conviction of her wifely duty to do so. She loved him and she feared for his immortal soul:

> I should say also there is a danger in giving up revelation which does not exist on the other side, that is the fear of ingratitude in casting off what has been done for your benefit as well as for that of all the world and which ought to make you still more careful, perhaps even fearful that you should not have taken all the pains you could to judge truly . . . I should be most unhappy if I thought we did not belong to each other for ever.

Darwin's response to this was rather poignant:

> When I am dead, know that many times I have kissed and cryed over this. C.D.[38]

We have no definite information, but it would seem that husband and wife were mutually concerned not to let their religious differences mar their domestic relations, and that they thenceforth avoided the topic, confining themselves to their respective spheres. Darwin continued with his science and his scepticism and Emma busied herself with his person and not with his distressing ideas and work, which she nevertheless loyally supported and promoted by her domestic arrangements and by her acquiescence in relinquishing the London society and theatre parties she had enjoyed so much. Darwin's increasing ill-health and absorption in his work dictated the latter necessity, and Emma's life narrowed to one of "watching and nursing . . . cut off from the world" (Henrietta's description; Litchfield 1915, 2; 56). She had her reward in his gratitude expressed in the fulsome tributes of Darwin's *Autobiography*. She was his "greatest blessing", his "wise adviser and cheerful comforter throughout life", so infinitely his superior in "every single *moral* quality" (my emphasis) (Barlow 1969, 96–97).

---

38 Barlow 1969. 235–39, 97. The one woman of Darwin's acquaintance who might have challenged his conventional notions of women was Harriet Martineau, a close friend of his brother Erasmus, who moved in more radical and literary circles than Charles. Martineau, when Charles knew her, was already a noted writer and intellectual, well-travelled and an ardent defender of women's rights. It is possible that it was through knowing Martineau, an acknowledged female sceptic, that Darwin decided to discuss his religious doubts with Emma. For Martineau's views on the emancipation of women, see Rossi 1973, 118–43.

This stereotype of Victorian feminine servitude, domesticity and piety, is given a bit of a jolt by Henrietta's ascription of "remarkable independence" to her mother's character and way of thinking (Litchfield 1915, 1: 61–62). True, there are glimpses of another Emma behind the facade of the perfect nurse. She was, for her time, a reasonably cultivated woman. She knew French and Italian, and her German was considerably better than Darwin's. Characteristically, she helped him with his translations. Her letters show her to have had humour and a wide general knowledge. If Darwin's taste dictated the choice of the popular, sentimental novels she read aloud to him (typically, he preferred happy endings and a lovable and pretty heroine), her own choice was wider ranging. In spite of her professed indifference to Darwin's work, she seems to have understood it and its implications pretty well. And how much of this indifference was really aversion on religious grounds? Again, for all her piety, she could, on occasion, dissent from conventional religious opinion, as when she defended the morality and ethics of "this new breed of agnostics". After Darwin's death, she took a great interest (although a decidedly conservative one) in politics, avidly following the election results and parliamentary debates. She knew she ought to care about the higher education of women, although she did not (Litchfield 1915, 2: 172, passim). Nevertheless, stereotype and historical person coincide fairly well. Whatever independence of mind Emma exhibited, it hardly appears remarkable even in Victorian terms, and it certainly did not extend to any notion of female equality. Her background, training and circumstances concurred to that end. Henrietta's account of her mother's early life is an unwitting testament to the powerful patriarchal conditioning of Victorian women.

Emma's maternal grandfather had been in the habit of thumping his fist on the table and ordering his daughters to talk when he wished to be entertained after dinner. His daughters all became good talkers but went in "nervous dread" of their father who made their home life utterly constrained and miserable. Not surprisingly, Emma's mother considered men as "dangerous creatures who must be humoured" and treated her husband accordingly. Emma's father, Josiah Wedgwood, son of the potter industrialist of the same name, also inspired nervous awe in most of his female relations, one of whom described him as "always right, always just, and always generous". Charles Darwin's sisters, who had their own household patriarch to placate in Dr Robert Darwin, were astounded at the ease and familiarity with which Charles treated Uncle Jos, "as if he was a common mortal" (Litchfield 1915, 1: 1–14).

The second, third and fourth generation Wedgwoods and Darwins who so often intermarried, may have inherited some unconventional theological and political notions, but they were entirely orthodox in their understanding and expectations of woman's domestic and social roles. These staunch supporters of negro emancipation would have been confounded by the suggestion that their wives, daughters, sisters, needed emancipating. The elaborate division of labour that underlay the successful pottery enterprise that founded the Wedgwood fortunes extended to the domestic sphere, where the respective roles of men and women were thoroughly understood and defined. A Wedgwood (Emma's father) required his wife to be

> sensible to his pains and his pleasures, participat[e] in his hopes,. . . [strengthen] his good dispositions and gently discourag[e] his harshness and petulance, and more than all . . . become flesh of his flesh and bone of his bone, by bearing him children . . . (Litchfield 1915, 1: 14).

Men might indulge in "philosophy", women were assumed to be bound by religious piety to their roles of moral preceptors of family life. A husband should guard his religious opinions lest he distress his wife. In all his life, Darwin's father had known only three women sceptics, and of one of these he was not certain (Litchfield 1915, 1: 95–96). A high premium was placed of feminine prettiness, vivacity and sweetness; little or none on feminine intellect, education or independence. In choosing his wife from his Wedgwood cousins, Darwin could be as comfortable in his expectations of her assumption of his male supremacy and importance, as he was of her substantial dowry.[39]

Not that Darwin was in any sense a typical Victorian patriarch. The historian, Gertrude Himmelfarb, who is one of Darwin's harshest critics, concedes:

> The most cynical reader of biographies would be hard put to it to dispute the genuineness of the love and respect borne him by his family, and his most determined enemies were unable to call into question his gentleness, modesty, and good nature. There may be much in his work and mind to criticise, but little in his character.

Nevertheless, Himmelfarb continues tartly, his character and mind were all of a piece: ". . . what was admirable in the one was not necessarily so in the other, tenderness of character sometimes showing itself as softness of mind".[40]

---

39 Emma's dowry was a bond of £5000 and an allowance of £400 a year. See Litchfield 1915, 2: 3.

40 Himmelfarb 1967, 142. Himmelfarb is one of the few Darwin historians to have subjected Darwin's concept of sexual selection to a searching critique and her pithy criticisms are often very apt: "(T)his standard of beauty that is so capricious among savages must have been even more so among prehistoric men, to favour a patch of hair around the chin of man and to discourage it on woman. To complicate matters, this capriciousness must have remained constant for an untold number of generations, if the species was to evolve at the slow pace Darwin set for it. It was a bold experiment to make so tenuous and hypothetical an idea as the aesthetic standards of our ape-like progenitors bear the burden of such weighty matters as the evolution of man from the animals and the distinctions of sex and race" (ibid., 366). Although she perceived the anthropomorphism of Darwin's discussion of sexual selection (ibid., 346), Himmelfarb failed to discern its androcentrism. Unfortunately, her lack of customary reverence for her subject and the gusto with which she set about mowing down this tall poppy provoked a storm of criticism from more conventional Darwin scholars, and her work (in many respects very good) was not well reviewed.

It is a curious contradiction, that the man whose writings have been credited with such revolutionary impact, should have clung so tenaciously to the familiar, cosy and innocuous after his arduous stint on the "Beagle" – to have made the shawl, sofa and feminine attendant a way of life. There has been a good deal of controversy about the nature of Darwin's ill-health and suggestions range from those of specific aetiology to the frankly Freudian. A more plausible explanation is that Darwin turned himself into an invalid simply to get on with his work (Pickering 1974). This would explain his acquiescence in the excessive care Emma bestowed on him, the advantage he consistently took of his semi-invalidism to avoid the strains of a social life which would have interfered with his work, and the enormous amount of scientific work, both experimental and literary, he managed to accomplish in spite of his chronic ill-health. He did not have to trouble himself about the management of house, garden or livestock. Emma "shielded him from every avoidable annoyance, and omitted nothing that might save him trouble, or prevent him becoming over-tired" (F. Darwin 1888, 1: 159). He was a loving, kindly and indulgent father, but his children "all knew the sacredness of working time" (F. Darwin 1888, 1: 136). For all his free and easy relations with them, he inculcated the Victorian virtues of respect and obedience: "Whatever he said was absolute truth and law to us".[41] The atmosphere of Down House has been so often evoked as affectionate and homely, but there is no question that Darwin's invalid status and work routine were dominant, and that his family patterned their lives around the demands of his twin occupations. Without departure from his consistent "gentleness, modesty and good nature", he nevertheless achieved what he wanted. His most diffident wishes were as much deferred to as the despotic demands of any fist-thumping, awe-inspiring patriarch, and his love and gratitude endorsed the narrow, entirely domestic lives he tenderly imposed on wife and daughters. The unacknowledged stresses of that cosy environment are suggested by Henrietta's prolonged and mysterious breakdown between the ages of thirteen and eighteen years, when she too assumed the role of invalid, a role she continued to exploit for much of her life. When Henrietta was eighty-six, she told her niece that she had never made a pot of tea in her life, that she had never been out in the dark alone, that she had never travelled without her maid, and that since the age of thirteen she had had breakfast in bed. It was the opinion of this niece that it was unfortunate that Aunt Etty had had no "real work" into which she might have channelled her unbounded energy and managerial talents: "As it was, ill-health become her profession and absorbing interest".[42]

---

41 Henrietta's words, in F. Darwin 1888, 1: 137.
42 Raverat 1954, 99. Darwin's other surviving daughter Bessy never married and was judged incompetent by her relations: "she was not good at practical things . . . and she could not have managed her own life without a little help and direction now and then" (ibid., 121). Not that I am suggesting that the sons emerged unscathed from the over-protective care of Charles and Emma. All had their share of the "family hypochondria" and "lived all their lives under [Darwin's] shadow" (ibid., 177). Nevertheless, all had professions of one kind or another and were not confined to the

The social nature of the epidemic of female illness among the Victorian middle and upper classes has been explored by a number of scholars who argue that illness was a socially acceptable retreat for those women unable to come to terms with the contradictions and limitations of their narrow and unproductive lives.[43] Whereas Darwin resorted to illness in order to get on with his work, Henrietta retreated to it because she had no work. Female invalidity conformed with Victorian notions of feminine frailty and dependency and reinforced society's strict and rigid definitions of sex roles and sexual differences. In Henrietta's case, these differences had marked her out from infancy. From their birth, Darwin observed and compared the development of his sons and daughters. To his fatherly eyes, his infant sons showed an innate aggressive aptitude for throwing things at anyone who annoyed them, while his daughters were more passive and demonstrated their feminine superiority at manual dexterity. It followed from this infantile recapitulation of primitive evolution, that his sons exhibited reason at a much earlier age than his daughters and were more intelligent.[44]

In conventional fashion the sons were educated at school and university, while Henrietta and her sisters were taught at home by a series of governesses chosen by Emma who was not overtly concerned with their educational qualifications. In later life, Henrietta regretted the poor quality of her education (Litchfield 1915, 2: 178). As might be expected, the daughters were conventionally religious, while the sons tended more towards the scepticism of their father.

It was feminine conventionality that overrode the wishes of the sons when Darwin's *Autobiography* was published with the deletion of his religious opinions. Henrietta went so far as to threaten legal proceedings to stop its publication altogether. She felt that on religious questions it was "crude and but half thought-out", a strongly worded criticism she never ventured to make of any other aspect of Darwin's writing (Barlow 1969, 12). It was Henrietta who proofed *The Descent*, in fact edited it, for Darwin thanked her profusely for her rephrasing of various sections. But she seems to have found nothing to cavil at in the section on woman's intellectual inferiority, which of course gave due recognition to the notion of feminine moral superiority. Similarly, Emma's only concern with *The Descent* was that she would "dislike it very much as again putting God further off"; otherwise she found it "very interesting" (Litchfield 1915, 2: 196). Apart from matters of syntax it would seem that religion was the one acceptable area in which a Darwin female felt competent to make an intellectual judgment, while asserting her moral authority.

---

domestic sphere like their sisters. One even married a "feminist" (ibid., 169) and Francis Darwin's second wife was Ellen Crofts, a Fellow and lecturer in English literature at Newnham (ibid., 162).

43 See Ehrenreich and English 1979, Ch. 4, "The Sexual Politics of Sickness"; see also papers by Ann Douglas Wood, Carol Smith-Rosenberg and Regina Morantz in Hartman and Banner 1974.

44 C. Darwin, "A Biographical Sketch of an Infant" in Gruber 1974, 465–74. This paper, published in 1877, was based on observations Darwin had made of his own children and notes he kept in a diary on the development of his oldest son, William.

Henrietta married shortly after *The Descent* was published and Darwin could give her no better advice on that occasion than the following formula, an amusing blend of sentiment and hypochondria:

> I have had my day and a happy life, notwithstanding my stomach; and this I owe almost entirely to our dear old mother, who, as you know well, is as good as twice refined gold. Keep her as an example before your eyes, and then Litchfield will in future years worship and not only love you, as I worship our dear old mother (Litchfield 1915, 2: 204–205).

It never seems to have occurred to Darwin to question the excessive maternal solicitude and protectiveness he evoked from wife and children who conspired to shield him from his over-sensitive self. He was eternally grateful, he was Emma's slave, he worshipped her, he was a selfish brute, but he could console himself with the reflection that woman was naturally more tender and less selfish than man. Emma was simply exhibiting her innate qualities, as he was. He was very likely referring to his own career when he wrote in *The Descent*:

> Man is the rival of other men; he delights in competition, and this leads to ambition which passes too easily into selfishness. These latter qualities seem to be his natural and unfortunate birthright (Darwin 1889, 563).

It was unfortunate, but it was the natural order of things. The thought that he might have attained his own high eminence at the expense of his beloved Emma would have been too painful to bear. The concept of the innate mental differences between the sexes was as psychologically indispensable as it was theoretically consistent. Emotional comfort could be distilled from theoretical necessity. Not that I am suggesting that this was in any way a conscious process on Darwin's part.

Emma herself once wrote of him: "He is the most open, transparent man I ever saw, and every word expresses his real thoughts" (Litchfield 1915, 2: 6). With due allowance for wifely sentiment, all Darwin's writings, published and private, bear this out. They may have been confused, at times inconsistent, certainly in some ways as we have seen they were biased, but they were remarkably open and unselfconscious. For Darwin, the differences between the sexes were as self-evident as the differences in beaks and plumage between the finches of the Galapagos Islands, and both sets of phenomena were reducible to the same causes. There was, after all, no inconsistency between his personal experience and his theoretical argument. The women he had known most intimately conformed entirely to Victorian conventions of femininity and domesticity. Of his own part in reinforcing those conventions he remained sublimely unaware.

That Darwin never managed to transcend these conventions and take seriously Mill's critique of them, should occasion no surprise. He had not Mill's advantage of a Harriet Taylor. Not that he would have been happy in the company of a liberated, intelligent and strong-minded woman. He had wanted a "nice soft wife" and in Emma he found one. The domestic relations of the Darwins are best understood as an expression of the class and sexual divisions of Victorian society, and to these I shall now turn. For before all, Darwin was a Victorian, "a gentlem[a]n and a family m[a]n, of complete financial, political and sexual respectability",[45] and while this was of great advantage in the promotion of unorthodox opinion, and Darwin, Huxley and the entire Darwinian party capitalized on it, in return it imposed its own orthodoxy.

III. Feminism, Darwinism and the Social Context

It is one of the characteristic prejudices of the reaction of the nineteenth century against the eighteenth, to accord to the unreasoning elements of human nature the infallibility which the eighteenth century is supposed to have ascribed to the reasoning elements. For the apotheosis of Reason we have substituted that of Instinct; and we call everything Instinct which we find in ourselves and for which we cannot trace any rational foundation. This idolatry, infinitely more degrading than the other, and the most pernicious of the false worships of the present day, of all of which it is now the main support, will probably hold its ground until it gives way before a sound psychology, laying bare the real root of much that is bowed down to as the intention of Nature and the ordinance of God. –
– J. S. Mill (and Harriet Taylor Mill), The Subjection
of Women (Rossi 1970, 128)

The nineteenth century was a period of extraordinary social and economic transformation and expansion, in which pre-industrial modes of legitimation, religion in particular, were giving way to a secular redefinition of the world. In the process, science increasingly took over from religion the task of defining and upholding the moral and social order. Evolution was central to this transition, and took on a newfound respectability.

The Origin was published, acclaimed and accepted within the body of scientific knowledge in the mid-Victorian era of capitalist enterprise, when industrial capitalism became a genuine world economy. In the prevailing mood of complacent confidence and general prosperity, the revolutionary notion of evolution no longer seemed to imply social upheaval.[46] On the contrary, the secular ideology of progress, assimilated to the capitalist requirements of industrial and economic

45 J. W. Burrow, "Introduction" to Darwin 1968, 41.
46 See Hobsbawm 1975, Ch. 14; 1973, Chs. 12, 13, 15; Mendelsohn 1980. Mendelsohn writes, "During the course of the nineteenth century the term evolution came to be contrasted directly with revolution." In the earlier part of the century, evolutionary speculations such as those of Erasmus Darwin and Robert Chambers were opposed for largely political reasons, because in a period of

growth, catch-cry of a rapidly advancing liberal and "progressive" bourgeoisie, proved amenable to the notion of biological evolution, particularly when it was so congenially expressed in the familiar terminology of classical political economy. Progress could now be scientifically sanctioned, for Darwinism guaranteed it where the utilitarians had only been able to hope that they could engineer it.[47] The "social Darwinism" forged by Spencer from his earlier social evolutionism and shored up with Darwinian biological concepts (themselves heavily dependent on social theory)[48] made unobstructed competition and the resultant "survival of the fittest" the guarantee of continuous social progress without revolutionary or radical change. It has been pointed out that Spencer's unique appeal lay in "his ability to support the foundations of the status quo while at the same time introducing to the middle class the revolutionary mechanism of evolutionary law and the discoveries of science" (Haller and Haller 1974, 61–62). Recent scholarship has emphasized the central role played by economic and political factors in the reception of evolutionary theory, and it is clear that it was in its social, rather than its biological form, that "Darwinism" was most widely known and popularized in the late nineteenth century.[49] In the process, the traditional radical component of evolutionary thinking was swamped by the rising tide of Social Darwinism, which went on to provide the intellectual underpinnings of imperialism, war, monopoly capitalism, militant eugenics and racism. Darwinism could and can mean many things to many people, but there is little doubt that its dominant nineteenth-century mode was that Social Darwinism that so well served late Victorian imperialist interests.[50]

Darwin's own part in this was not insignificant, as has been so often asserted. He did not have to endorse the activities of "every cheating tradesman"[51] for his work to have a profound impact on nineteenth-century social and political theory. Darwin's neutrality can hardly be asserted and sustained in the face of his own application of his theory of evolution to the interpretation and justification of existing economic and social relations and his insistence that social progress

---

great social and industrial upheaval, they were perceived as threatening social stability and morality (Garfinkle 1955; Millhauser 1959).

47 Young 1974, 28. See also Burrow 1966, Chs. 3 and 4.

48 Two recent studies have demonstrated the importance of the writings of social theorists and political economists such as Comte, Adam Smith, Dugald Stuart and James McIntosh, as well as Malthus, in the genesis of Darwin's concept of natural selection: Schweber 1977, and Manier 1977. But the classic study remains Young's "Malthus and the Evolutionists: The Common Context of Biological and Social Theory", 1969.

49 See for instance the papers in Glick 1974; Mendelsohn 1980.

50 See Young 1974. The classic study is Hofstadter [1944], 1955. There have been some attempts to revise Hofstadter's thesis on the grounds that it was possible to encompass several meanings – including a non-competitive model of society – within the spectrum of Social Darwinism (notably Wilson 1967). But see the comments by Jones 1978, 19; also Rogers 1972; Greene 1977.

51 Darwin wrote to Lyell in 1860: "I have received, in a Manchester newspaper, rather a good squib, showing that I have proved 'might is right', and therefore that Napoleon is right, and every cheating tradesman is also right". F. Darwin 1888, 2: 262 [CCD 8: 189].

could only occur through severe and sustained competitive struggle. When he incorporated contemporaneous social thought in support of this belief in *The Descent*, he opened up his work to its reciprocal appropriation as Social Darwinism.[52] Young has argued persuasively for a "common context" of biological and social thought associated with the themes of struggle and adaptation which was the main interpretative resource for both nineteenth-century evolutionists like Darwin and social theorists like Spencer (Young 1969). When the problem of human evolution had finally to be faced, Darwin was as dependent upon Spencer and others of the social evolution tradition for the larger social and political generalizations by which to make evolution explicable to his audience, as they were, in a scientifically minded age, on his biological ratification of their social evolution. From the alliance of Darwinian biology and Spencerian social evolutionism, which *The Descent* consolidated, came Social Darwinism.

It was an alliance that made for success. As Darwin reported to Henrietta:

> Murray reprinted 2000 [of *The Descent of Man*] making the edition 4500, and I shall receive £1470 for it. That is a fine big sum. . . . Altogether the book, I think, as yet, has been very successful, and I have been hardly at all abused (Litchfield 1915, 2; 202).

The atmosphere of general assent and goodwill that greeted *The Descent* is a notable indication of the change in opinion that had taken place since the publication of *The Origin* (Himmelfarb 1967, 354–59). It is all the more notable in view of the fact that *The Descent* was published on the eve of the suppression of the Paris Commune. When *The Times,* stirred to fever pitch by the events in Paris, invoked the Commune to attack the dangerous and immoral "disintegrating speculations" of *The Descent*, it found itself out of step with the more general anxiety to dissociate Darwinism from political revolution and absorb it into the traditional sphere of natural theology and conservative politics and morality.[53]

From the 1870s on, it became possible for those who found it expedient, to look to evolution rather than religion for the corroboration of their social values.

---

52 Jones (1978, 17–19) has suggested that the alliance of Darwinian biology and Spencerian social evolutionism profited Darwin in a period when his concept of natural selection was facing "formidable problems" posed by the lack of a satisfactory theory of heredity. It "secured the survival of his theory as a major part of British scientific and intellectual tradition in the later nineteenth and early twentieth century before its reintegration with the theory of heredity in the 1920s". The discrimination of Darwin from Social Darwinism that is so frequently urged by historians is not as simple as they suggest. See Barnes and Shapin 1979.

53 Hobsbawm (1975, 167–69) writes that the Paris Commune was "more formidable as a symbol than as a fact". Although it did not seriously threaten the bourgeois order, its brief period of existence created a wave of panic and hysteria, as the international press accused it variously of "instituting communism, expropriating the rich and sharing their wives, terror, wholesale massacre, chaos, anarchy and whatever else haunted the respectable classes". After the Commune, what their "betters" feared was not social revolution in general, but proletarian revolution.

The more theologically minded could make a "subtle accommodation with the theory . . . adopting an attendant natural theology which, while it made God remote from nature, made his rule grander", thus securing at a stroke the double ratification of God and science (Young 1974, 23). It was a double ideological ratification that also appealed strongly to American "robber barons", reaching its apotheosis in the well-known Sunday School Address by J.D. Rockefeller, where he defended the morality of the monopolistic practices of Standard Oil as "not an evil tendency in business" but "merely the working-out of a law of nature and a law of God" (Hofstadter 1955, 45).

Contradictory as it may seem, in certain respects (as a number of scholars have stressed[54]) Darwinism represents not so much a revolutionary break as an underlying continuity with natural theology, which, by the time *The Origin* burst on the scene, had made its own accommodation with Malthusian social theory and the ideology of progress and was moving cautiously towards a more naturalistic or scientific interpretation of earth's history. As suggested above, Darwinism was simply one aspect of a much broader movement that can be traced back to the end of the eighteenth century, and embraced not only directly evolutionary writings, such as those of Erasmus Darwin and Robert Chambers, but the population theory of Malthus, utilitarianism and laissez-faire doctrine, feminism and natural theology. All aimed at reinterpreting more naturalistically, traditional views of nature and society, while assuming a basically theistic view of both. Where they differed was in where to draw the line, the evolutionists insisting that *all* of nature including humanity and mind was under the domain of natural law and therefore a legitimate object of scientific inquiry, the natural theologians disputing the inclusion of humanity, or at least mind, in the course of material nature. Viewed in this light, the Darwinian controversy becomes a "demarcation dispute within natural theology" (Young 1974, 24), and the ability of theology ultimately to accommodate Darwinism, when faced with the necessity for doing so, becomes explicable.

This interpretation also helps us to understand why, having triumphed and made men's and women's minds subject to natural law, many leading Darwinians became so rigidly determinist in their views on human social and economic arrangements. To reiterate, the Darwinian debates were merely the focus of the more general controversy that preoccupied nineteenth-century intellectuals as secular naturalism challenged traditional theological modes of explanation: are human affairs governed by fixed laws or are they the result either of chance or of supernatural interference? To put it another way, if human actions are intelligible, it can only be because they, like the rest of nature, can be subsumed under fixed and immutable laws (Burrow 1966, 106–107). The whole spectrum of nineteenth-century progressive thought (including feminism) was influenced by this naturalistic assumption, which stemmed partly from conscious opposition to conventional wisdom and authority, partly from an ever-increasing confidence in

54 See Cannon 1961; Young 1969; 1974; Bowler 1977.

the "certainties" of science and the universality and inevitability of natural law. Harriet Martineau, one of the founders of British sociology and an ardent defender of women's rights, wrote enthusiastically of Comte's *Positive Philosophy*:

> We find ourselves suddenly living and moving in the midst of the universe – as a part of it, and not as its aim and object. We find ourselves living, not under capricious and arbitrary conditions, unconnected with the constitution and movements of the whole, but under great, general, invariable laws, which operate on us as part of a whole.[55]

Thus Darwin, in pushing his case against the divine origin of human mind and conscience, argued for their evolution according to the same processes that had produced all living things. His refusal to concede any but naturalistic explanations of human intelligence and morality, hardened into a biological determinism that rejected all social and cultural causation other than that which could be subsumed under the natural laws of inheritance and thus become innate or fixed.[56]

We can trace this process through Darwin's writings. There is an early Notebook emphasis on the significance of education to a materialist view of morality: "Believer in these views will pay great attention to Education" (Gruber 1974, 390; OUN 28; Barrett et al. 1987, 608). At this stage, he was even willing to concede that the education of women could play a definite role in social evolution, both through women's own [inherited] intellectual and moral improvement and through their general influence as moral preceptors:

> Educate all classes, avoid the contamination of castes, improve the women (double influence) & mankind must improve (C 220; Barrett et al. 1987, 309).

It is to be noted, however, that he stressed the deleterious effects of miscegenation. By the time of *The Descent*, Darwin's confidence in the improving power of education and other environmental agencies was waning before his increasing emphasis on the biological basis of mental and moral differences, and his insistence on the necessity of continuous competitive struggle for human mental and moral improvement. In *The Descent* he advocated eugenics as a means of social advancement, and not long before his death he wrote:

---

55 Burrow 1966, 106–7. The Seneca Falls Declaration on Women's Rights of 1848 began with the words: "When, in the course of human events, it becomes necessary for one portion of the family of man to assume among the people of the earth a position different from that which they have hitherto occupied, but one to which the laws of nature and of nature's God entitle them . . ." (Rossi 1973, 415–16).

56 Primarily through the law of inheritance of acquired characters; see Note 20 above. Nurture thus merged into nature. As Greene (1977, 24) observes: "The 'Lamarckian' principle of the inheritance of acquired characters, far from constituting a rival principle of explanation, was viewed as cooperating with the law of natural selection in bringing about the gradual improvement of the human race".

I am inclined to agree with Francis Galton in believing that education and environment produce only a small effect on the mind of anyone, and that most of our qualities are innate.[57]

The contradiction was that such rigid exclusion of environmental explanation led full circle back to the Wise Designer and Law Giver who ultimately sanctioned the social order which men and women could not change by their own efforts. Mill summed it all up in the extract from the powerful opening chapter of *The Subjection of Women* that heads this section. It was the "intention of nature and the ordinance of God" that men and women should occupy their socially and culturally sanctioned positions, and it made little practical difference whether one attributed the cause primarily to the designing hand of providence or evolution by natural and sexual selection.

From the 1870s on, the dominant Darwinian tradition was characterized by a moralizing naturalism, to which *The Descent* gave a powerful boost (Weber 1974, 279–82). Huxley, Romanes, Galton, Lubbock and Spencer all produced popular writings of this kind. Their language sometimes assumed an inspired evangelical tone. Galton wanted to "elicit the religious significance of the doctrine of evolution". Huxley, the self-designated agnostic, saw in anthropology a "religion of man", whom he pictured as potentially raised upon his accumulated and organized collective experience as "on a mountain top, far above the level of his humble fellows, and transfigured from his grosser nature by reflecting, here and there, a ray from the infinite source of truth" (cited in Weber 1974, 280). For many Darwinians, playing churchman merely required translation of ecclesiastical into scientific language. What had been sin, became biologically and therefore socially injurious (Doyal 1979, 148). While it was the intent of many leading Darwinians like Spencer and Vogt to bring political legislation and social procedure into harmony with human biology, not antiquated notions of natural reason or Christian morality, it was surprising how often the new "truths" of science affirmed the traditionally sanctioned stereotypes of men and women.

Huxley, distinguished for his celebrated stand against the deduction of ethical "oughts" from biological "ises" that characterized Social Darwinism, wrote sweepingly that women were "by nature, more excitable than men – prone to be swept by tides of emotion . . . naturally timid, inclined to dependence, born conservative . . ." (Huxley [1865] 1898, 71). Yet his liberal principles of democracy and individualism could not deny a better education to women, for all their natural inferiority. Let us have "sweet girl graduates" by all means: "They will be none the less sweet for a little wisdom; and the "golden hair" will not curl less

---

57 Barlow 1969, 43. Galton's influence on Darwin in this respect was considerable. See Darwin's letter to Galton of December 1869: "You have made a convert of an opponent in one sense, for I have always maintained that, excepting fools, men did not differ much in intellect, only in zeal and hard work; and I still think (this) is, an eminently important difference" (F. Darwin and Seward (ed.) 1903, 2: 41; *CCD* 17: 531). See also Greene 1977.

gracefully outside the head by reason of there being brains within". Let women become merchants, barristers, politicians; Huxley could reassuringly assert that it would make no difference to the status quo:

> Nature's old salique law will not be repealed, and no change of dynasty will be effected. The big chests, the massive brains, the vigorous muscles and stout frames of the best men will carry the day, whenever it is worth their while to contest the prizes of life with the best women. . . . The most Darwinian of theorists will not venture to propound the doctrine, that the physical disabilities under which women have hitherto laboured in the struggle for existence with men are likely to be removed by even the most skilfully conducted process of educational selection.

Huxley's liberal "oughts" could not help but come into conflict with what was commanded by biological "ises". Nevertheless, justice must prevail, and law and custom should not add to the biological burdens that weigh woman down in the "race of life":

> The duty of man is to see that not a grain is piled upon that load beyond what Nature imposes; that injustice is not added to inequality (ibid., 73–75).

Huxley's prediction was correct. Those Darwinian theorists (and they were many, including Darwin) who pronounced upon the "woman question," raised insuperable evolutionary barriers against feminine intellectual and social equality. Where they did not argue directly against the extension of the franchise and higher education to women on biological grounds, as did Spencer and Cope, they followed Huxley's liberal line of conceding to women their right to the vote and education, but imposing strict evolutionary limitations on the outcome, as did Romanes or Geddes and Thompson.[58] In order to obliterate the innate intellectual and emotional differences between men and women it would be necessary to have all evolution over again on a different basis, a patent absurdity:

> What was decided among the prehistoric Protozoa cannot be annulled by Act of Parliament (Geddes and Thompson 1889, 271).

Huxley's "higher moral tone" and the biologically-based moral guidance offered by other Darwinians were factors in the struggle they were waging to establish science as a profession worthy of middle-class status and rewards (Figlio 1978), and fed into the current economic and political climate. By the 1870s, the cold winds of change were beginning to blow about the ears of the British middle-classes, as the limits of the steam-based technology of the first Industrial Revolution become

---

58 See Sleeth Mosedale 1978; Alaya 1977; Conway 1970; Fee 1974.

visible, and the "Great Depression" of 1873–1896 undermined the foundations of mid-nineteenth-century liberalism. After its glorious advances of the '50s and '60s, the economy stagnated, and Britain's industrial and economic global dominance was increasingly challenged by Germany and the U.S.A. When this competition became acute, the only major escape left for British capital was the traditional one of the economic (and increasingly the political) conquest of hitherto unexploited areas of the world – that is, imperialism – a route which was also quickly adopted by the competing powers. This period was also characterized by urban and industrial unrest, and saw the emergence of mass socialist working-class politics all over Europe.

With the end of the age of unquestioned expansion, the growing doubts about the economic prospects of Britain, and the abiding fear of working class insurrection, the optimistic and confident liberalism of the boom period hardened into an entrenched conservatism. The bourgeois social order of the 1870s was more than ever anxious to consolidate and justify its class and racial superiority and to preserve that basic bourgeois institution, the family – the cornerstone of the bourgeois social order:

> The "family" was not merely the basic social unit of bourgeois society but its basic unit of property and business enterprise, linked with other such units through a system of exchange of women-plus-property (the "marriage portion"). . . . Anything which weakened the family unit was impermissible (Hobsbawm 1979, 127–32).

By the 1870s, feminism was beginning to be perceived as a direct threat to the bourgeois family. Nineteenth-century feminism, from Mary Wollstonecraft on, was thoroughly bourgeois in its derivation and aspirations. Its demands for women's suffrage, higher education and entrance to middle-class professions and occupations grew out of that progressive middle-class liberalism for which John Stuart Mill was the leading spokesman. By 1870, not only had Mill's powerful voice been raised in the service of feminism, but women were already attending courses at London and Cambridge (although not as official members of the universities). A few had even managed with great difficulty to gain entrance to medicine and qualify as doctors, while many others were being prepared to compete with boys for the university lower examinations. In 1870, Oxford University decided to open its lower examinations to women also. It seemed only a matter of time before middle-class women not only gained the franchise, but would be able to take out degrees and compete professionally with men, thus acquiring not only intellectual but economic and political independence of the family (Burstyn 1973). Moreover the possibility of family limitation was discreetly beginning to be raised by some feminists – a prospect that struck at the heart of a growing middle-class concern with its reproductive potential versus that of the teeming, irresponsible and potentially insurrectionary lower orders. Inevitably, in the context of a general hardening of attitudes, the increasing intensity and urgency of the demands of feminism fostered a strong reaction against the gains it had made during the confident and prosperous '50s and '60s.

The traditional sexual division of labour, which had been characteristic of the pre-industrial and pre-capitalist period, where women had a clearly defined domestic role, was accentuated by the new organization of labour demanded by industrial capitalism. This was particularly so for bourgeois women:

> For them the division between public life and the private world of the home was absolute, and most became mere symbols by which their husband's financial and social status was evaluated. They were embodiments of conspicuous consumption and remained in their homes to provide their husbands and children with the tenderness, sensitivity and devotion to the arts which was so conspicuously lacking in the factories and mines of Victorian industry. . . . Women worked inside the home and men outside it, and this strict differentiation between the spheres of men and women lay at the heart of Victorian society (Doyal 1979, 151).

It was woman's responsibility to guard the values inherent in the "family" and the "home", where her maternal virtues of love, patience and compassion were to temper the savagery of capitalist competition. The feminists' demand for their liberal "rights" was thoroughly at odds with this renewed emphasis on the sexual division of labour. As in other areas of social concern, during the 1870s science was increasingly invoked to reinforce the traditional religion-sanctioned belief in the essential domesticity of women. With the timely appearance of *The Descent* at the beginning of the decade, Darwin's growing authority and prestige were pitted against the claims by women for intellectual and social equality. This was carried out primarily through the medium of the "new" anthropology of the '70s, which was also the purveyor of the scientific racism that dominated late-Victorian science and social theory.

> There was scarcely an anthropologist who did not take up the moral problem of the evolution of the family and who did not on that basis pronounce upon the emancipation of women.[59]

The massive upsurge of anthropological and medical writings endorsing traditional conceptions of woman and her role that began around the 1870s has now

---

59 Weber 1974, 279; see also Fee 1974. Burstyn (1973, 81) makes the point that "medicine was the first occupation to be assailed by women in their attempts to enter the professions, and it was medical practitioners who made the strongest attack against higher education for women". In an age of extreme reticence about sex, it was considered by many that women would make more appropriate gynaecologists and obstetricians than men. The majority of nineteenth-century anthropologists and biologists were doctors by training, and a persuasive case could be made that they had a professional interest in warding of feminine competition that lent itself readily to anthropological and biological endorsements of the status quo.

been thoroughly documented and explored. The bias at the root of this "scientific" refutation of the claims of feminism has been exposed, and its key social and political role in the anti-feminist backlash of the late-Victorian period demonstrated.[60] The profound dislocation of late nineteenth-century feminism in the face of this scientific onslaught has been less thoroughly explored and understood. However, in the light of the above analysis, Flavia Alaya's suggestion of a crisis of feminist ideology is persuasive. Alaya argues that the "impact of nineteenth-century science . . . gave such vigorous and persuasive reinforcement to the traditional dogmatic view of sexual character that it not only strengthened the opposition to feminism but disengaged the ideals of feminists themselves from their philosophic roots [of Enlightenment egalitarianism]" (Alaya 1977, 261–62). Nineteenth-century feminists became entrapped within the same framework of biological determinism as Darwin. The earlier alliance the feminists had forged with science in the opposition of naturalistic interpretations of human nature and society to conventional wisdom and authority, ultimately betrayed them when science, particularly Darwinism, gave a naturalistic, scientific basis to the class and sexual divisions of Victorian society. The only recourse for feminism to this concerted scientific drawing of naturalistic limits to its claims, was to assert that woman was "different but equal": to claim for woman a biologically based "complementary genius" to man's – a "genius" which was rooted in her innate maternal and womanly qualities.

Thus Antoinette Brown Blackwell, the American feminist and evolutionist, in her critique of Darwin's evolutionary argument for woman's physical and intellectual inferiority, offered an evolutionary argument for the equality of men and women. She did not dispute Darwin's view that the mental differences between men and women were biologically based and the product of evolution; rather she disputed whether woman's innate mental differences could properly be called inferior to man's.[61] She balanced man's greater strength, reasoning powers and sexual love against woman's greater endurance, insightfulness and parental love, and concluded with a final evolutionary endorsement of Victorian values:

> If Evolution, as applied to sex, teaches any one lesson plainer than another, it is the lesson that the monogamic marriage is the basis of all progress. Nature, who everywhere holds her balances with even justice, asks only that every husband and wife shall co-operate to develop her most diligently-selected characters. . . . No theory of unfitness, no form of conventionality, can have the right to suppress any excellence which Nature has seen fit to evolve. Men and women, in search of the same ends, must co-operate in as many heterogeneous pursuits as the present

---

60 See papers by Alaya 1977; Sleeth Mosedale 1978; Fee 1974; Burstyn 1973; Ehrenreich and English 1979; Smith Rosenberg and Morantz in Hartman and Banner 1974.
61 A. Brown Blackwell, *The Sexes Throughout Nature* (New York, 1875), extract reprinted in Rossi 1970, 356–377. I am indebted to Randall Albury for this point.

development of the race enables them both to recognise and appreciate (Rossi 1970, 376–77).

Such argumentation could only reinforce traditional stereotypes and cater to the drawing of biological limits to human potentiality.[62]

The refusal by Harriet Taylor and Mill to ground human nature in Nature stands out against this overwhelming nineteenth-century trend; but it is to be noted that Mill himself was not immune from contemporary ideology. He too put his faith in science, in a "sound psychology", which would lay bare the "real root of much that is bowed down to as the intention of Nature and the ordinance of God".

IV. Conclusion

I sometimes marvel how truth progresses, so difficult is it for one man to convince another, unless his mind is vacant.
— Darwin *to* Wallace *on Sexual Selection,* 1868[63]

Darwin's consideration of human sexual differences in *The Descent* was not motivated by the comtemporary wave of anti-feminism (as can be said of most late-Victorian biologists who dealt so exhaustively with the attributes of women), but was central to his naturalistic explanation of human evolution. It was his theoretically directed contention that human mental and moral characteristics had arisen by natural evolutionary processes which predisposed him to ground these characteristics in nature rather than nurture – to insist on the biological basis of mental and moral differences as the raw material on which natural and sexual selection might operate. This brought him into opposition with Mill and others who argued for an environmental or cultural explanation of such differences, and into line with the biological determinism of Galton, Vogt, Spencer and others, whose related

---

62 The socialist and visionary, Eliza Burt Gamble, who offered the other major nineteenth-century rebuttal of Darwin's arguments for the continuing inferiority of women, was an even more thoroughgoing Darwinian than Brown Blackwell. She entirely accepted and endorsed Darwin's account of the differentiation of the sexes, but held to the view that it confirmed woman's innate superiority. According to Gamble all "progressive" moral and social principles stem from woman's maternal instincts. Man is innately egoistic and selfish, concerned primarily with the gratification of his animal instincts" and to this end he has dispossessed woman of her "fundamental prerogative" of aesthetic choice. Women have become "economic and sexual slaves . . . dependent upon men for their support". Gamble looked forward to the time when women would emerge from the "murky atmosphere of a sensuous age", regain their rightful power of sexual selection and through the transmission of their more refined instincts and ideas peculiar to the female organism" (such as altruism, sympathy, etc.) to their offspring, found a "new spiritual age": "society advances just in proportion as women are able to convey to their offspring the progressive tendencies transmissible only through the female organism". See Gamble [1894], 1916.

63 F. Darwin and Seward 1903, 2: 77 [*CCD* 16, 1: 452].

but more explicit social and political conceptions he borrowed and built into *The Descent*. In return he proffered additional support and the prestige of his name, which entered into social theory as "social Darwinism" and was widely used to endorse late-Victorian assumptions of while middle-class male supremacy. In this fashion, Darwin endorsed the anti-feminist arguments of those "Darwinians" like Huxley, Spencer, Romanes, Geddes and Thompson, who drew biological limitations to woman's political and social potentiality. His own foray into social justification and prescription in *The Descent* was a specific contribution by Darwin to the scientific anti-feminism that characterized this period.

Further, through his concept of sexual selection, Darwin promoted an androcentric account of human evolution, which rationalized Victorian conceptions of male dominance and importance and confirmed Victorian sexual stereotypes. An examination of his early Notebook entries demonstrates that Darwin consistently held to these values and by a process of circularity fed them into his conceptions of human biological and social evolution.

Darwin's feminist critics are therefore correct in asserting the bias at the root of Darwin's characterization of women as innately domestic and intellectually inferior to men, and in pointing to the cultural and social values implicit in his concept of sexual selection. They are also correct in asserting the political effects of Darwin's argument for woman's continuing inferiority in the contemporary struggle by feminists for higher education, and the general political role of Darwinism in scientifically endorsing anti-feminism through late nineteenth-century biology and anthropology.

However, to do Darwin historical justice, it must be acknowledged that Darwin's personal experience did not lead him to question Victorian sexual stereotypes and the sexual division of labour, and his bourgeois class position reinforced them. Nor was he primarily motivated by anti-feminism, but by the defence of his theory of evolution. Apart from the social and political constraints within which Darwin operated, there were powerful intellectual ones which led not only Darwin but many feminists into biological determinism in their joint effort to replace traditional theological modes of explanation with scientific ones.

Nor did Darwin engage actively in sexual discrimination as did Huxley, when this long-time "supporter" of higher education for women fought hard to exclude them from ordinary meetings of the Geological and Ethnological Societies, on the grounds that their "amateur" presence would jeopardize the professional status of those institutions.[64] True, it would have been quite out of character for Darwin to engage in political struggle, and with his handsome income from his solidly invested inherited capital,[65] he could remain comfortably outside the struggle

64 See Huxley's letter to Lyell of 1860; L. Huxley 1900, 1: 211–12; see also 387, 417; and Burstyn 1973, 88.
65 By Darwin's death, his estate amounted to £282,000, a sum compounded of money inherited from his father, Emma's dowry and income from investments. His annual income from investments

for scientific professionalization and keep his liberal principles intact. He wrote approvingly of the "triumph of the Ladies at Cambridge" when women were finally accorded the right to present themselves for the "Little-Go" and Tripos Examinations in 1881.[66]

To suggest, therefore, that Darwin's theory of sexual selection was primarily a political ploy, is simply not correct.[67] Moreover, in spite of its potential for exploitation for anti-feminist purposes, it was very little called upon by those Darwinians who pronounced upon woman's abilities and potential. Only Romanes, Darwin's direct intellectual heir, took it up and applied it to the "woman question" where he used it to support the notion of woman's complementary genius (Romanes 1887). Geddes and Thompson, in their influential and widely read work *The Evolution of Sex* (1889), took pains to separate themselves from Darwin on the influence of sexual selection upon secondary sexual characteristics. Spencer, who wrote most voluminously upon woman's biological limitations, made very little use of sexual selection. With typical tenacity he shunted along his own intellectual railways tracks of "survival of the fittest" and Neo-Lamarckian and recapitulatory explanation of women's evolutionary inferiority.[68] Most Darwinians seem to have concurred with Wallace who wrote to Darwin on reading *The Descent:*

> There are . . . difficulties in the *very wide* application you give to sexual selection which at present stagger me . . . (F. Darwin and Seward 1903, 2: 93 [*CCD* 19: 46]).

With sexual selection, Darwin had tried to explain too many aspects of evolution, which his fellow Darwinians could explain as well as or better through natural selection aided by use-inheritance. Ironically, it was Wallace's views on the primacy of natural selection in sexual dimorphism that were to prevail.[69]

The recent attempts by Ghiselin and others[70] to resurrect the theory of sexual selection in all its androcentric glory in the context of the current wave of scientific anti-feminism are therefore doubly ironic, and feminists have a legitimate concern to expose the Victorian roots of the theory. However there are dangers

---

alone (apart from royalties on his books) was £8000, on which he paid £40 income tax. See Himmelfarb 1967, 134.

66 Darwin to his son George, 1881, in Litchfield 1915, 2: 245.

67 See Note 7 above.

68 See Darwin's letter to Spencer, in F. Darwin and Seward 1903, I: 351–52. It could be argued that Darwin's endorsement of Vogt's recapitulatory argument was far more pernicious in its effects. Most nineteenth-century arguments for the lower evolutionary status of women sooner or later resorted to recapitulation theory.

69 See the papers by Simpson, Dobzhansky and Mayr in Campbell 1972.

70 See the papers by Ehrman (esp. p. 127, shades of Galton!), Trivers and Fox in Campbell 1972; see also E. O. Wilson 1975; Wilson 1978, Chs. 2, 4, 6; Wickler 1973; Dawkins 1976, Ch. 9.

in the wholesale extrapolation of nineteenth-century events to the twentieth, and vice versa. The attribution of Victorian values to twentieth-century biologists is not only historically incorrect but politically meaningless. Twentieth-century biologists are patently *not* conducting their arguments in a late Victorian social, political and intellectual context, but very much in the present, and only a thorough analysis of the present context can clarify the ideological role of such biological arguments in our society and lay bare their political ramifications.

Similarly, Darwin cannot be personally judged by twentieth-century yardsticks, any more than his work can be assessed by twentieth-century standards and concepts. To label him a sexist may be technically correct and emotionally satisfying to those who oppose all manifestations of sexual discrimination, but is mere rhetoric in the context of a society in which almost everyone was a sexist – who held discriminatory views of woman's nature and social role. Those men and women who managed to transcend these socially induced conventions to live their personal lives and locate their theoretical constructs outside them were rare indeed. This was not achieved by most feminists, nor by that other great theoretician of the Victorian era – Karl Marx.

Rather, from the historical analysis of Darwin's theoretical constructs, we may gain some valuable insights into the complex on-going interplay between theories of nature and theories of society. They are insights that have eluded Ghiselin who thinks we can still "reasonably hope to develop ethical standards consistent with biological reality".[71] They have also eluded those feminist biologists and anthropologists who have opposed the androcentric evolutionary constructions of Ghiselin and his kind with oestrocentric[72] ones infused with feminist values, who scour ethology and anthropology for data to support their views and scurry down the old determinist pathways to Nature's laws.

Even Darwin could occasionally rise above the positivist distinction between facts and values and concede the impossibility of bringing a "vacant mind" to bear on scientific "truth".

Acknowledgements

An earlier draft of this paper benefited considerably from the comments and criticisms of Randall Albury, Ian Langham, David Oldroyd and John Schuster. Needless to say, the present version is entirely the author's own responsibility.

---

71 Ghiselin 1974, 263. Ghiselin, like most sociobiologists, engages in some rhetoric on the distinction of "ought" from "is" (248). Nevertheless he erects a "new theory of moral sentiments" based on reproductive competition ("we have evolved a nervous system that acts in the interests of our gonads, and one attuned to the demands of reproductive competition") on the grounds that through "self-discipline" we may "perceive the world as it really is" and that "truth has ethical significance" (263).

72 Female-centred theories. See for instance Morgan 1973; Reed 1975.

Bibliography

Alaya, Flavia. 1977. "Victorian Science and the "Genius" of Woman." *Journal of the History of Ideas* 38: 261–80.

Albury, Randall W. 1975. "Darwinian Evolution and the Inferiority of Women." *GLP! – A Journal of Sexual Politics* 8: 10–19.

Barlow, Nora (ed.) 1969. *The Autobiography of Charles Darwin, with Original Omissions Restored*. New York: W. W. Norton.

Barnes Barry and Steven Shapin, eds. 1979. *Natural Order; Historical Studies of Scientific Culture*. Beverly Hills, London: Sage Publications.

Barrett, Paul H. et al. eds. 1987. *Charles Darwin's Notebooks, 1836–1844*. Cambridge: Cambridge University Press.

Bowler, Peter J. 1977. "Darwinism and the Argument from Design: Suggestions for a Re-evaluation." *Journal of the History of Biology* 10: 29–43.

Burkhardt, Frederick, et al., eds. 1985-. *The Correspondence of Charles Darwin*, 23 vols. Cambridge: Cambridge University Press.

Burrow, John W. 1966. *Evolution and Society: A Study in Victorian Social Theory*. Cambridge: Cambridge University Press.

Burstyn, Joan N. 1973. "Education and Sex: The Medical Case Against Higher Education for Women in England, 1870–1900." *Proceedings of the American Philosophical Society* 117: 79–89.

Campbell, Bernard, ed. 1972. *Sexual Selection and the Descent of Man, 1871–1971*. London: Heinemann.

Cannon, Walter F. 1961. "The Bases of Darwin's Achievement: A Revaluation." *Victorian Studies* 5: 109–34.

Conway, Jill. 1970. "Stereotypes of Femininity in a Theory of Sexual Evolution." *Victorian Studies* 14: 47–62.

Crooke, John H. 1973. "Darwinism and the Sexual Politics of Primates." *Social Science Information* 12: 7–28.

Darwin, Charles. 1871. *The Descent of Man, and Selection in Relation to Sex*. 2 vols. London: John Murray.

——. 1889. *The Descent of Man, and Selection in Relation to Sex*. 2nd ed. London: John Murray.

——. 1909. *Foundations of the Origin of Species: Two Essays Written in 1842 and 1844 by Charles Darwin*. Edited by Francis Darwin. Cambridge: Cambridge University Press.

——. [1859] 1968. *The Origin of Species*. Reprint of First Edition, ed. J. W. Burrow. Harmondsworth: Penguin Books.

Darwin, Francis, ed. 1888. *The Life and Letters of Charles Darwin*. 3 vols. London: John Murray.

Darwin, Francis and A. C. Seward, eds. 1903. *More Letters of Charles Darwin*. 2 vols. London: John Murray.

Dawkins, Richard. 1976. *The Selfish Gene*. Oxford: Oxford University Press.

Doyal, Lesley. 1979. *The Political Economy of Health*. London: Pluto Press.

Durant, John R. 1979. "Scientific Naturalism and Social Reform in the Thought of Alfred Russel Wallace." *British Journal for the History of Science* 12: 31–58

Eiseley, Loren. 1979. *Darwin and the Mysterious Mr. X*. New York: E. P. Dutton.

Ehrenreich, Barbara and Diedre English. 1979. *For Her Own Good: 150 Years of the Experts' Advice to Women*. London: Pluto Press.

Fee, Elizabeth. 1974. "The Sexual Politics of Victorian Social Anthropology." In *Clio's Consciousness Raised*, edited by Mary S. Hartmann and Lois Banner, 86–102. New York, Evanston, San Francisco, London: Harper Torchbooks.

Figlio, Karl. 1978. "Chlorosis and Chronic Disease in Nineteenth Century Britain: The Social Constitution of Somatic Illness in a Capitalist Society." *Social History* 3: 167–97.

Gale, Barry G. 1972. "Darwin and the Concept of Struggle for Existence: A Study in the Extrascientific Origins of Scientific Ideas." *Isis* 63: 321–44.

Gamble, Eliza Burt. (1894) 1916. *The Sexes in Science and History: An Inquiry into the Dogma of Woman's Inferiority to Man*. Rev. ed. New York and London: G. P. Putnam.

Garfinkle, Norton. 1955. "Science and Religion in England, 1790–1800: The Critical Response to the Work of Erasmus Darwin." *Journal of the History of Ideas* 16: 376–88.

Geddes, Patrick and J. Arthur Thompson. 1889. *The Evolution of Sex*. London: Walter Scott.

Ghiselin, Michael T. 1972. *The Triumph of the Darwinian Method*. Berkeley: University of California Press.

——. 1974. *The Economy of Nature and the Evolution of Sex*. Berkeley: University of California Press.

——. 1976. "Two Darwins: History versus Criticism." *Journal of the History of Biology* 9: 121–32.

Glick, Thomas F., ed. 1974. *The Comparative Reception of Darwinism*. Austin and London: University of Texas Press.

Gould, Stephen Jay. 1977. *Ontogeny and Phylogeny*. Cambridge, MA: Harvard University Press.

Greene, John C. 1975. "Reflections on she Progress of Darwin Studies." *Journal of the History of Biology* 8: 243–73.

——. 1977. "Darwin as a Social Evolutionist." *Journal of the History of Biology* 10: 1–27.

Gruber, Howard E. 1974. *Darwin on Man: A Psychological Study of Scientific Creativity*. London: Wildwood House.

Haller, John S. and Robin M. Haller 1974. *The Physician and Sexuality in Victorian America*. Urbana: University of Illinois Press.

Hartman, Mary S. and Lois Banner, eds. 1974. *Clio's Consciousness Raised: New Perspectives on the History of Women*. New York, Evanston, San Francisco, London: Harper Torchbooks.

Herbert, Sandra. 1974, 1977. "The Place of Man in the Development of Darwin's Theory of Transmutation, Parts 1 and 2." *Journal of the History of Biology* 7: 217–58; 10: 155–227.

Himmelfarb, Gertrude. 1967. *Darwin and the Darwinian Revolution*. Gloucester, MA: Peter Smith.

Hobsbawm, Eric J. 1973. *The Age of Revolution*. London: Cardinal.

——. 1975. *The Age of Capital: 1848–1875*. London: Wiedenfeld and Nicolson.

——. 1979. *Industry and Empire*. Harmondsworth: Penguin.

Hofstadter, Richard. [1944] 1955. *Social Darwinism in American Thought*. Rev. ed. Boston: Beacon Press.

Hubbard, Ruth. 1979. "Have Only Men Evolved?" In *Women Look at Biology Looking at Women: A Collection of Feminist Critiques*, edited by Ruth Hubbard et al., 7–35. Boston: G. K. Hall.

Huxley, Leonard, ed. 1900. *Life and Letters of Thomas Henry Huxley*. 2 vols. London: Macmillan.

Huxley, Thomas Henry. 1898. "Emancipation – Black and White"(1865). In *Collected Essays*, 3: 66–75. New York: Greenwood Press.

Johnston, Ron. 1976. "Contextual Knowledge: A Model for the Overthrow of the Internal External Dichotomy." *Australian and New Zealand Journal of Sociology* 12: 193–203.

Jones, Greta. 1978. "The Social History of Darwin's *Descent of Man*." *Economy and Society* 7: 1–23.

Litchfield, Henrietta. [1904] 1915. *Emma Darwin, A Century of Family Letters, 1792–1896*, 2 vols. London: John Murray.

Macleod, Roy M. 1977. "Changing Perspectives in Social History of Science." In *Science, Technology and Society: A Cross-Disciplinary Perspective*, edited by I. Spiegel-Rosing and D. de Solla Price, 189–95. Beverley Hills and London: Sage.

Manier, Edward. 1977. *The Young Darwin and His Cultural Circle*. Dordrecht: D. Reidel Pub. Co.

Mendelsohn, Everett. 1980. "The Continuous and the Discrete in the History of Science." In *Constancy and Change in Human Development*, edited by O. G. Brim and J. Kagan, Ch. 3. Cambridge MA: Harvard University Press.

Millhauser, Milton. 1959. *Just Before Darwin: Robert Chambers and* Vestiges. Middletown CT.: Wesleyan University Press.

Morgan, Elaine. 1973. *The Descent of Woman*. London: Bantam Books.

Mulkay, Michael. 1979. *Science and the Sociology of Knowledge*. London: Allen & Unwin.

Oppenheimer, Jane M. "An Embryological Enigma in the *Origin of Species*." In *Forerunners of Darwin: 1745–1859*, edited by Bentley Glass, Owsei Temkin and William L. Straus, 292–322. Baltimore: The Johns Hopkins Press.

Pickering, George White. 1974. *Creative Malady*. London: Allen & Unwin.

Prichard, James Cowles. (1813) 1973. *Researches into the Physical History of Man*. Ed. By George W. Stocking. Reprint, Chicago: University of Chicago Press.

Raverat, Gwen. 1954. *Period Piece, a Cambridge Childhood*. London: Faber & Faber.

Reed, Evelyn. 1975. *Woman's Evolution: From Matriarchal Clan to Patriarchal Family*, New York: Pathfinder Press

Richards, Evelleen. 1976. *The German Romantic Concept of Embryonic Repetition and its Role in Evolutionary Theory in England up to 1859*. Ph.D. Dissertation, University of New South Wales.

Rogers, James A. 1972. "Darwinism and Social Darwinism." *Journal of the History of Ideas* 33: 265–80.

Romanes, George J. 1887. "Mental Differences between Men and Women." *The Nineteenth Century* 21: 654–71.

Rossi, Alice S., ed. 1970. *Essays on Sex Equality: John Stuart Mill and Harriet Taylor Mill*. Chicago and London: The University of Chicago Press.

——, ed. 1973. *The Feminist Papers: From Adams to de Beauvoir*. New York: Bantam Books.

Ruse, Michael. 1975. "Charles Darwin and Artificial Selection." *Journal of the History of Ideas* 36: 339–50.

Sahlins, Michael. 1977. *The Use and Abuse of Biology*. London: Tavistock.

Sandow, Alexander. 1938. "Social Factors in the Origin of Darwinism." *Quarterly Review of Biology* 13: 315–26.

Schweber, Sylvan. 1977. "The Origin of the *Origin* Revisited." *Journal of the History of Biology* 10: 229–316.

Sleeth Mosedale, Susan. 1978. "Science Corrupted: Victorian Biologists Consider 'The Woman Question'." *Journal of the History of Biology* 11: 1–55.

Smith, Roger. 1972. "A. R. Wallace: Philosophy of Nature and Man." *British Journal for the History of Science* 6: 177–99.

Stocking, George W. 1968. *Race, Culture and Evolution: Essays in the History of Anthropology*. New York: Free Press.

——. 1971. "What's in a Name? The Origins of the Royal Anthropological Institute." *Man* 6: 369–90.

Vogt, Carl. 1864. *Lectures on Man: His Place In Creation, and in the History of the Earth*. London: Anthropological Society of London.

Vorzimmer, Peter J. 1970. *Charles Darwin: The Years of Controversy*. Philadelphia: Temple University Press.

Wallace, Alfred R. 1864. "The Origin of Human Races and the Antiquity of Man Deduced from the Theory of Natural Selection." *Anthropological Review* 2: clvii-clxxxvii.

——. 1869. "Sir Charles Lyell on Geological Development and the Origin of Species." *Quarterly Review* 126: 379–94.

Weber, Gay. 1974. "Science and Society in Nineteenth Century Anthropology." *History of Science* 12: 260–83.

Wickler, Wolfgang. 1973. *The Sexual Code: The Social Behaviour of Animals and Men*. Garden City: Anchor Books.

Wilson, Edward O. 1975. *Sociobiology: The New Synthesis*. Cambridge, MA: Harvard University Press.

——. 1978. *On Human Nature*. Cambridge, MA: Harvard University Press.

Wilson, Raymond Jackson, ed. 1967. *Darwinism and the American Intellectual*. Homewood: Dorsey Press.

Young, Robert M. 1969. "Malthus and the Evolutionists: The Common Context of Biological and Social Theory." *Past and Present* 43: 109–45.

——. 1971a. "Darwin's Metaphor: Does Nature Select?" *The Monist* 55: 442–503.

——. 1971b. "Evolutionary Biology and Ideology – Then and Now." *Science Studies* 1: 177–206.

——. 1973. "The Historiographic and Ideological Contexts of the Nineteenth Century Debate on Man's Place in Nature." In *Changing Perspectives in the History of Science*, edited by M. Teich and R. M. Young, 344–438. London: Heinemann.

——. 1974. "The Impact of Darwin on Conventional Thought." In *The Victorian Crisis of Faith*, edited by A. Symondson, 13–35. London: Society for the Promotion of Christian Knowledge.

6

HUXLEY AND WOMAN'S PLACE IN SCIENCE

The "woman question" and the control of Victorian anthropology

If then we are not to speak of grave or scientific things in "society", & are shut out from almost all scientific "societies", how are we to learn?

— Eliza Linton to T.H. Huxley (1868)[1]

As John Greene has stressed, it was Thomas Henry Huxley rather than Herbert Spencer who became the "chief expounder and champion of Darwinism" as a "world view" after the publication of the *Origin of Species*. It was Huxley who in 1860 undertook to depict "the picture which science draws of the world" and throughout the decade he elaborated a Darwinian world view of "harmonious order governing eternally continuous progress" in which Nature was always "fair, just, and patient" but "without remorse" in enforcing the universal struggle for existence. One of Huxley's more colourful metaphors pictured the world as "Nature's university" in which all "mankind" are enrolled. Those who will not learn and obey her laws are ruthlessly "plucked", while those who learn to live in harmony with Nature are the liberally educated: "They will get on together rarely; she as his ever beneficent mother; he as her mouth-piece, her conscious self, her minister and interpreter."[2]

In the 1860s, therefore, Huxley constituted himself as Nature's leading Darwinian "mouth-piece", and his better-known role as the interpreter of "man's place in nature" was subsidiary to his overarching "moralizing naturalism". The picture of the world that Huxley projected in his popular essays and lectures of this period was one in which "natural knowledge" would lay the foundations of a "new morality"; a world where the "man of science" through his access to reliable natural knowledge would guide the conduct and organization of society. Those men of science best fitted to bring about the "New Reformation" envisaged by Huxley were of course the

---

1 E. Lynn Linton to T. H. Huxley, 11 November 1868, Imperial College Archives, Huxley Papers (HP), 21.223–6.
2 T. H. Huxley, "A Liberal Education" (1868), quoted in Greene 1981, 141–3.

"young guard" Darwinians, and throughout the 1860s he worked tirelessly towards their interrelated professional and social advancement.[3] Much of this effort was channelled into the socially sensitive science most closely concerned with the study of "man" – anthropology – and its divided and divisive theoretical models and institutions. For in this crucial period of the Huxley-led drive for Darwinian dominance of nineteenth-century science, professional anthropology was riven by the conflict between the older-established and religious-oriented Ethnological Society and the short-lived, but extremely influential, Anthropological Society of London (1863–71), the institutional stronghold of the physical anthropologists who were led by the charismatic racist, James Hunt. Although, as I shall show, Huxley's anthropological position was more congruent with that of the naturalistic and anti-clerical oriented Anthropologicals, he and the leading Darwinians allied themselves with the Ethnological traditionalists, and Hunt and Huxley became locked in an extended and acrimonious struggle for the control of Victorian anthropology (Stocking 1971; Rainger 1978; Richards 1989). It is my intention in this essay to explore the significance of the "woman question" to their conflict and its resolution.

Although it has received scant attention from historians, Hunt himself pointed up the decision of the Ethnologicals to follow the example of the Royal Geographical Society and admit "Ladies" to their meetings, as one of the major reasons for his secession from the Ethnological Society in 1863 and the foundation of the Anthropological Society. In addition to its racism, Hunt's Anthropological Society was characterized by an overt anti-feminism, and it played a key role in the "scientific" refutation of the claims by nineteenth-century liberal feminists for social and intellectual equality (Fee 1974, 1979). By contrast with Hunt, Huxley has been conventionally depicted as adopting an "enlightened" stand on women's issues (Young 1985, 617). However, closer scrutiny makes his publicly reiterated support for female education and entry to the professions somewhat problematic. Moreover, Huxley played a little-known but leading role in actively excluding women from scientific societies, specifically the Ethnological Society. Huxley and Hunt, it would seem, were at one in their opposition to female admission to the Ethnological Society.

Most accounts of the Huxley/Hunt conflict have argued the basic intellectual incompatibility of the Darwinians and the Anthropologicals, but as John Greene has reminded us: "The lines between science, ideology, and world view are seldom tightly drawn" (Greene 1981, 2). In this essay I offer a reinterpretation that stresses the crucial role of ideological and social factors in the Huxley/Hunt dispute. It thus provides a framework within which Huxley's position on the woman question and its implications for the professional and scientific aspirations of the Darwinians may be clarified.

Finally, I have sought to relate the politics of female admission to the complexities of nineteenth-century feminism through the controversial figure of Eliza

---

3 Ibid. See also Turner 1978; Weber 1974.

Lynn Linton, one-time radical, successful woman journalist, and latterly, committed Darwinian. Lynn Linton offered the only documented resistance to Huxley's exclusion of women from the Ethnological Society, and her passionate eight-page petition on behalf of the women "visitors" stands as a unique (and symbolically neglected) testament to the relations of women to mid-Victorian organized science and Darwinism in particular.

"I perceive . . . that Ladies are to come–More's the pity"

The admission of women to the Ethnological Society was first formally proposed at the council Meeting of 17 October 1860, when John Crawfurd, the recently elected president, gave notice that at the next meeting he would move a resolution that "Ladies be admitted as visitors".[4] Crawfurd's move was in line with recent liberal developments in female education, where, following the establishment of Queen's College and the Ladies College in Bedford Square, women had shown an increasing interest in attending public lectures on science at a time when these constituted almost the only form of scientific education available to them. However, Crawfurd seems to have been less motivated by a liberal concern with furthering female education than by his desire, as president, to increase attendance at the meetings of the moribund Society by making them, as Hunt sneeringly put it, "fashionable and popular".[5]

The attendance of the "fair sex" at public scientific lectures had been generally endorsed by their male patrons on the grounds that it "doubled the enjoyment" of the men by providing a sort of decorative backdrop to the occasion (Barber 1980, 132; Alic 1986, 178–81). Crawfurd and a large and powerful section of the Ethnological Society evidently wished to capitalize on this feminine drawing-power in the same way the Geographical Society had recently done. However, while it was socially acceptable for women to decorate the more frivolous scientific occasions, when it came to their actual membership of learned societies and attendance at society meetings, even avowed liberal "advocates" of female education like Huxley drew the line. Some months previously he had forcefully stated and explained his opposition to female admission to Charles Lyell.

Lyell had written to Huxley stating that the admission of women to the Geological Society might aid the cause of "geology" by exposing them to the non-creationist viewpoint and counteracting the unfavourable influence of organized religion. That "jesuit in disguise", the Bishop of Oxford, was going about and "inculcating the doctrine that no woman should ever be allowed in any of the authorised places of education to hear both sides discussed". But Lyell saw no reason why "the power of the tongue & its influence on one half of

---

4 Archives, Royal Anthropological Institute, "Council Minute Book, 1844–1869," Ethnological Society of London (ESL) Minutes.
5 James Hunt, "Dedication to Broca," in Vogt 1864, viii.

society" should be left exclusively to those "60,000 sworn teachers of endorsed opinions".[6]

However, by return of post, Huxley quashed Lyell's notion of recruiting the forgotten half of society to the Darwinian cause. He reacted against it as vehemently as the Bishop of Oxford, Samuel Wilberforce, if for somewhat different reasons:

> [T]he Geological Society is not, to my mind, a place of education for students but a place of discussion for adepts: and the more it is applied to the former purpose the less competent it must become to fulfil the latter – its primary and most important object.

Women were necessarily amateurs, and their presence at serious scientific discussions would jeopardize the professional status of the Society. It was not, Huxley hastened to assure Lyell, that he wished to place "any obstacle" in the way of the intellectual advancement and development of women. On the contrary, he did not see how society could progress as long as one half of the human race remained sunk in the ignorant superstitions inculcated by parsondom. But he did not believe that others would follow his plans for educating his own daughters in basic science to the extent that the next generation of women might become "fit . . . companions of men in all their pursuits" (not that Huxley thought men had anything to fear from their "competition");

> [Y]ou know as well as I do that other people won't do the like, and five sixths of women will stop in the doll stage of evolution, to be the stronghold of parsondom, the drag on civilization, the degradation of every important pursuit with which they mix themselves – "intrigues" in politics and "friponnes" in science.[7]

If Huxley's "claws and beak" were "good for anything", he assured Lyell that such "dolls" would be kept from hindering the progress of any science he had "to do" with.

Even by early 1860, as this exchange makes explicit, Huxley had put a Darwinian gloss on the woman question and prescribed its limitations. A minority of women, suitably educated, might become the "fit companions" of men, but not their "competitors". Like Henrietta Huxley (that paragon of scientific wives), they might assist their husbands – exhibit an intelligent interest in their work, illustrate or proofread their manuscripts, even occasionally accompany them to the more popular scientific meetings. Their proper role was to be more concerned with the scientist than his science. For careerists like Huxley, who had a "good deal of fighting to do in the external world", it was essential to have "light and warmth and confidence within the four walls of home".[8] It was inconceivable that

6 C. Lyell to T. H. Huxley, 16 March 1860, HP, 6.32.
7 T. H. Huxley to Lyell, 17 March 1860, HP, 30.34; published in L. Huxley 1900, 1: 211–12.
8 T. H. Huxley to E. Haeckel (L. Huxley 1900, 1: 289).

women might actually engage in the "fighting" that Huxley found so invigorating, and advance the professional status and rewards of science. On the contrary, their inexpert and unprofessional presence would "hinder [its] progress". As for the great majority of women, the "five sixths" whose lot was to remain stunted at the "doll stage" of evolution in the thrall of parsondom, their reactionary and frivolous presence was to be excluded at all costs from the forums of professional science.

Huxley's prejudices were founded on his own experiences and expectations. Mid-Victorian science was an all-male preserve, which women entered, if they entered at all, only as spectators – at the most as fashionable dabblers, not to be taken seriously. Thus Huxley could be jovially impressed by Mrs Buckland's newly discovered fossil "Echinoderm", but even the "very jolly" Mrs Buckland brought out his only semi-facetious "unutterable fear of scientific women".[9] The majority of women who dabbled in natural history were "Naturalists of the Boudoir" who kept a shell or mineral collection, a fern case or aquarium, and rarely ventured into the serious study of their hobby. The exclusion from the learned societies of those few women who went beyond the dilettante pursuit of their interest was in itself a strong disincentive to research, for it was usually only through the various *Transactions* that findings could be published (Barber 1980, 125–38).

Victorian feminine fragility also obstructed serious research. Girls were discouraged from real intellectual application on the grounds that it was unhealthy and fatiguing. While Huxley persistently overworked himself to the point of mental exhaustion, churning out publications and lectures until he was persuaded to recoup his forces by energetically clambering up and down various European peaks, Henrietta carefully inculcated the Victorian canon of feminine intellectual and physical frailty. "I am sure you are right", wrote her acquaintance the Countess of Portsmouth, "in not allowing your girl to do much work–I am grown very nervous about any forcing of the brain with young growing girls. I think it horribly capable of addling the brain instead of filling it . . ."[10]

Another major obstacle to the feminine pursuit of the natural sciences was the crippling "delicacy" of the age, which made it necessary for middle-class women to shy in prudish alarm from any publicly expressed hint of sex or reproduction. It took a strong motivation indeed for women to flout this Victorian sensibility. The few male naturalists who consented to address women on such issues, expurgated their lectures to such an extent that it is doubtful whether their audience grasped the essential physiology. Publications likely to be read by women were also suitably censored, or the offensive matter might be rendered in Latin. In 1856 Richard Owen negotiated carefully with John Murray over the respectability of his proposed inclusion of the "reproductive economy and apparatus" of a bee in his article on parthenogenesis for the Tory *Quarterly Review*; while around the same time, his arch-rival, Huxley, sniggeringly refused an invitation to lecture to ladies

9   T. H. Huxley to E. Dyster (L. Huxley 1900, 1: 125).
10   Eveline, Countess of Portsmouth, to Henrietta Huxley, 21 October 1874, HP, 28.107.

at London University: "What on earth should I do among the virgins, young and old in Bedford Square? . . . I should be turned out . . . for some forgetful excursions into the theory of Parthenogenesis or worse" (Barber 1980, 133, 134).

It was not only on sexual matters that Huxley had to guard his tongue in female company. The overwhelming majority of Victorian women did not share Henrietta Huxley's faith in the ennobling revelations of science and her open-mindedness on religious issues. Huxley could not speak without repercussion even to his fellow naturalists' wives, using the same openness with which he evidently discussed his agnosticism with Henrietta. Some years earlier, Huxley's "unusually plain manner (of speaking) of his want to faith" had so "alarmed" Andrew Ramsay's young wife Louisa that she "worked herself into a fever" after an "intellectual evening" at the Huxley's. On this occasion, as Ramsay had feared, Huxley's plain speaking broke up the projected joint visit of the Ramsays and the Huxleys to Switzerland.[11] Such incidents undoubtedly reinforced Huxley's conventional Victorian view of the pious frailties of middle-class women: they were plainly unequal to the give and take of robust "intellectual" discussion and thus better excluded from it.

It is fascinating to note how this female stereotype manifested itself on the legendary occasion of Huxley's celebrated confrontation with Wilberforce at the Oxford Meeting of the British Association for the Advancement of Science, only a few months after his exchange of views with Lyell. This was the highpoint of the Darwinian debates of the 1860s (now regarded by historians as more apocryphal than apocalyptic), when Huxley supposedly routed the reactionary and anti-scientific forces of parsondom in the person of Wilberforce with his devastating and apposite response to the bishop's fatal quip about Huxley's simian ancestry: "Was it through his grandfather or his grandmother that he claimed his descent from a monkey?" Huxley's son Leonard (who had not then been born) authoritatively describes the "ladies" who packed the windows of the lecture room to urge on the bishop's attack on natural selection with a dainty "waving and fluttering" of their white handkerchiefs.[12] The full horror of the social solecism of Huxley's unprecedented public counter-attack on a man of the cloth is epitomized in the "lady" who fainted and had to be carried from the room. This often-recounted incident also, of course, encapsulates the intellectual inappropriateness of the response and, more generally, of the presence of women at the debate. In one contemporary version, the bishop's slur on Huxley's ancestry is interpreted as a misguided attempt to trade on the antipathy to degrading woman (presumably in the person of Huxley's grandmother) to the level of the quadrumana. Huxley's reply is then represented as a scholarly and dignified eschewing of such vulgarity,

---

11 Andrew C. Ramsay, 15 March 1856, Imperial College Archives, Lett's Diary no. 1, Ramsay Papers/1/24, 42V. I am indebted to James Secord for this reference.
12 L. Huxley 1900, 1: 181. Women, primarily as wives and daughters of British Association members, had been admitted to all sections of the meetings in 1839, but they were confined to separate galleries or railed-off areas. The first woman member of the Association was admitted in 1853, but even as late as 1876 women were not permitted to hold office. See Alic 1986, 179–81.

and Wilberforce, abashed, is forced to recognize that he "had forgotten to behave like a perfect gentleman".[13] The piquancy of this account lies in its reversal of roles, with the bishop emerging as no gentleman, and the upstart marginal middle-class Huxley winning the day through his higher moral tone and implicit defence of Victorian values and Victorian womanhood. It gains some support from historians who have pointed out the extent to which the leading Darwinians capitalized upon their collective gentlemanly image – their solid financial, political and sexual respectability and general Victorian conventionality – in the promotion of unconventional scientific opinion (Stocking 1971, 380–1; Burrow 1968, 4; Hodge 1974, 11; Ruse 1979, 251–2).

Whatever the reality behind the mythology that has accreted around the Oxford debate, it serves as a cliché of the conventional relation of women to mid-Victorian science and to the Darwinian debates in particular. It was not a convention that Huxley cared to flout, and indeed, in a number of ways, he subscribed to it. He was, moreover, as I shall show, not averse to deploying it for institutional and social ends, and to the detriment of his liberal views. In contradiction of his public stance, Huxley's personal antipathy to the presence of women at scientific lectures was made explicit in a note he wrote in 1862 to Edward Perceval Wright, who had invited him to address the Dublin University Association on the question of the "common origin of men and apes". Wright had expressed some concern that Huxley's controversial topic might provoke a religious backlash, but Huxley in reply made it clear that he was far more concerned at the emasculating prospect of female attendance than by any mere "blackguarding" by "your Irish Holy Willies": "I perceive I misunderstood the tenor of your former note – and that Ladies are to come–More's the pity–I shall have to emasculate my discourse or else be unintelligible–I think I prefer the latter alternative."[14]

By 1864 Huxley had found anatomical evidence of the feminine inferiority that he and his Victorian contemporaries took for granted. In his Hunterian Lectures for that year he described the structural differences he had supposedly observed between the brains of men and women: "On the whole [the cerebral convolutions] are simpler in women than in men, and in the lower races the convolutions have a greater simplicity and symmetry than in the higher" (Di Gregorio 1984, 169). It was this anatomical and intellectual ranking of women and blacks below white European males that, for Huxley, made it "simply incredible" that women and blacks could ever endanger the supremacy of men like himself, and it was this scientific certainty that underpinned the reassuring message of his essay, "Emancipation–Black and White", published the following year. Here Huxley eschewed the "new woman-worship" on scientific grounds and confronted the "irrepressible" woman question with the unshakable Victorian conviction that "in every excellent character, whether mental or physical, the average woman is inferior to the average man, in the sense of

---

13 L. Huxley 1900, 1: 183–4. This account is attributed to Professor Farrar, Canon of Durham.
14 T. H. Huxley to E. P. Wright, 8 March 1862, HP, 29.115.

having that character less in quantity and lower in quality". History proves that man is more intelligent, responsible, passionate, artistic, and even more beautiful than woman. But although Nature has not made men and women equal, these "facts" do not afford the "smallest ground" for refusing to educate women as well as men, or giving them the same civil and political rights. In the name of justice, moralized Huxley, law and custom should not add to the biological burdens that weigh women down in the "race of life": "The duty of man is to see that not a grain is piled upon that load beyond what Nature imposes; that injustice is not added to inequality." Let women achieve their liberal rights: let them compete with men; give them a fair field but no favour, and let Nature judge the outcome. "So far from imposing artificial restrictions upon the acquirement of knowledge by women, throw every facility in their way"; let us have "sweet girl graduates"; let women even become merchants, barristers, politicians; it would make no difference to the status quo:

> Nature's old salique law will not be repealed, and no change of dynasty will be effected. The big chests, the massive brains, the vigorous muscles and stout frames of the best men will carry the day, whenever it is worth their while to contest the prizes of life with the best women. . . . The most Darwinian of theorists will not venture to propound the doctrine, that the physical disabilities under which women have hitherto laboured in the struggle for existence with men are likely to be removed by even the most skilfully conducted process of educational selection.
>
> (Huxley [1865] 1968, 3: 73–4)

But not even such newly erected Darwinian biological barriers were strong enough to keep women in their proper place and out of serious scientific discussions. Although as secretary and then president, Huxley had vigilantly kept the "friponnes" from the door of the Geological Society, he had not been able to perform the same service to ethnology, for he was not then a member of the Ethnological Society. In 1860, under the aegis of Crawfurd and in Huxley's absence, women had been formally admitted to the meetings of the Society. In 1868, when Huxley was elected president, he immediately flew in the face of his own liberal platitudes by initiating the move to exclude them. Huxley was now the acknowledged self-constituted Darwinian spokesman and anatomical expert on the central and most contentious issue of the Darwinian debates: "man's place in nature". His popular lectures to workingmen on the relations of man to the lower animals, and his 1863 book with this title, had assured him of professional and public recognition.[15] But this "question of questions", which precipitated Huxley into the fight for Darwinian control of Victorian anthropology, was not to be debated by women. The new science of "man" might pronounce upon "woman" (indeed, this

---

15 For an assessment of Huxley's anthropological writings and their motivations, see Di Gregorio 1984, 129–84 and Desmond 1982.

was shortly to become one of its major concerns), but it was defined and applied by Huxley specifically to exclude the very object's participation. The symbolic Huxleyan "beak and claws" were not only used to rend parsondom in defence of the Darwinian programme, but also, as he had forewarned Lyell, turned against women's scientific aspirations. The Ethnological Society, which had so briefly (and disastrously in Huxley's view), opened its doors to women, was to be professionally closed to them. No longer would it be viewed disparagingly as a "ladies' Society", but henceforth, like its institutional competitor, the all-male Anthropological Society, it would be relieved of the emasculating presence of women. At a stroke, Huxley thereby sought to upgrade the professional status of the Darwinian-led Ethnologicals and to remove one of the major impediments to their amalgamation with the recalcitrant and vociferous Anthropologicals. By 1868, the continuing schism between the two Societies had become a serious obstacle to Darwinian dominance of this key discipline, and Huxley was intent upon their unification under the "proper direction" of the Darwinians.[16]

"We are the students and interpreters of nature's laws"

In 1860, when the contentious issue of female admission was first broached in the Ethnological Society, ethnology had not been admitted to the Darwinian pantheon. It was not one of the sciences with which Huxley had as yet "to do", and the Society was not, generally speaking, a grouping likely to attract his iconoclastic attention. The ethnologists, who had their roots in Quaker and evangelical philanthropy, conducted their researches within a religious framework and sought to account for racial diversity in terms consistent with the Bible. They were primarily "monogenists" who accepted some modification over time as races diverged from their original unity of type.

However, in recent years, the religious conservatism of the Society had been challenged by a small but growing membership of physical anthropologists, including James Hunt. The physical anthropologists were primarily "polygenists" who advocated the ultimate diversity of the human races and opposed the theological concern of the ethnologists to derive all races from a single stock. They were generally men like Hunt himself, with a background in medicine, and their method was strictly anatomical. They placed great emphasis on describing, measuring and classifying the physical types of humanity, forming rigid categories that maximized racial differences and justified the polygenist belief in essential human diversity and inequality. On the whole, they lacked the benevolent, protectionist racial attitudes of the ethnologists, and they were inflexibly determinist in their interpretation of racial differences. In Hunt's case, racial determinism extended to an extreme and virulent racism that eventually precipitated his final break with the Ethnologicals (Stocking 1971, 376). But by that stage Hunt had established the Anthropological Society,

16 T. H. Huxley to J. Lubbock, 18 October 1867, British Library, Avebury Papers (Correspondence of Sir John Lubbock, V), 49642.63.

which met for the first time on 6 January 1863. And as Hunt and others represented it, the issue of female admission was crucial to the formation of the new Society.

Hunt conceived the Anthropological Society as a platform for his racial/political opinions, which he could not voice within the confines of the Ethnological Society. He and his supporters wanted a forum where they could pursue their version of anthropology, untrammelled by theological or social restraints. And the admission of women to meetings of the Ethnological Society was viewed by Hunt as the single greatest threat to the objects and duties of a truly scientific society. The "grave, erudite, and purely scientific study" of anthropology required the "most free and serious discussion, especially on anatomical and physiological topics", and this was totally at odds with the admission of women. As a dedicated anthropologist, Hunt had been the most vigorous opponent of this "fatal mistake", but in vain:

> You will, doubtless, smile at the strange idea of admitting females to a discussion of all Ethnological subjects. However, the supporters of the "fair sex" won the day, and females have been regularly admitted to the meetings of the Ethnological Society during the past three years. Even now the advocates of this measure do not admit their error, nor do they perceive how they are practically hindering the promotion of those scientific objects which they continue to claim for their society.
>
> (Hunt in Vogt 1864, viii; see also Hunt 1868, 433)

Hunt's account is supported by the sparse evidence of the Ethnological Society minutes. As they record, Crawfurd's proposal that women should be admitted as "visitors" was discussed at the meeting of 27 November 1860 and clearly provoked considerable opposition, not only from Hunt and his followers. "After some discussion", the motion was amended to read that "Ladies be admitted to the Meetings on all occasions specified by the Council." "This amendment", the minutes tersely state, "was carried", and the secretaries were authorized to invite "strangers" to the meetings. Hunt was then joint secretary, and such was his indignation and chagrin at the decision that at the meeting following his defeat, he tendered his resignation on the ostensible grounds of "health" and "too much Society business".[17] This, however, he was persuaded to withdraw, and he continued as secretary until his final rupture with the Ethnological Society, when he could devote the whole of his considerable energy and talents to strengthening his flourishing male-only stronghold, the Anthropological Society. One of the earliest motions adopted by the new society, with Hunt in the chair, was that "Ladies" might become financial members of the Anthropological Society, but "shall on no occasion whatever be allowed to attend any of the meetings of the Society".[18] Their cash was welcome, but not their persons.

17 ESL Minutes, 27 November 1860; 6 and 20 February 1861.
18 Royal Anthropological Institute Archives, Anthropological Society of London (ASL), "Council Minutes," 5 August 1863.

From the start, Hunt made clear that he was not merely founding a new society, but a "new science", and that the overwhelming significance of the new anthropology devolved upon its political implications. Race was for Hunt, as it had been for his mentor, the controversial transcendental anatomist and racial determinist Robert Knox, the key to "scientific" political legislation and social procedure (Richards 1989). From the platform of the Anthropological Society meetings and the pages of its prolific publications, Hunt waged ferocious and unceasing war on the "unnatural" notions of those liberals and radicals who suffered from the "rights-of-man mania". On behalf of the Anthropologicals and their new science, Hunt contested Huxley's liberal Darwinian bid to be nature's "mouthpiece". As Hunt saw it, the Anthropologicals rather than the Darwinians were the "interpreters of nature's laws", and it was their duty to deliver their expert opinions on the practical applications of their science. According to Hunt, this meant that John Stuart Mill's claim for black and female suffrage was a scientific absurdity, contradicted by the "facts of human nature" as revealed by the researches of the anthropologist (Hunt 1863, 1864, 1866, 1867).

Hunt and the Anthropologicals were quickly infamous for their anthropological endorsement of slavery and the more racist manifestations of British imperialism, such as Governor Eyre's bloody suppression of black revolt in Jamaica. They prided themselves on creating a forum for "liberty of thought and freedom of speech" unequalled by any other scientific society, and their members (who included the notorious Richard Burton and the Duke of Roussilon), went out of their way to confront middle-class morality. Nevertheless, the provocative, topical and often salacious discussions of the meetings, which dwelt obsessively on such essential anthropological topics as female circumcision, phallic symbolism and the anatomy of "the Hottentot Venus", initially attracted a large and enthusiastic membership. Within two years of their foundation, the Anthropologicals numbered over five hundred members (almost twice the size of the Ethnological Society); they were engaged in an active publication and translation programme; and were attempting to displace the Ethnologicals from the meetings of the British Association (Stocking 1971, 377, 380). The significance of the exclusion of women for the phenomenal success of the Anthropological Society was crystal-clear to Alfred Russel Wallace, who spelt it out for Huxley's benefit after Huxley had tried to dissuade the backsliding Wallace from attending the Society's meetings:

> I cannot agree with you that "there was not the slightest reason for [the Society's] existence". It seems to me that its establishment is a good protest against the absurdity of making the Ethnological a ladies' Society. Consequently many important and interesting subjects cannot possibly be discussed there; – & as the Geographical is also a ladies' Society the Anthrop. is the only place where they can be discussed.[19]

19 A. R. Wallace to T. H. Huxley, 26 February 1864, HP, 28.91. Wallace continued sporadically to attend the meetings of the Anthropological Society until 1868, when Huxley excluded women from the ordinary meetings of the Ethnological Society.

Huxley's failure to convince Wallace that his Darwinian duty lay exclusively with the Ethnologicals undoubtedly reinforced his determination to put an end to this "absurdity" of a "ladies' Society" when he was in a position to do so.

Huxley and the leading liberal Darwinians were quick to express their outrage at the racist pronouncements and political polemics of the all-too-successful Anthropologicals. But the Darwinians were not so outspokenly critical of the anthropological excursions of their competitors into that other major socio-political topic of the day, the woman question. Here, apparently, they could find common "scientific" ground with the Anthropologicals in the writings of Carl Vogt, whose *Lectures on Man* was translated by Hunt and published by the Anthropological Society in 1864.[20]

According to Vogt (currently one of the best-known European exponents of Darwinism, which he linked with a militant materialism and racism), the crania of men and women differed to such an extent that they could be classified "as if they belonged to different species". Moreover, he claimed, "they differ in their proportions more than many typical or race skulls". It was the polygenist Vogt's opinion that the human races were actually different species whose separate lines of evolution might be traced back into the very remote past to a common ancestry, but whose current differences were so great as to be virtually unbridgeable. As for the sexes, comparative anatomy had demonstrated for him that the crania of adult women were more childlike than those of men, thus referring them to the inferior development of the lower races. Indeed, Vogt claimed that this anatomical difference increased with the development of the race, "so that the European excels much more the female than the Negro the Negress". This meant, as he interpreted it, that there could be no possibility of sexual equality among "progressive" civilizations:

> Just as, in respect of morals, woman is the conservator of old customs and usages, of traditions, legends and religion; so in the material world she preserves primitive forms, which but slowly yield to the influence of civilization. We are justified in saying, that it is easier to overthrow a government by revolution, than alter the arrangements in the kitchen, though their absurdity be abundantly proved. In the same manner woman preserves, in the formation of the head, the earlier stage from which the race or tribe has been developed, or into which it has relapsed. Hence, then, is partly explained the fact, that the inequality of the sexes increases with the progress of civilization.
>
> (Vogt 1864, 81)

Among the lower races, the occupations of the two sexes are similar – Bushmen and women share the same tasks – but among civilized nations there is a sexual

---

20 Not all of the Darwinians shared Huxley's more liberal views on racial issues and the Governor Eyre incident. See Lorimer 1978, 131–200.

division of labour, both physical and mental, which can be bridged only at the cost of a common degeneracy.

Hunt clearly found Vogt's anatomical endorsement of female inferiority and its political implications as congenial as his racism and polygenism. It was no coincidence that Hunt aired his version of the conflict over the issue of female admission to the Ethnological Society and its significance for the formation of the Anthropological Society in the preface to his translation of Vogt's *Lectures*. The "fatal mistake" of the ethnologists lay not only in the constraints the presence of women might place on the free and open discussion of crucial anthropological topics, but in the laughable notion that women might actually engage in serious anthropological debate. Hunt and his fellow Anthropologicals were as scientifically certain of the intellectual and cultural inferiority of the female as they were of the Negro. As Vogt had shown, sexual anatomical and physiological differences were as indicative of intellectual and cultural differences as racial ones. Hunt intimated that some readers might find parts of the work offensive, but he confidently appealed to their masculine solidarity and superior scientific understanding: "The Fellows of the Anthropological Society of London are happily neither women nor children" and should, as men of science, be quite ready to accept such of Vogt's opinions as can be "logically deduced from well-ascertained facts".[21]

These "well-ascertained facts" proved equally acceptable to the Darwinians. Darwin himself in the *Descent of Man*, would reproduce Vogt's anatomical evidence in support of his evolutionary argument for the innate and continuing intellectual inferiority of women (Darwin 1871, 2: 329–30, n24). But, more immediately, Hunt's translation of Vogt's Lectures was clearly the inspiration of Huxley's anatomical and intellectual equation of women and blacks in his Hunterian Lectures of 1864. A year later, echoes of Vogt's claim that women were insusceptible to radical change could be discerned in Huxley's "Emancipation–Black and White", where Huxley asserted that women were "born conservatives" (Huxley [1865] 1968, 3: 71). As his earlier comments to Lyell indicate, Huxley was in full agreement with Vogt on the impossibility of educating women in unconventional religious and scientific opinions.

Despite their differing political positions, the reactionary Hunt and the "enlightened" Huxley were in fundamental agreement on the "natural" inferiority of women and on a "natural" hierarchy of race, and both men put their respective anthropologies to socio-political use. Nor was Hunt opposed to theories of development, providing these were located within an acceptable polygenist framework, like Vogt's.[22] Given the degree of coincidence between their anthropological systems and their shared naturalistic and anti-clerical orientation, it is not surprising

---

21 Hunt, "Dedication to Broca", and "Editor's Preface," in Vogt 1864, xii–xiii. Paul Broca was the acknowledged French leader of craniometry who used his researches to support his assumption of the intellectual inferiority of women and blacks. See Gould 1981, 73–112.

22 Hunt consistently maintained his commitment to a vaguely defined naturalistic developmentalism. See Richards 1989.

that initially Hunt and his followers had tried to make common cause with the Darwinians against the more conservative and theologically oriented Ethnologicals. With Lyell and Darwin, Huxley was one of the first five Honorary Fellows to be elected to the newly formed Anthropological Society.[23] It is significant that at this stage Huxley had still not joined the Ethnological Society. It was only after the Anthropologicals (who had their own ideological and professional axes to grind) showed unequivocally that they were not to be recruited to the Darwinian cause by thoroughly alienating Huxley with a "coarse attack" on his *Man's Place in Nature*, that Huxley resigned from the Anthropological Society and threw in his lot with the Ethnologicals. The leading Darwinians then rallied to take over the Ethnological Society and establish it as their institutional power base in the human sciences. Thereafter, relations between the Anthropologicals and the Darwinians deteriorated to the point of open hostility and conflict, and Hunt's rhetoric became increasingly anti-Darwinian (Richards 1989).

The conflict was less theoretical in character than ideological and professional, and Hunt's "anti-Darwinism" must be interpreted in this vein. His was a hegemonic struggle with Huxley and the Darwinians. The object was to define the ideological role of anthropology in Victorian society. In a period when traditional theological modes of explanation were giving way before a secular redefinition of the world, the Anthropologicals and the Darwinians offered two competing versions of a legitimating scientific naturalism. From their institutional stronghold of the flourishing Anthropological Society, Hunt and his followers were able, for some considerable time, to resist incorporation into the Darwinian anthropological model proffered by Huxley, and to offer formidable professional opposition to the takeover of London science by the Darwinians.[24] Hunt's vehemently proclaimed opposition to the "monogenism" of the Darwinians was therefore largely strategic, and served rhetorical and political purposes. It was a convenient peg on which he could hang their ideological differences and demarcate the Anthropologicals from the competing but institutionally weaker "Darwinian club".

Both the Darwinians and the Anthropologicals were well aware that Darwinism was not incongruent with polygenism, and the leading Darwinians (including Huxley, Wallace and Francis Galton), were instrumental in bringing this more forcefully to the notice of the Anthropologicals. All of these Darwinians had, by the close of the 1860s, demonstrated how monogenism and polygenism might be reconciled in evolutionary biology. Thus in the process of liberating Darwinism from the charge of "monogenism" with which Hunt and his cohort persisted in identifying it, the Darwinians made a number of significant concessions to the polygenist platform. This interpretation explains why so many prominent Darwinians came to incorporate so much specifically polygenist thinking into their interpretations of human history and racial and sexual differences. Their

---

23 The others were the polygenist Crawfurd and the developmentalist Richard Owen. See ASL Council Minutes, 18 February 1863.

24 On the takeover of London science by the Darwinian "young guard", see Turner 1978.

anthropological writings were designed not only to promote Darwinism as the key to the scientific study of humanity and society, but also to bridge the theoretical and institutional gap between the rival societies. When the two societies at last merged in 1871, a "new" evolutionist anthropological model had been formed, shaped by the confrontations and negotiations between the Darwinians and the Anthropologicals.[25]

By 1868 the Darwinians were in the ascendancy in their power struggle with the Anthropologicals, who were in serious financial difficulties through their overly ambitious publishing activities and a decline in membership brought about by internal dissension and their increasingly disreputable image. Huxley, grown impatient with the situation, took upon himself the task of putting an end to this "scientific scandal". He accepted the presidency of the Ethnological Society on condition that its Council support his efforts towards unification. With typical energy and ruthlessness, he set about reorganizing and strengthening the Ethnological Society and making the Anthropologicals more amenable to amalgamation. He pushed inept office-bearers off the Council, stepped up the Society's publications, and recruited the membership and support of those leading Darwinians, such as Joseph Hooker, who had not already rallied to the cause.[26] Finally, he tackled the pressing political problem of the Ethnological Society's derogatory image as a "ladies' Society".

Huxley's campaign was certainly political, and not only in the narrow institutional sense. By 1868 the issue of women's suffrage and women's rights had come to the fore with Mill's magnificent championing of the cause in the House of Commons, and the Anthropologicals had temporarily abandoned their pursuit of racial issues in order to confront this latest *reductio ad absurdum* of Mill's outmoded and unscientific political economy. They knew of "no subject upon which [anthropology] ought to give a more authoritative decision than upon the claims of women to political power".[27] Predictably, the consensus of the "humble Anthropologists" on this topical and threatening issue was that sexual differences and capacities had arisen from the "widespread action of natural laws, and are not to be annihilated by a merely human decree". As Vogt had established, the mental and moral differences between men and women corresponded with their anatomical differences, and the latter were the "true, irrevocable, everlasting, natural source of the practical and beneficial division of duties between men and women". What then was the natural mission of woman? Nature (as interpreted by the Anthropologicals) answered the woman question with one resounding word – "Maternity":

> It is woman's great function, and it should be her proud privilege, that she can bear and rear children to be men. . . . Is it possible to conceive a more

25 For characterizations of this "new" evolutionist anthropology, see Stocking 1971; Weber 1974; Lorimer 1978, 131–61.

26 T. H. Huxley to J. Hooker, 24 January 1868, HP, 2.140. See Stocking 1971, 383.

27 See Pike 1869, xlvii; Harris 1869. For historical background, see Liddington and Norris 1984.

contemptible and deplorable spectacle than that of the female (I will not profane the beautiful name of woman) who, having undertaken, and having appointed to her, by nature, those functions, in the proper fulfilment of which consists the charm and glory of the sex, deliberately neglects and abdicates the sacred duties and privileges of wife and mother, to make herself ridiculous by meddling in and muddling men's work?

(Allan 1869, ccxii)

A certain professional self-interest may be detected in these anthropological refutations of women's claim to do the "work" of men. As indicated above, many of the Anthropologicals were medical men, and medicine was the first occupation to be assailed by women in their attempts to enter the professions. For once, Victorian prudery worked to women's advantage here, for in an age of extreme reticence about sexual matters, many considered that women would make more fitting obstetricians and gynaecologists. Women's rights advocates, such as Elizabeth Blackwell, often deployed this argument (which was based on a full acceptance of Victorian delicacy and was not without its ideological hazards), in order to promote the entry of women into medicine (Burstyn 1980, 85; Morantz 1974, 48). The extent to which the professedly anti-Darwinian Anthropologicals co-opted evolutionary arguments to ward off this feminine threat is impressive, and demonstrates yet again the compatibility of Anthropological and Darwinian thought. Women, for instance, possessed less than men of that "combativeness which is necessary not only in political life, but even in the ordinary struggles for existence". Woman's subordination to man was "natural and eternal" and any attempt to "revolutionize the education and status of woman on the assumption of an imaginary sexual equality" would induce a "perturbation in the evolution of races".[28]

What difference there was between such arguments and Huxley's denial to women of any "natural equality", existing or potential, lay in Huxley's extension to women of their right to legal and political emancipation on the understanding that they would not be able to overcome their biological limitations and compete with men. The Anthropologicals, on the contrary, saw women's "competition" as a real professional threat and predicted social and biological upheaval from such a violation of the laws of nature. Huxley and Hunt may not have been able to resolve all their differences in negotiating the amalgamation of the two Societies, but they could at least reach full agreement on the professional unfitness of women anthropologists. Nor had Huxley any need to expect much opposition on the issue from within the Darwinian-dominated Ethnological Society. The timely death of the aged Crawfurd had fortuitously eliminated the leading proponent of female admission, and without his patronage the "ladies" themselves, having

---

28 Pike 1869, liii, lix; Allan 1869, ccxiii. Allan is here quoting Paul Broca's criticism of the feminists; see Broca 1868, 50.

only the status of "visitors" without the full entitlements of membership, could scarcely press their case.

However, at least one of them tried. When Huxley gave notice of the intended "expulsion" of the ladies from the Society, he received an impassioned petition on their behalf from Eliza Lynn Linton, who commended herself to him as a "representative woman of the bread-winning class".[29]

"Darwin . . . opened a new world to me"

The relation of Eliza Lynn Linton (1822–98), a minor novelist and journalist, to Victorian feminism is controversial. Her early life might have served as a blueprint for the emancipated woman of the time, for she built a reputation as an upholder of women's rights in education, family property and divorce. But in early 1868 came a *volte face*: in a series of much-talked about essays in the *Saturday Review*, she became the great literary opponent of the nascent women's movement.[30]

As a young woman, Eliza Lynn demonstrated an interest in ethnology – it was her article on "Aborigenes" that led to her becoming the first salaried professional woman journalist in Britain. With the failure of her marriage to the radical artisan William Linton (who was prominent in the National Chartist movement), she turned to science for consolation. All her life Eliza was torn by a contradiction she never managed to resolve, a contradiction that is reflected in all her writings on the woman question. On the one hand, she earned her own living, associated with leading radicals and intellectuals, and lived the life of an independent, strong-willed woman. Yet on the other, she clearly yearned for the more conventional Victorian role of wife and mother and its conservative social rewards. After 1865 the public humiliation of her failed marriage and her violation of her idealized womanly role precipitated one of those Victorian crises of faith so characteristic of the period. Eliza overcame her despair and found spiritual and social redemption in the certainties of science and the scientific meetings she began to frequent:

> Those Friday Evening Lectures at the Royal Institution, when Tyndall experimented or Huxley demonstrated . . . what evenings in the Court of Paradise those were! How I pitied the poor wretches who did not come to them! . . . I do not think there was one in the audience who drank in the wine of scientific thought with more avidity than I. . . . [I]t strengthened, warmed, exhilarated and almost intoxicated me.[31]

Eliza's new creed was scientific naturalism: "in science were FACTS, and these were of the kind to make a new mental era – a new departure of thought for the whole world". In the substitution of the scientific method for the theological, she

29 Lynn Linton to T. H. Huxley, 11 November 1868, HP, 21.223–6.
30 Biographical details on Eliza Lynn Linton are available in: Layard 1901; Van Thal 1979; Smith 1973.
31 Linton 1976, 3: 96. This work is a dramatization of Lynn Linton's own life in a male persona.

saw the emancipation of the human intellect from superstition, and she pinned her faith in human progress and her own moral redemption on Darwinism. "Darwin", she later wrote, "opened a new world to me. . . . The unity of Nature was the core of the creed to which I owe my subsequent mental progress – the Doctrine of Evolution that by which I have come to peace" (Linton 1976, 3: 79). In Darwinism Eliza found a substitute for William Linton's republican idealism that was more soothing to her lonely and difficult path through life, and more consistent with her growing political and social conservatism. All this found expression in the articles she published anonymously in the *Saturday Review*, which became known collectively by the title of one of them: "The Girl of the Period".

With attention-catching titles and vivid prose, Eliza vehemently fought university education for women, birth-control, women's suffrage and women's entry to the professions: "The Girl of the Period" ("a creature who dyes her hair and paints her face . . . who lives to please herself . . . bold in bearing . . . masculine in mind"); "Modern Mothers" ("this wild revolt against nature, and specially this abhorrence of maternity . . ."); "What is Woman's Work?" ("professions are undertaken and careers invaded which were formerly held sacred to men; while things are left undone which, for all the generations that the world has lasted, have been naturally and instinctively assigned to women to do"); "The Shrieking Sisterhood"; "Wild Women"; "Modern Man Haters"; etc., etc. (Linton 1883).

Fundamental to Eliza's reiterated opposition to the goals of the "shrieking sisterhood" was her insistence (which she held to the end of her life) that "the sphere of human action is determined by the fact of sex, and that there does exist both natural limitation and natural direction" (Linton 1883, "Preface"). Like Huxley, she assumed naturalistic limits to women's aspirations, and she acquired these directly from the Victorian stereotype of femininity. There is a striking coincidence between Eliza's ideal woman and Huxley's "fit companion". Eliza's ideal was the domestically competent (like Huxley, she deplored "dolls" who were "hopelessly useless" and could do nothing with their brains or their hands), but inherently modest, maternally oriented girl who, "when she married, would be her husband's friend and companion, but never his rival; one who would consider his interests as identical with her own, . . . who would make his house his true home and place of rest" (Linton 1883, 1).

Eliza's essays caused a sensation. They inspired cartoons, fashions in clothing, a satirical journal and several other "Girl of the Period" publications. Eliza did not announce her authorship until 1883, when the essays were republished in book form, but since in certain circles it was an open secret, she recorded that her attacks on the "sacred Sex" caused some "ill-blood" among her literary acquaintances. After her death, a leading advocate of female emancipation summed up what might serve as the feminist consensus on the essays: Eliza Lynn Linton found it "more profitable" to attack rather than defend her Sex (cited in Van Thal 1979, 76).

While this assessment has a good deal going for it, it would be a great mistake to dismiss Eliza Lynn Linton simply as a gifted writer, one-time radical and emancipated woman, who sold out to a reactionary anti-feminism through personal disappointment and for professional gain. Beneath the superficialities of her mawkish concern for the proprieties and her exaggerated expressions of distaste for those whom she denounced as "shrieking" extremists and "manhaters", Eliza consistently stated her allegiance to those three issues she regarded as the "core of this question of woman's rights". They were women's right to an education "as good as, . . . but not identical with, that of men" (what form this was to take remained unclear, but Eliza regarded some training in science as absolutely essential[32]); their right to property; and their right to divorce and custody of their children. These "rights" were "just and reasonable" and, above all, did not conflict with Eliza's insistence on the "natural limitation of sphere . . . included in the fact of sex". Her belief in this fundamental biological determinism was paramount and absolute; it was grounded in her unshakable materialism and Darwinism. Women could no more emancipate themselves from the laws of biology than the earth could free itself from the law of gravitation (Linton 1976, 3: 2–4), and it was this rigid scientific certainty that underpinned (and undermined) Eliza Lynn Linton's position on the woman question – and her petition to Huxley.

"It is not fair to exclude us"

Under the patronage of "dear old Mr. Crawfurd", whom she had known as a child, Eliza had become a regular "visitor" to the meetings of the Ethnological Society (Linton 1976, 3: 171). Its meetings were "real meat and bread" to her mind, obsessed as she was with "aborigines" and the liberating "FACTS" of science. There she came in contact with "clever men" like Huxley himself and into the presence of the "living thought" instead of the "deader reading" to which women like herself were generally restricted. She went home from the meetings feeling "cleverer, enriched, ennobled". In 1868, however, the prospect of her expulsion by the ruthless Huxley from this essential "communion" with learned men provoked Eliza to step outside her paid role of deriding and attacking the "girl of the period", to plead passionately on behalf of her right to a better education and better opportunities.

For one who subscribed so vehemently to the ideology of natural limitation and separate spheres, Eliza, when hard pressed, exhibited a very clear (and somewhat bitter) perception of the crippling effects of the social and cultural restrictions conventionally imposed upon Victorian women. "I have been and am a newspaper writer to some extent", she told Huxley in a letter, "but my area is limited and my powers are unequal to the best kind of work, owing to the want of those advantages and opportunities which men have." Men could meet together in clubs and societies, talk, discuss, "strike out new thoughts, hear a multiplicity of views", but "we women sit at home and spin from the one poor brain unaided. Hence the

32 See Lynn Linton to Mr Benn, 1881, in Layard 1901, 203.

comparative poverty of woman's work." Well aware of the Victorian convention that knowledge in women was unfeminine, "and scientific knowledge especially so", Eliza urged on Huxley the attendance of women at scientific meetings as one of the very few opportunities they had for serious discussion with men. Here they could set aside their social obligation to charm and amuse:

> You know how few opportunities we women have for getting any seri-ous or valuable talk with men. We meet you in "society" with crowds of friends about & in an atmosphere of finery & artificiality. Suppose I, or any woman – let her be as fascinating as possible – were to bombard you with scientific talk – would you not rather go off to the stupidest little girl who had not a thought above her pretty frock, than begin a discussion on the Origin of Species?[33]

If women were not to talk of science in society, and were excluded from scientific societies, how were they to learn about science?

In words that might have served as a paraphrase of Huxley's own "Emancipation–Black and While", Eliza reminded him of his liberal Darwinian obligations:

> [W]hat are the facts of woman's personal condition? We are thrown into an active hand to hand struggle for existence all the same as men. . . . The battle of life is a very serious matter to some of us, and we are frequently hindered and heavily weighted. . . . [I]t is not fair to exclude us from the means of knowledge & of active thought, of extended views – such as we get from attending learned discussions – on the simple plea of our womanhood.

Eliza, being Eliza, did concede the ostensible reason, that "in the interests of science (paramount of every other consideration)", there might be "necessary" discussion on "special subjects" that would render the presence of women "hin-dering or indelicate". But on such "rare" occasions, "we can always be got rid of" by advance warning, or a message left with the porter. "I pray you with all my strength", Eliza wrote to Huxley, "to keep us as attendants at the Ethnological meetings, & when you are going to discuss hazardous papers, give us warning, & we will stay away. Else let us be free still to attend".[34]

Although Eliza, fearing to put herself forward and seem "presumptuous", had made her case according to her own precepts, with proper regard for the proprie-ties and without "shrieking", in the form of a personal letter to Huxley (even "muddling up" her reasons – "like a woman!"), her powerful but womanly plea

---

33 Lynn Linton to T. H. Huxley, 11 November 1868, HP, 21.223–6. This passage has also been pub-lished by Burstyn (1980, 44) to evince the Victorian convention that women were never to talk "business" with men.

34 Lynn Linton to T. H. Huxley, 11 November 1868, HP, 21.223–6.

did not meet with the anticipated "fair" treatment from the relentless Huxley. The problem was that her offer of voluntary female exclusion from "hazardous" papers did not meet the real point at issue. If women controlled their own occasional exclusion, the Ethnological would still be a "ladies' Society", and Huxley was adamant that it must cease to be. He and the restructured Council came up with the ingenious compromise of demarcating between "Ordinary Meetings" that would be for "scientific" discussions to which "ladies will not be admitted", and larger, public "special Meetings" where "popular" topics could be discussed, and to which "ladies will be admitted" by "special invitation".[35] With one timely stroke, this admirable (and typically Huxleyan) solution reconstituted the Ethnological as a "gentlemen's society", paid lip service to the liberal principle of female admission, and retained the decorative drawing power of the "fair sex". All must have been well satisfied – except Eliza (and those she represented), who was now inexorably relegated to the more frivolous "popular element" she deplored and exiled from the serious scientific discussions she craved. But, as a leading public advocate of the "separate spheres" ideology, she was hardly in a position to complain.

As for those "hazardous" but crucial "scientific" subjects deemed unsuitable for female ears and, presumably, beyond the intellectual comprehension of such amateurs: the very first all-masculine "Ordinary Meeting" convened by Huxley featured a series of reports on "customs connected with childbearing amongst the natives of Australia and New Zealand".[36] The irony of excluding women from a male discussion of such an indelicate topic as childbirth is compounded by the fact that the "expert" author of the reports was Joseph Hooker, the leading Darwinian botanist, who had been hastily recruited to the ranks of the "professional" ethnologists by the self-same Huxley who was so anxious to expel the "amateur" ladies. The perversity of this would have been lost on Huxley, who, having demonstrated his concurrence with Hunt on the issue of female admission, was now busily pushing the goal of amalgamation with the Anthropologicals and Darwinian dominance of this key discipline to an ultimately successful conclusion. With Hunt's sudden death in mid-1869, the Anthropologicals lost their charismatic leader, and the remaining dissidents offered only token resistance to the forceful Huxley. By the beginning of 1871, the Ethnological and Anthropological Societies were amalgamated as the Anthropological Institute of Great Britain and Ireland, and the Darwinians were firmly in control of this politically sensitive science of "man".[37]

35 "Report of the Council," *Journal of the Ethnological Society of London, n.s.,* 1 (1868–9), vii–xv (xiv).
36 "Ordinary Meeting" (23 February 1869, with Huxley in the Chair), *Journal of the Ethnological Society of London,* n.s., 1 (1868–9), 68–75.
37 On the domination of the Anthropological Institute by the Darwinian "ethnologicals", see Stocking 1971, 383–6.

Figure 6.1 Huxley in Control, as depicted by his son-in-law John Collier in 1891. A very masculine Huxley, right hand trousered, leans casually but authoritatively on his accumulated wisdom, a symbolic cranium in hand. Courtesy of Wellcome Library, London.

"I am at a loss to understand"

Darwin's *Descent of Man*, published in the same year, consolidated the Darwinian endorsement of many features of Hunt's polygenist platform (Stocking 1971). Racial issues aside, much of the book's discussion of the mental and moral differences between men and women had previously been rehearsed by the Anthropologicals. Darwin was as insistent as any Anthropological on the biological basis of the continuing intellectual inferiority of women, and as much opposed to Mill's environmentalist interpretations. Like the Anthropologicals (and Huxley) before him, he brought Vogt's anatomical "observations" and their built-in social implications to his aid. By asserting the instinctively maternal and inherently modest traits of the human female, and the male's innate aggressive and competitive characteristics, Darwin provided naturalistic corroboration of deeply entrenched Victorian values and proffered an evolutionary justification of woman's narrow domestic role and contemporary social inequalities. Above all, he followed Huxley's lead by arguing on evolutionary grounds that the higher education of women could have no long-term impact on their social evolution and was, strictly speaking, a waste of resources (Richards 1983).

Recent scholarship has begun to undermine the traditional historiographic distinction between Darwinism and Social Darwinism, arguing that "Darwinism was 'social' from the start", and that "social Darwinism" is the "artifact of a professional discourse that increasingly pretended to divorce science from ideology" (Moore 1986, 39; Young 1985). The process, it now seems clear, began with the "young guard" Darwinians themselves, who, under Huxley's leadership, sought to advance their interconnected professional and social interests by relating science and Darwinism in particular to the social stability of the nation. It was Huxley's great achievement as a propagandist that he so skilfully distanced the scientist from the ideologue, while promoting science and the scientist as the independent and neutral arbiters of pressing social and political problems. He was the "master of concealed debate", and while he took the heavy-handed Spencer to task, Huxley made his own subtle accommodation of ethical "oughts" to biological "ises". His winning "social stratagem" of the 1860s was to bring Darwinian anthropology and biology to the aid of a rapidly advancing liberal bourgeoisie who, with the decline of religion, lacked a compelling ideological defence against equality and democracy. It was primarily with Huxley's dexterous hands that ideologically neutral Darwinism erected the necessary barriers, by proving that the inferior could not compete in an open society (Desmond 1982, 158–64; Helfand 1977; Hobsbawm 1975, 268).

The triumph of the Darwinians over the Anthropologicals must therefore be seen as much an ideological as an institutional victory. It was not merely a question of "style" – that Hunt and the Anthropologicals alienated scientific and social support by their tasteless and destructive confrontation with Victorian morality and organized religion, and their total lack of "gentlemanly" comportment in debate (Stocking 1971, 380–1; cf. Richards 1989). It was more the fact that the more "stylish" (i.e., more socially conformist) Darwinians, and Huxley in particular, occupied the higher moral ground politically and ideologically as well. Rather than directly counterposing his anthropology to liberal bourgeois politics and ideology in the manner of Hunt, Huxley's Darwinian anthropological model indirectly accommodated them. Where Hunt biologized a range of reactionary political and social positions on the interrelated Negro and woman questions, Huxley argued the more socially appealing and insidious liberal line that these groups might achieve "emancipation", while imposing strict biological limitations on the consequences.

It was a two-edged ideological weapon that Huxley deployed with considerable finesse and success in the face of his own demonstrated violation of the liberal principles he professed. When Sophia Jex-Blake turned to Huxley for assistance in her fight to gain medical qualification at the University of Edinburgh, she got no practical help whatever from this certified "emancipator" who had rejected women's egalitarian claims in the name of science, but who had publicly pledged to "work heart and soul" towards the "practical ends" of women's emancipation.[38] Jex-Blake and a handful of pioneering women students had been reluctantly accepted into the medical faculty under new regulations for the admission of women, which stipulated that they be instructed in separate classes. Matters came to a head in 1872 when their attempts to comply with the regulations and receive their separate instruction in anatomy were blocked by the University Court, which denied the legitimacy of their instructor's qualifications but refused to permit him to prove them by examination. Jex-Blake begged Huxley to examine the women's instructor (who had already successfully filled the position of anatomical demonstrator at the Surgeon's Hall), and provide him with a certification that would be recognized by the University Court. But the eminent anatomist and dedicated "emancipator" proved singularly reluctant to involve himself in this feminine "storming of the citadel". Instead, Huxley doled out a carefully measured *soupçon* of ideological support. He "sympathized" with Jex-Blake's cause (even though he did not think that women were on average as intelligent as men), but expressed his professional solidarity with those professors of anatomy, physiology and obstetrics who objected to teaching mixed classes of young men and women. He himself had consistently refused on moral grounds to admit women to his own lectures on comparative anatomy. Nevertheless, the same Huxley who had categorically refused to teach the "virgins" of Bedford Square, assured Jex-Blake that he would not hesitate to teach anything he knew to a class of women. It was, therefore, with "great regret" (and a totally uncharacteristic lack of resourcefulness) that Huxley was "compelled to refuse" her plea.[39] The women students were forced to bring an action against the University, which had recommended that they give up their claim to graduation (the only legal passport to practice) and accept informal (and professionally useless) certificates of "proficiency".

Yet again, when Jex-Blake's papers were referred to him for his professionally disinterested scrutiny, Huxley endorsed her failure at the final examinations. These had been taken under the combined stress of the women's contested legal action against the University (which they eventually lost), and continual

38 Huxley 1968, 3:71. See also Todd 1918, 383–4, 415–18; Bell 1953, 62–109.
39 T. H. Huxley to S. Jex-Blake, 28 October 1872, in L. Huxley 1900, 2: 74, 387.

intimidation and harassment from the medical staff and male students. Huxley found that "certain answers were not up to the standard". Huxley and Nature had sat in judgement, and Jex-Blake was "plucked", if not from "Nature's university", most definitely from the University of Edinburgh. She had failed to compete, presumably in a "fair field", and most assuredly with "no favour". Once more Huxley manipulated the situation to his ideological advantage, on this occasion with a letter to *The Times* (which had publicized Jex-Blake's failure and Huxley's part in it). Lest Miss Jex-Blake might think that his decision was "influenced by prejudice against her cause", for the last time Huxley flourished his tarnished liberal credentials: without seeing any reason to believe that women were on average as strong physically, intellectually or morally as men, he could not shut his eyes to the fact that many women were much better endowed in these respects than many men. To exclude such women from the profession of medicine went against all justice and the best interests of society, and Huxley (with the exclusion of women from the Ethnological Society and the newly amalgamated Anthropological Institute safely behind him), was "at a loss to understand" it.[40]

It is notable that when the London School of Medicine for Women was founded by Jex-Blake in 1874, Huxley was not among those qualified lecturers who gave their time and help to the School. Nor did this famous Darwinian, who made the tag of a "liberal education" synonymous with his name, ever use his unparalleled opportunities, professionally or publicly, to advance the cause of women's higher education, to which he was supposedly so devoted, in any practical way. With professional anthropology now a male preserve, Huxley's interest in the woman question in science was over. It had served its purpose, and he quietly dropped it when, in the more conservative climate of the 1870s, the advocacy of women's emancipation came to be seen as more of a liability than a strategic advantage to the Darwinians, by then effectively running Victorian science from the epicentre of the influential X Club (an exclusively masculine enterprise, although the select "x's" occasionally brought their "yv's" to the more frivolous group events).[41] In any case, a woman's successful graduation from "Nature's university" meant that she had learned to live in harmony with Nature's laws, and, as liberally interpreted by Nature's dominant Darwinian "mouthpiece", these all but precluded her graduation from any Victorian institution of higher education.

That Huxley continues to be historically evaluated as an "enlightened" advocate of women's emancipation is a tribute to the manoeuvrability of his position on the woman question. The historical reality is that he suborned women's emancipation to the Darwinian control of anthropology. He not only used his

---

40 T. H. Huxley, letter to *The Times*, 8 July 1874, in L. Huxley 1900, 1: 417.

41 L. Huxley 1900, 1: 258. On the influence of the X-Club, see MacLeod 1970, 305–22. A rough correlation may be drawn between Darwinian dominance of a learned society and its resistance to the admission of women. Thus the Royal Society, the Geological and Linnean Societies, and the Royal Microscopical Society (all of which had a heavy preponderance of Darwinians), did not admit women until the twentieth century. See Alic 1986, 181.

considerable professional powers to exclude women from organized science, but, in conjunction with the leading Darwinians, he also subtly reinforced late-Victorian assumptions of white male supremacy and contributed to the scientific anti-feminism that characterized evolutionary biology and anthropology in this period (Weber 1974). In effect, Huxley excluded women from science in the name of science and redefined that science to ratify their exclusion. It could be argued that the impact of his two-edged ideological position on the woman question was even more damaging to nineteenth-century feminist ideology than that of Hunt's total opposition. For while Huxley appeared to proffer ideological support to the feminists in the name of liberalism, he paved the way for the scientific subversion of their liberal egalitarian roots through his rejection of their egalitarian claims in the name of Darwinism. For many feminists, themselves deeply committed to naturalistic scientific explanations and to the new Darwinism, the only recourse from the concerted Darwinian drawing of naturalistic limits to their claims, was to retreat from the egalitarian ideal, and to assert that woman was "different but equal". They claimed a biologically based "complementary genius" for woman, a "genius" rooted in her innate maternal and womanly qualities. Eliza Lynn Linton, the contradictions notwithstanding, is best located among those advanced women of the nineteenth century whose confidence in the liberating powers of science, and whose opposition of naturalistic interpretations of human nature and society to conventional theological wisdom and authority, ultimately betrayed them, when science, especially Darwinism, gave a naturalistic, scientific basis to the class and sexual divisions of Victorian society (Alaya 1977; Richards 1983).

It is only comparatively recently, in the wake of the second wave of feminism, that feminist scholars have been able to transcend Eliza's dilemma and emancipate themselves from the sociobiological laws of the latter-day "Darwinians", by shifting the focus from the "woman question in science", to the "science question in feminism" – by asking how a science "so deeply involved in distinctively masculine projects can possibly be used for emancipatory ends" (Harding 1986, 29).

Acknowledgements

I should like to thank Jim Moore and James Secord for discussions and criticism, John Greene for his earlier encouragement of my work in this area, and the following institutions and libraries for permission to study manuscript material: the Imperial College of Science and Technology, the British Library, and the Royal Anthropological Institute.

Bibliography

Alaya, Flavia. 1977. "Victorian Science and the 'Genius' of Woman." *Journal of the History of Ideas* 38: 261–80.
Alic, Margaret. 1986. *Hypatia's Heritage: The History of Women in Science from Antiquity to the Late Nineteenth Century*. London: Women's Press.

Allan, J. McGrigor. 1869. "On the Real Differences in the Minds of Men and Women." *Journal of the Anthropological Society* 7: cxcv–ccxix.

Barber, Lynn. 1980. *The Heyday of Natural History, 1820–1870*. London: Jonathan Cape.

Bell, E. Moberly. 1953. *Storming the Citadel: The Rise of the Woman Doctor*. London: Constable.

Broca, Paul. 1868. "On Anthropology." *Anthropological Review* 6: 35–52.

Burrow, John W. 1968. "Introduction." In *The Origin of Species*, edited by Charles Darwin, reprint edn. Harmondsworth, Middlesex: Penguin Books.

Burstyn, Joan. 1980. *Victorian Education and the Ideal of Womanhood*. London: Croom Helm.

Darwin, Charles. 1871. *The Descent of Man, and Selection in Relation to Sex*. 2 vols. London: John Murray.

Desmond, Adrian. 1982. *Archetypes and Ancestors: Palaeontology in Victorian London, 1850–1875*. London: Blond and Briggs.

Di Gregorio, Mario A. 1984. *T.H. Huxley's Place in Natural Science*. New Haven, CT: Yale University Press.

Fee, Elizabeth. 1974. "The Sexual Politics of Victorian Social Anthropology." In *Clio's Consciousness Raised: New Perspectives on the History of Women*, edited by Mary S. Hartman and Lois Banner, 86–102. New York: Harper Torchbooks.

———. 1979. "Nineteenth-Century Craniology: The Study of the Female Skull." *Bulletin of the History of Medicine* 53: 415–33.

Gould, Stephen Jay. 1981. *The Mismeasure of Man*. New York: W. W. Norton.

Greene, John C. 1981. *Science, Ideology, and World View: Essays in the History of Evolutionary Ideas*. Berkeley: University of California Press.

Harding, Sandra. 1986. *The Science Question in Feminism*. Ithaca, NY: Cornell University Press.

Harris, G. 1869. "On the Distinctions, Mental and Moral, occasioned by the Difference of Sex." *Journal of the Anthropological Society* 7: clxxxix–cxcv.

Helfand, Michael. 1977. "T.H. Huxley's 'Evolution and Ethics': The Politics of Evolution and the Evolution of Politics." *Victorian Studies* 20: 157–77.

Hobsbawm, Eric J. 1975. *The Age of Capital, 1848–1875*. London: Weidenfeld and Nicolson.

Hodge, M. J. S. 1974. "England." In *The Comparative Reception of Darwinism*, edited by Thomas F. Glick, 3–31. Austin: University of Texas Press.

Hunt, James. 1863. "On the Negro's Place in Nature." *Memoirs of the Anthropological Society* 1: 1–64.

———. 1864. "Anniversary Address to the Anthropological Society of London, January 5, 1864." *Journal of the Anthropological Society of London* 2: lxxx–xciii.

———. 1866. "Race in Legislation and Political Economy." *Anthropological Review* 4: 113–35.

———. 1867. "Anniversary Address, January 1, 1867." *Journal of the Anthropological Society* 5: xliv–lxx.

———. 1868. "On the Origin of the 'Anthropological Review' and Its Connection with the Anthropological Society." *Anthropological Review* 6: 431–42.

Huxley, Leonard, ed. 1900. *Life and Letters of Thomas Henry Huxley*. 2 vols. London: Macmillan.

Huxley, Thomas Henry. [1865] 1968. "Emancipation – Black and White." In T. H. Huxley *Collected Essays*, 9 vols., reprint edn., 3, 66–75. New York: Greenwood Press.

Layard, George S. 1901. *Mrs. Lynn Linton: Her Life, Letters, and Opinions*. London: Methuen.

Liddington, Jill, and Jill Norris. 1984. *"One Hand Tied Behind Us": The Rise of the Women's Suffrage Movement*. London: Virago Press.

Linton, Eliza Lynn. 1883. *The Girl of the Period and Other Social Essays*. London: Richard Bentley and Son.

———. 1976. *The Autobiography of Christopher Kirkland*. 3 vols., reprint edn. New York: Garland.

Lorimer, Douglas A. 1978. *Colour, Class and the Victorians: English Attitudes to the Negro in the Mid-Nineteenth Century*. Leicester: Leicester University Press.

MacLeod, Roy M. 1970. "The X-Club: A Social Network of Science in Late-Victorian England." *Notes and Records of the Royal Society of London* 24: 305–22.

Moore, James. 1986. "Socializing Darwinism: Historiography and the Fortunes of a Phrase." In *Science as Politics*, edited by Les Levidow, 38–80. London: Free Association Books.

Morantz, Regina. 1974. "The Lady and her Physician." In *Clio's Consciousness Raised: New Perspectives on the History of Women*, edited by Mary S. Hartman and Lois Banner, 38–53. New York: Harper Torchbooks.

Pike, Luke Owen. 1869. "On the Claims of Women to Political Power." *Journal of the Anthropological Society* 7: xlvii–lxi.

Rainger, Ronald. 1978. "Race, Politics, and Science: The Anthropological Society of London in the 1860s." *Victorian Studies* 22: 51–70.

Richards, Evelleen. 1983. "Darwin and the Descent of Woman." In *The Wider Domain of Evolutionary Thought*, edited by David Oldroyd and Ian Langham, 57–111. Dordrecht and Holland: Riedel.

———. 1989. "The 'Moral Anatomy' of Robert Knox; The Interplay Between Biological and Social Thought in Victorian Scientific Naturalism." *Journal of the History of Biology* 22: 373–436.

Ruse, Michael. 1979. *The Darwinian Revolution*. Chicago: University of Chicago Press.

Smith, F. B. 1973. *Radical Artisan: William James Linton, 1812–97*. Manchester: Manchester University Press.

Stocking, George W. 1971. "What's in a Name? The Origins of the Royal Anthropological Institute (1837–71)." *Man* 6: 369–90.

Todd, Margaret. 1918. *The Life of Sophia Jex-Blake*. London: Macmillan.

Turner, Frank M. 1978. "The Victorian Conflict Between Science and Religion: A Professional Dimension." *Isis* 69: 356–76.

Van Thal, Herbert. 1979. *Eliza Lynn Linton: The Girl of the Period*. London: George Allen and Unwin.

Vogt, Carl. 1864. *Lectures on Man: His Place in Creation, and in the History of the Earth*. London: Anthropological Society.

Weber, Gay. 1974. "Science and Society in Nineteenth Century Anthropology." *History of Science* 12: 260–83.

Young, Robert M. 1985. "Darwinism *is* Social." In *The Darwinian Heritage*, edited by David Kohn, 609–38. Princeton, NJ: Princeton University Press.

7

REDRAWING THE BOUNDARIES

Darwinian science and Victorian women intellectuals

In *The Descent of Man, or Selection in Relation to Sex*, his long-awaited work on human evolution of 1871, Charles Darwin wrote:

> The chief distinction in the intellectual powers of the two sexes is shown by man's attaining to a higher eminence in whatever he takes up, than can woman – whether requiring deep thought, reason, or imagination, or merely the use of the senses or hands.

Those aspects of intelligence conventionally attributed to women, such as intuition, rapid perception and imitation, Darwin dismissed as "characteristic of the lower races, and therefore of a past and lower state of civilization". For Darwin, the intellectual differences between the sexes, like their physical differences, were entirely predictable on the basis of a consideration of the long-continued action of natural and sexual selection aided by use-inheritance. Male intelligence, he argued, would have been consistently sharpened through the struggle for possession of the females (sexual selection) and through hunting and other male activities such as the defence of the females and young (natural selection). "Thus", he concluded, "man has ultimately become superior to woman" (Darwin 1871, 2: 326–9).

By 1871, Darwinism, in the capable hands of its leading popularizers and propagandists, the scientist Thomas Henry Huxley and the social theorist Herbert Spencer, was well on its way to becoming the new orthodoxy in Victorian science and society. Darwin's theory of evolution (first publicly presented in *The Origin of Species* in 1859) was accepted into the body of scientific knowledge in a period of extraordinary social and economic transformation, in which pre-industrial modes of legitimation, religion in particular, were giving way to a secular, naturalistic redefinition of the world. In the process, the natural sciences increasingly took over from religion the task of defining and upholding the moral and social order. Darwinism was central to this transition.

Darwin took the biological "struggle for existence", the basis of his theory of natural selection, from Malthusian social theory, and it has been compellingly argued that the image of nature presented in Darwin's work was contingent upon his own social context of mid-Victorian capitalist enterprise (Young 1985; Desmond and Moore 1991). Earlier evolutionary doctrines had been closely associated with political radicalism (Desmond 1989). Natural selection, with its emphasis on progress through competition and the elimination of the less well adapted, dissociated evolution from revolution, and, at the same time, brought it into line with the competitive, free-trading ideals of the newly powerful industrialists and reform-oriented professionals who constituted a ready-made receptive audience for Darwin's views.

Huxley, in particular, capitalized on the opportunity thus provided to promote his claims for social progress through scientific advance. Darwinism was his lever for shifting power from an old privileged, ecclesiastical elite to a new technocratic elite of professional scientists whose authority to guide the conduct and organization of society rested in right reasoning and reliable natural knowledge, not mythical "truths." Throughout the sixties, he actively popularized and institutionalized a form of evolutionary naturalism that recruited support from a wide spectrum of society. In an increasingly secular and scientifically minded age, "progressives" of all kinds, including many feminists, rallied to a scientifically credentialed creed that its leading advocates overtly opposed to outmoded theological modes of explanation and linked with social and technological progress (Desmond 1994, 310–63).

But, in certain respects, the new Darwinism represented less a revolutionary break than an underlying continuity with the natural theology tradition it displaced. For all their differences, both doctrines were concerned to justify much the same set of underlying assumptions about economic and social relations, to preserve the status quo (Young 1985). In a context of imperial expansionism, economic uncertainty, urban and industrial unrest, the emergence of mass socialist working-class movements all over Europe, and the increasing urgency of the demands by women for the suffrage, higher education and entrance to middle-class professions, the origin of "man" by natural law rather than divine creation was made more palatable for its Victorian audience by Darwinian concepts of "natural" and inevitable white, middle-class male supremacy.

Huxley led the way with his widely read "Emancipation–Black and White" of 1865. Here he steered a carefully calculated middle course, advocating votes and education for both women and blacks, but invoking Darwinian natural law of fair competition and no favours to reassure their oppressors and his threatened fellow-professionals that "Nature's old salique law will not be repealed, and no change of dynasty will be effected." Women, like blacks, were the natural inferiors of white men and would remain so. Not "even the most skilfully conducted process of educational selection", Huxley asserted, could remove the "physical disabilities under which women have hitherto laboured in the struggle for existence with men" (Huxley [1865] 1968, 74).

With the publication of *The Descent of Man*, Darwin put his imprimatur on such evolutionary ratification of Victorian values. His reconstruction of human evolution is pervaded by Victorian racial and sexual stereotypes and assumptions of the inevitability and rightness of the sexual division of labour. By asserting the instinctively maternal and inherently modest traits of the human female and the male's innate aggressive and competitive characteristics, Darwin provided naturalistic corroboration of woman's narrow domestic role and contemporary social inequalities (Richards 1983; Rosser and Hogsett 1984; Jann 1994). Following this, there was scarcely an evolutionist who did not take up and pronounce upon the woman question.

The more conservative, even reactionary, position was hammered out by Spencer, chief architect of "Social Darwinism." Spencer opposed the extension of the franchise and higher education to women primarily on the grounds that they were less highly evolved than men and constitutionally less fit to handle political or social and professional responsibilities. Other prominent Darwinians such as George J. Romanes, Francis Galton and Patrick Geddes joined forces with anthropologists, psychologists and gynaecologists to forge a formidable body of biological determinist theory that purported to show that women were inherently different from men in their anatomy, physiology, temperament, and intellect – that women, like the "lower" races, could never expect to match the intellectual or cultural achievements of men, nor obtain an equal share of power and authority. Victorian science (and evolutionary science in particular), as feminist scholars have documented, was strongly gendered (Conway 1970; Fee 1974; Rosenberg 1975; Mosedale 1978; Russett 1989).

The response by the women concerned to counter the concerted Darwinian reinforcement of traditional views of their social and cultural roles has also begun to be charted (Alaya 1977; Love 1983; Tedesco 1984; Egan 1989; Erskine 1995). But few studies that locate individual Victorian women in precise relation to the institutional and wider socio-political contexts of Darwinian science and its practitioners have been undertaken (Richards 1989). By unpacking specific instances of the engagements of particular representative women with Darwinism and its institutions, we may better our understanding of the ways in which the discursive categories of gender, sexuality and science were constructed and contested by some of the scientists and women concerned, and how that discourse was rooted in an historically specific set of ideas and practices about gender, sexuality and science (Outram 1987; Haraway 1989; Hall 1992).

In keeping with such contextual, comparative approaches, I want here to examine the contrasting responses of two Victorian women intellectuals, Eliza Lynn Linton and Frances Power Cobbe, to Darwinian science and its institutions. My two case studies are intended to uncover something of the diversity and complexity of Victorian feminism and the contradictions inherent in its general reliance on Victorian stereotypes of femininity upon which Victorian science was also contingent. They also offer a means of exploring the various strategies adopted by the dominant Darwinians in redrawing the boundaries of organized science against the incursions by women into this most masculine profession.

The women I have chosen were born in the same year – 1822. Both became self-supporting writers with a keen interest in science and the related social and political issues of their time. Both were political conservatives who fully endorsed the Victorian conventions of womanhood. But there the similarities end.

Eliza Lynn Linton (1822–1898) was a Victorian paradox, an "emancipated woman opposed to women's emancipation" (Anderson 1987, x). Successful journalist and ardent Darwinian, Lynn Linton represents the extreme pole of the biological determinist position on the woman question as advocated by the Darwinians. She campaigned vehemently against women's higher education, birth control, suffrage, and entry into the professions, largely on Darwinian grounds. But she also confronted Huxley over the exclusion of women from the Ethnological Society – a confrontation that brings to the fore all the contradictions of her position as a woman and an evolutionist in Victorian society and exposes Huxley's own manipulations of the woman question in pursuit of his interrelated goals of the professionalization of science and the Darwinian control of anthropology.

Frances Power Cobbe (1822–1904) represents another response to Darwinism – the theistic alternative that accepted the evolution of the body but not of the mind. Cobbe was well known in middle-class circles as a leading advocate of women's rights and an antivivisectionist. Her antivivisection crusade brought her into conflict with the professional and social aspirations of the Darwinians, notably Huxley and Darwin, who strongly defended the right of the scientist to animal experimentation. Their conflict also illustrates the ways in which women like Cobbe tested and extended the limits of the sphere of femininity and constructed political identities for themselves on a terrain different from that of the scientists.

I. Eliza Lynn Linton and the Masculine "New World" of Darwinized science

> Darwin opened a new world to me . . . The Unity of Nature was the
> core of the creed to which I owe my subsequent mental progress –
> the Doctrine of Evolution that by which I have come to peace.
>
> (Linton [1885] 1976, 3: 79)

The young Eliza Lynn was a poorly educated, strong-willed, but sensitive girl, who in the 1840s went to London from the obscurity of a country vicarage and in defiance of a patriarchal father to become the first salaried professional woman journalist in Britain. An intense, bookish woman, Eliza moved in the circles of the radical intelligentsia. In 1858, when she was thirty-six years old, she married the radical artisan William Linton. Linton was an engraver of considerable artistry who subordinated his talents to the active promotion of his political views. He was prominent in the national Chartist movement, but in some respects, particularly in his views on education, the liberating powers of science, and female emancipation, Linton shows the influence of the Owenite socialists (Smith 1973; Taylor 1983).

When Eliza first met him, he was living with the consumptive Emily Wade and their seven children, all of whom, girls and boys alike, were dressed in long blue flannel blouses, with shoulder-length hair and identical broad-brimmed hats. Initially, Eliza was charmed by Linton, his republicanism, his "moral purity," and his strange "Bohemian" household. She took over the family, helping the impractical Linton and Emily financially and imposing a certain middle-class order on the chaotic household. The children's diet, table manners, hair-length and accents were reformed to Eliza's exacting standards. After Emily's death, she decided to "legalize" her position with the children by marrying Linton (Anderson 1987, 72–81).

All her life, Eliza Lynn Linton was riven by a contradiction she never managed to resolve and which is reflected in all her writings on the woman question. She earned her own living, associated with leading radicals and intellectuals, and lived the life of an independent, strong woman; yet she clearly yearned for middle-class respectability and the more conventional Victorian role and rewards of wife and mother. Anderson's recent psychological portrait of Lynn Linton makes a strong case for her life-long inability to transcend the difficulties of her formative years, her subsequent conflicted self-hatred, and the intense male identification that fuelled her criticisms of female character and women's rights (Anderson 1987). Anderson, however, fails to recognize the extent to which Eliza's Darwinism sustained these tensions in her personal life and her relation to Victorian feminism.

Eliza's attempt to acquire a ready-made family and live out her idealized Victorian role of wife and mother was an abysmal failure. Her choice of husbands is an instance of the contradiction that dominated her life. She married a committed radical activist and tried to turn him into a conventional Victorian husband, a "capable & successful *doer*" (Anderson 1987, 80). Her failure to achieve this precipitated one of those Victorian crises of faith so characteristic of the period.

By 1865 she and Linton had separated, and the intelligent, hard-working writer could not accept the public humiliation she attached to her own violation of her socially assigned womanly role. After a period of despair, she found spiritual and social redemption in the certainties of science and the scientific meetings she eagerly attended at every opportunity:

> Those Friday Evening Lectures at the Royal Institution, when Tyndall experimented or Huxley demonstrated . . . what evenings in the Court of Paradise those were! How I pitied the poor wretches who did not come to them! . . . I do not think there was one in the whole audience who drank in the wine of scientific thought with more avidity than I. . . . [I]t strengthened, warmed, exhilarated and almost intoxicated me.
> (Linton 1976, 3: 83–4)

Eliza's new creed was the doctrine of scientific naturalism: "in science were FACTS, and these were of the kind to make a new mental era – a new departure of thought for the whole world, as well as for myself individually." In the "substitution of the scientific method for the theological," she saw the emancipation of the human intellect from superstition, and she pinned her faith in human progress and her own moral redemption on the new Darwinism (Linton 1976, 3: 79–81).

Eliza's association with Linton and the radicals probably prepared the ground for her ready conversion to Darwinism. Transmutationism was widely popular among radical artisans who believed that it served their republican and materialist platform (Desmond 1989). Around the time of her marriage to Linton, Eliza favourably reviewed the pre-Darwinian arguments of Robert Chambers and Herbert Spencer for the "progressive improvement" of life and society (Linton 1858). Spencer's later role as foremost Social Darwinist was to defuse the revolutionary appeal of transmutationism with the more socially acceptable mechanism of continuous social progress through elimination of the "unfit". The regeneration of society was now guaranteed by the "fixed laws" of capitalist competition, and here Eliza found a substitute for William Linton's republican idealism that was more consistent with her growing political and social conservatism. Further, she palliated her unconventional agnosticism and materialism and her self-perceived anomalous social situation with an obsessive outward conformity with Victorian prudery and propriety. All this found expression in the famous series of articles she published in the *Saturday Review* in early 1868, which became known collectively by the title of one of them: "The Girl of the Period."

With attention-catching titles and vivid prose, she vituperatively attacked and caricatured just about everything nineteenth-century feminism represented in articles such as "The Girl of the Period" ("a creature who dyes her hair and paints her face . . . who lives to please herself . . . bold in bearing . . . masculine in mind"), "Modern Mothers" ("This wild revolt against nature, and specially this abhorrence of maternity"), "What is Woman's Work?" ("professions are undertaken and careers invaded which were formerly held sacred to men; while things are left undone which, for all the generations that the world has lasted, have been naturally and instinctively assigned to women to do . . ."), "Wild Women," "Modern Man Haters," and so on (Linton 1883).

Eliza's essays caused a sensation and ensured her professional success. They inspired cartoons, fashions in clothing, a satirical journal, *The Girl of the Period Miscellany*, and several other publications (Anderson 1987, 120–5). The "Girl of the Period" or "GOP" became fixed in the Victorian vocabulary as a catchphrase or acronym for a "modern" or "fast" girl. Against this unnatural hussy of her creation, Eliza held up the ideal of the inherently modest, domestically-oriented girl who "when she married, would be her husband's friend and companion, but never his rival; one who would consider his interests as identical with her own . . . who would make his house his true home and place of rest" (Linton 1883, 1). The fact that her own practice of these feminine virtues had driven Linton from the marital home was beside the point, and the contradiction between this ideal and her own circumstances was generally ignored.

Obviously, Eliza's reassertion of traditional Victorian values was highly marketable in a context of middle-class antipathy towards the threatening economic and political independence of women. But it would be simplistic to dismiss her as a gifted writer, one time radical and emancipated woman, who sold out to a reactionary anti-feminism through personal disappointment and for professional gain. Despite her exaggerated concern for the proprieties and her diatribes against the "shrieking sisterhood," Eliza consistently held to three issues she regarded as the "core of this question of woman's rights." They were women's right to an education "as good as . . . but not identical with, that of men" (this, she thought should include some science education), their right to property, and their right to divorce and custody of their children. These "rights" were "just and reasonable" and, above all, did not conflict with Eliza's insistence on the "natural limitation of sphere . . . included in the fact of sex" (Linton 1870, 224–38, 1976, 3: 2–4).

Like the Darwinians, Eliza assumed naturalistic limits to women's aspirations and based these firmly within a traditional rendering of Victorian femininity. Her insistence on the essential domesticity and modesty of women was grounded in her unshakeable materialism and her Darwinism. Women, she held, could no more emancipate themselves from the laws of biology than the earth could free itself from the law of gravitation, and it was this rigid scientific certainty which underpinned and undermined her stance on the woman question. All the contradictions of that stance – personal, professional and scientific – are manifest in her eight-page petition to Huxley over the issue of the exclusion of women from the meetings of the Ethnological Society in 1868 (Huxley Papers 21.223–6).

Three years earlier, Huxley had publicly declared himself on the "irrepressible" woman question by asserting his Darwinian certainty that women would remain the natural inferiors of men. Nevertheless, he had argued, it was the liberal man's duty to see that "not a grain is piled upon that load beyond what Nature imposes; that injustice is not added to inequality" (Huxley [1865] 1968, 74).

Huxley's "Emancipation–Black and White" of 1865 served a number of purposes, but it was aimed primarily at the rival and rabidly racist Anthropological Society, which had broken away from the Ethnological Society in 1863 over the ostensible issue of the admission of women. The Anthropologicals quickly built up a large, enthusiastic and exclusively masculine membership devoted to the dissemination of anti-feminist and racist propaganda in the guise of physical anthropology. Their anatomical method of describing, measuring and classifying racial and gender differences allegedly proved the "natural" inferiority of women and blacks. John Stuart Mill's claim for black and female suffrage was therefore a scientific absurdity, contradicted by the "facts of human nature" as revealed by the researches of the anthropologist. The Anthropologicals endorsed slavery and the more racist manifestations of British imperialism. Primarily medical men who felt themselves particularly threatened by the professional aspirations of middle-class feminists, they also vented their spleen against those unnatural women who sought to deny their natural mission of motherhood and make themselves ridiculous by "meddling in and muddling men's work." Professedly anti-Darwinian, the Anthropologicals even on occasion harnessed evolutionary rhetoric to their anti-feminist stance: Women possessed less than men of that "combativeness which is necessary not only in political life, but even in the ordinary struggles for existence." Woman's subordination to man was "natural and eternal," and any attempt to "revolutionize the education and *status* of woman on the assumption of an imaginary sexual equality" would induce a "perturbation in the evolution of the races" (Richards 1989, 261–70).

There was little difference between such arguments and Huxley's denial to women of any "natural equality", existing or potential. His extension to women of their right to legal and political emancipation was offered on the understanding that they would not be able to overcome their biological limitations and "compete" with men on equal terms. Huxley was as certain as any Anthropological of the crucial cerebral differences between men and women and ranked women's intelligence with that of the "lower races." He was also forcefully opposed to the admission of women to scientific societies. For all his liberal rhetoric, Huxley's personal views of women were remarkably consistent with the publicly expressed opinions of the GOP author. With few exceptions, he viewed women as mostly frail, religious creatures, stuck at the "doll stage of evolution." To the careerist Huxley, women were *ipso facto* amateurs, fit for the classroom, but utterly out of place in the cut and thrust of professional scientific forums where their amateur presence threatened that Darwinian expertise and status to which he was so committed (Huxley 1900, I: 211–12; Richards 1989, 225–61; Desmond 1994, 310–63).

In 1865 the Anthropologicals were in the ascendant over the Darwinians who had colonized the moribund Ethnological Society and were vying with the Anthropologicals for control of the strategically significant science of "man." "Emancipation–Black and White," with its prohibition on denying their liberal rights to blacks and women on anatomical or any other grounds, was Huxley's attempt to refuse scientific authority to the Anthropologicals while asserting it for the Darwinians through the subordination of black and female equality to the inescapable struggle for existence. When the unruly Anthopologicals proved recalcitrant to Darwinian control, Huxley's strategy became one of amalgamation of the two societies under the "proper direction" of the Darwinians. One of the tactics he deployed as its newly elected president was to initiate the exclusion of women from Ethnological Society meetings (Richards 1989, 267–70).

This then was the context in which Eliza Lynn Linton, who, through her interest in human evolution and need for communion with "clever men," had become an assiduous attender of Ethnological Society meetings, was forced to step outside her paid professional role of deriding and attacking the "girl of the period" to plead passionately on behalf of her right to a better education and opportunities. "You know how few opportunities we women have for getting any serious or valuable talk with men," she told Huxley in 1868.

> We meet you in "Society" with crowds of friends about & in an atmosphere of finery & artificiality. Suppose I, or any woman – let her be as fascinating as possible – were to bombard you with scientific talk – would you not rather go off to the stupidest little girl who had not a thought above her pretty frock, than begin a discussion on the Origin of Species . . .?
>
> (Huxley Papers 21.223–26)

If women were not to talk of science in society, and were excluded from scientific societies, how were they to learn about science? There were very few scientific meetings open to women, and it was not easy to obtain the favour of an invitation to these more popular and fashionable events – "I have been [to the Royal Institution] only thrice in my life."

In paraphrase of Huxley's own "Emancipation–Black and White", Eliza sought to remind him of his liberal Darwinian obligations:

> What are the facts of woman's personal condition? We are thrown into an active hand to hand struggle for existence all the same as men – we of the middle classes have to earn our own bread – with very badly trained hands & brains it must be sorrowfully confessed. . . . The battle of life is a very serious matter to some of us, and we are frequently hindered and heavily weighted . . . it is not fair to exclude us from the means of knowledge & of active thought, of extended views – such as we get from attending learned discussions – on the simple plea of our womanhood.
>
> (Huxley Papers 21.223–6)

Although Eliza made her case according to her own precepts, with proper regard for the proprieties and without shrieking, in the form of a personal letter to Huxley, even in (unconscious?) parody of feminine intellectual incompetence, "muddling up" her reasons ("like a woman!"), her powerful but womanly plea did not meet with the anticipated "fair" treatment from the ruthless Huxley. He and the Darwinian-dominated council came up with the ingenious compromise of demarcating "Ordinary Meetings" which would be for "scientific" discussions to which "ladies will not be admitted," from larger, popular "Special Meetings" to which "ladies" might be admitted "by special invitation" (Richards 1989, 275).

With one timely stroke, this admirable (and typically Huxleyan) solution reconstituted the Ethnological as a "gentlemen's Society" and paid lip service to the liberal principle of female admission. All must have been well satisfied except Eliza (and those she represented), who was now inexorably relegated to the more frivolous "popular element" she deplored and exiled from the serious scientific discussions she craved. But, as a leading public advocate of the "separate spheres" ideology, she was hardly in a position to complain.

The exclusion of women served Huxley's purposes: it upgraded the professional status of the Ethnologicals and it removed one of the major impediments to their amalgamation with the Anthropologicals, which he achieved in 1871, the same year in which Darwin's *Descent of Man* consolidated the Darwinian endorsement of many aspects of the Anthropologicals' platform. Darwin was as insistent as any Anthropological (and Huxley) on the biological basis of the continuing intellectual inferiority of women and blacks, and as much opposed to Mill's environmentalist explanations. Above all, he followed Huxley's lead, by arguing on evolutionary grounds that the higher education of women could have no long-term impact on their social evolution and was, strictly speaking, a waste of resources (Darwin 1871, 2: 326–9; Richards 1983, 1989, 276).

II. Frances Power Cobbe, the "duties of women" and the dissent from Darwinism

> To those amongst us who have not bowed to the new moral system of Darwin and Spencer, there is something almost pathetic in the ignorance both of the passions and of the spiritual part of human nature which these philosophers unconsciously betray.
>
> (Cobbe 1881a, 70n)

Frances Power Cobbe provides an illuminating contrast to Eliza Lynn Linton. She believed absolutely that women were more chaste, generous and moral than men and was adamant that they must nurture their offspring, look after the home, and not succumb to selfishness and loose, "Bohemian" manners and morals (Cobbe 1881a, iv). But unlike Lynn Linton, Cobbe did not dissociate herself from the professional and other liberal goals of the women's movement. Asserting that the "cause of the emancipation of women is identical with that of the purification of society," Cobbe actively sought the extension of woman's existing social role and a proper recognition of its importance (Cobbe 1881a, 11; Caine 1992, 103–49).

Cobbe also experienced the prevalent crisis of faith and was an early convert to evolutionary naturalism. But after a short period of agnosticism, the intolerable sense of severance caused by the death of her much-loved mother persuaded her to opt for a form of theism based on the idea of a just and rational God whose moral law was evident to all people through their own intuition, not through revelation. It also provided a basis for her rebellion against the moral authority of her domineering father, although she continued to serve him as a dutiful daughter. It was her theism, her absolute belief in the moral autonomy of women, and her strong sense of their mental and moral difference from men that constituted the core of Cobbe's feminism (Caine 1992, 115–19, 131–2).

While she paid lip service to the Victorian imperative of marriage for women, Cobbe was highly critical of the institution. She inveighed vigorously against the married woman's loss of legal identity, of property or earnings, and the domestic tyranny and misery to which so many were subjected (Cobbe 1862, 1863). She never married and, after her father's death, travelled widely, pursued an independent, hard-working writing vocation, and lived in domestic harmony and comfort for thirty-four years with Mary Lloyd, a painter. It seems to have been one of those Victorian "female marriages" that provided affection and commitment for its participants without any necessary sexual involvement (Caine 1992, 120–5).

Cobbe's attitude towards Darwinism was shaped by her feminism and her theism. She knew Darwin personally and, initially at any rate, greatly admired him. In her autobiography she recounts how, on encountering Darwin while out walking, they held a shouted discussion across a bramble patch about the significance of Mill's views on women for Darwin's forthcoming *Descent of Man*. Mill, Darwin asserted, could learn some things from science. Women's nature, like men's, was rooted in their biology, and it was through the "struggle for existence and (especially) for the possession of women that men acquire their vigour and courage" (Cobbe 1894, 2: 124–5). Cobbe got the opportunity to set him right when she reviewed Darwin's *Descent*. She had no theological difficulties with tracing "Man to the Ape." But, while she did not dissent from Darwin's views on the evolution of the physical differences between men and women, Cobbe forcefully opposed his "Simious Theory of Morals," his "most dangerous" utilitarian interpretation of the evolution of the human mind and morality from animal instincts (Cobbe 1872, 14, 1894, 2: 127). To the "Atheistic Morals" and materialism of the evolutionists, she opposed her doctrine of "Theistic Ethics," which asserted the existence of free will and the immortality of the soul, championed love over knowledge, and stressed the special duties of women to "those who have no free-will – the lower animals." This neatly removed the mental and moral differences between men and women from the biological to the spiritual domain and so, in Cobbe's view, guaranteed the moral autonomy and authority of women (Cobbe 1881a, 17, 67).

The *real politic* as far as Cobbe was concerned was less the campaigns for the suffrage and higher education than her antivivisectionist crusade. It was into this highly controversial movement that she channelled most of her abundant energy, intellect, and consummate political skills, and it was her passionate involvement in this cause that brought her into direct conflict with the Darwinians, Huxley and Darwin in particular.

Cobbe publicly became involved in the antivivisection movement in 1875 when she circulated a memorandum urging the Royal Society for Prevention of Cruelty to Animals to mobilize to restrict the practice of experimentation on live animals. Among the many eminent persons she approached was Darwin. Darwin refused to sign Miss Cobbe's "foolish paper" but reacted with alarm to the "many powerful names" who had indicated their support for her proposal. He alerted Huxley to the need for action lest the House of Commons, "being thoroughly non-scientific," should pass some "stringent law, enough to check or quite stop the revival of physiology in this country" (Darwin Archive 97: 37.8). Huxley backed Darwin's suggestion of a counter-proposal from "eminent physiologists and biologists" for "reasonable" legislation "as the best method of taking the wind out of the enemy's sails." He added for good measure, "My reliance as against that 'foolish fat scullion' & her fanatical following is not in the wisdom and justice of the House of Commons, but in the large number of fox-hunters therein" (Darwin Archive 166: 338).

For Darwin and Huxley, antivivisectionists were the "enemy" who threatened the progress and prestige of British science. They were characterized by their foolishness (i.e., irrationality) and their fanaticism (i.e., emotionalism), and they were also, as Darwin and Huxley well understood, female. Not only were they led by the redoubtable Cobbe, who came to personify the antivivisection movement, but, as Darwin put it, it was women "who from the tenderness of their hearts and from their profound ignorance" were the "most vehement opponents" of vivisection (Darwin 1994, letter 10546). On the home front, Darwin found himself called to account for his pro-vivisectionist stance by his daughter Henrietta (Darwin 1888, 3: 202–3). When the physiologist Romanes came to visit, Darwin warned him not to talk about experiments on animals "when in presence of my ladies" (Darwin 1994, letter 9916). Women, then, were the enemy of the progress of rational science, and they were so because of those very feminine traits that otherwise made them such desirable and conforming wives and daughters. As antivivisectionists they were doubly subversive, trading on their femininity to unman the scientist on his own ground.

The commitment of so many women to the antivivisection movement is another illustration, like that of spiritualism, of how concepts of femininity and moral superiority could be used to legitimate a range of public and quasi-public activity not usually associated with the traditional female role (French 1975, 240–1; Elston 1987; Owen 1990). Women like Cobbe could thus adopt a leadership role and lay claim to special skills and knowledge without doing damage to those qualities that constituted the Victorian ideology of femininity. At the same time, paradoxically, their involvement in the antivivisection movement provided an opportunity for women to use their very femininity to achieve and wield power, especially over male scientists and doctors, to subvert female subservience with the conventional feminine tools of sentimentality and womanly concern for suffering. But this interpretation does not account for the extraordinary identification of Cobbe and many other Victorian women with animals and their sufferings, particularly at the hands of medical scientists.

Lansbury's intriguing thesis is that for these women the vivisected animal was the surrogate of woman, humiliated, exposed and threatened by the gynaecologist's knife and by the pornographer's whip, stirrups and other paraphernalia taken from the stables and kennels. She argues that the full impact of the antivivisection movement on Victorian culture cannot be understood without recognition of the fusion of imagery from three major areas: gynaecological, pornographic, and literary. The pornographic literature of the period dwelt repetitively on the ritual bestialization of women. Women were "broken to the bit," "mounted," made to "show their paces," collared, chained, bound, flogged and seduced into grateful submission to their "masters." Lansbury points to the "uneasy similarity" between the devices made to hold women for sexual pleasure in such male fantasies and the gynaecological table and "stirrups" which came into general use around 1860. Women doctors like Elizabeth Blackwell and Anna Kingsford, both committed antivivisectionists, deplored the "degrading cruelty" with which poor women were treated in the major hospitals of the day. Antivivisection literature routinely conflated the plight of such women, who allegedly were made the victims of cruel and unnecessary experimental surgery, with that of the dogs, cats and monkeys who were the piteous, defenceless objects of the merciless vivisector. The symbols evoked by women antivivisectionists were "all the more potent because they were drawn from a muffled context of reticence and ambiguity" (Lansbury 1985; see also Moscucci 1990, 112–27; Elston 1987).

Lansbury's thesis is given greater plausibility by Ritvo's explorations of the political and social resonances of Victorian relations with animals. Their relation to dogs, for example, exemplified many of the tensions in Victorian class and sexual ideologies. Dogs, like lower-class women and prostitutes, possessed dangerous sexualities that necessitated control and were a potential source of contagion to middle-class humans. The identification of woman's sexuality and nature with dogs and other domestic animals was made most explicit in the discourse of breeders. When, for instance, they discussed the difficulty of getting a prize bitch to mate exclusively with a selected male, their discussion was transparent to assumptions about the sexuality of human females (Ritvo 1987, 3–4, 180–6; 1988). The dog, above all, signified the loved but subservient being that thoroughly understood and accepted its inferior position. Even its body proclaimed its profound submission to humanity. It was the most malleable of all man's domestic productions, its shape and size responding most readily to the caprice of the breeders (Ritvo 1987, 20–3).

There is more than a degree of coincidence between these traits associated with the domesticated dog and those accorded to women within the terms of Victorian domestic ideology. Women were to serve, to obey, to be pliant to masculine whim and will, to stay where men commanded or follow where they led. The young Darwin made the direct connection: marriage was analogous to pet keeping; a wife was an "object to be beloved & played with. – better than a dog anyhow" (Darwin [1838] 1986, 2: 444). Darwin's semi-facetious musings were not innocent. The stereotyping of women as domestic animals was deeply entrenched in Victorian culture.

Darwin's immersion in the literature of the breeders guaranteed his full exposure to such metaphorical discourse. In any case, his argument for the evolution of human mind and morality by the same natural agencies as the struggle for existence and for mates was dependent on breaking down the traditional theological distinctions between animal and human mentality. The pages of *The Descent of Man* bristle with anthropomorphic descriptions of animals, with loyal dogs and brave monkeys, proud peacocks, coy bitches, aggressive, promiscuous stags and cocks – tropes that when analogically reapplied to human behaviour and social institutions provided naturalistic corroboration of Victorian values. His concept of sexual selection, largely dependent on Darwin's studies of the observations and activities of contemporary animal breeders, was inescapably anthropomorphic, transferring Victorian social values and stereotypes back onto his conception of human biological and social evolution (Richards 1983).

When the aesthetic choice he attributed to female animals could not be made to fit this proper Victorian's conception of the submissive sexuality and "inferior" intelligence of human females, Darwin simply overturned it and put into men's hands the modifying and shaping power of human sexual selection. Man, he claimed, being "more powerful in body and mind," had seized the power of selection from woman. The differing standards of beauty of the various races offered the explanation, via male aesthetic preferences, of racial and sexual differentiation. "Monstrous" as it might seem, Darwin was convinced that the "jet-blackness of the negro" had been gained through the process of male selection, just as had the supposedly more pleasing secondary sexual characteristics of European women – sweeter voices, long tresses and greater beauty (Darwin 1871, 2: 368–84). For Darwin, the human male was the analogue of the animal breeder who exercised his caprice in varying the appearance of the breed; and woman's body, like the dog's, was pliant to male manipulation (Richards 1983, 76–9).

Talking about animals, therefore, offered those like Darwin who "would have been reluctant or unable to avow a project of domination directly a way to enact it obliquely" (Ritvo 1987, 6). The other side of the coin is Lansbury's claim that for women, the subjugated, to protest against vivisection "was to challenge a world of male sexual authority and obscenity which they sensed unconsciously, even if they had no direct experience of it" (Lansbury 1985, 422).

For Cobbe, the parallels were obvious. Women, like animals, were subject to the power of doctors and scientists and to the brutality of many men. Her powerful article, "Wife Torture in England" (1878a), was written on the crest of her involvement in the antivivisection campaign. It was an indictment of the endemic domestic violence that Cobbe recognized as not being confined to the working class; of the culture that expected female service in the home, that condoned and even drew entertainment from wife-beating; and of the system in which women lacked legal and political rights and were regarded as the property of their husbands (Cobbe 1878a; Caine 1992, 135–8). At the same time Cobbe attacked the arrogance and cruelty of the male medical profession whom she saw as exerting an increasingly oppressive control over women's lives, turning healthy women into a "whole sex of Patients" constantly liable to illness and dependent on the incompetent and callous attentions of doctors (Cobbe 1878b). Doctors were "doubly treacherous" to women, a reference Cobbe made between the way doctors treated women as patients and their fight to exclude women from the profession of medicine (Cobbe 1881b, 325).

Doctors and medical scientists were of course her targets in the antivivisection campaign, and she believed that those who engaged in vivisection were brutalized by the pain and suffering they inflicted so that, like those husbands who beat and assaulted their wives, they came to find excitement and even pleasure in the infliction of pain. Medicine, like marriage, exacerbated women's oppression. The unspeakable ambivalences of sexuality and cruelty that Cobbe evoked in her denunciations of vivisectors, medical men and wife torturers found expression in the outrage and distress that many women, inhibited from articulating them on their own behalf, evinced on behalf of vivisected animals. The extreme behaviour, the "emotionalism, floods of tears, and fainting," that characterized antivivisection meetings is legendary (French 1975, 248; Lansbury 1985).

How the sadosexual associations of vivisection meshed with the reactions of Darwin and Huxley to Cobbe and her "fanatical following" is not easy to judge.

It is unlikely that even these highly respectable family men were unaware of the pervasive masculine world of Victorian pornography. However, even if they perceived it, it is inconceivable that either Huxley or Darwin would have flouted Victorian convention to contest openly the darker sexual imagery conjured up by Cobbe's allusions. Their overt reaction to the emotionalism of the anti-vivisectionists was to deride and dismiss it as so much female hysteria and irrationality. Darwin thought the antivivisectionists to be "half mad" (Darwin 1888, 3: 210), while Huxley ridiculed their "fanaticism of philozoic sentiment" and contrasted it with the "rational" basis of experimental physiology (Huxley 1900, I: 434).

The ostensible issue for both men was the progress of British science. The reiterated danger was that the attacks of the antivivisectionists would undermine the already precarious status of the new science of experimental physiology. This was the issue over which Darwin was prepared to go public, and, for the first and only time in his long career, engage in direct political action. It was an issue that concerned Huxley, the professional scientist, even more. While he was busily promoting the social standing and moral responsibility of medical scientists, one of Cobbe's tactics was to downgrade their status and morality. Medical men, the upper class Cobbe alleged, were not gentlemen; they tended to come from the lesser ranks of society and were a "parvenu profession, with the merits and the defects of the class." Their social origins explained their defective morality, their trade unionism, and the sordid materialism that pervaded their ranks and made them unwontedly ambitious, motivated by monetary gain, and careless of suffering (Cobbe 1881b). Huxley strongly contested such charges, while Darwin was greatly offended by Cobbe's "monstrous" attribution of sadistic pleasure in animal suffering to leading scientists of the day (Darwin 1888, 3: 200, 203; Huxley 1900, 1: 427–34). A cruel scientist was an anomaly, a contradiction of the gentlemanly image and high moral standards they claimed for British science and its practitioners. Cobbe's accusations of scientific cruelty and assumption of the higher moral ground on behalf of women antivivisectionists were, therefore, in direct conflict with Huxley's professionalization strategy and his promotion of the scientist as the appropriate moral arbiter of important social questions.

For almost twenty years, from her headquarters of the Victoria Street Society, Cobbe fought a sustained but inevitably losing campaign against the growing power and authority of a science and medicine that, in her view, oppressed both women and animals. She came to abhor the "priest-like arrogance of some representatives of the modern scientific spirit". Her disillusion was complete when the "great naturalist who has revolutionized modern science" became the centre of an "adoring *clique* of vivisectors." She clashed with Darwin in the *Times* and pursued him through the pages of her journal, the *Zoophilist* (Cobbe 1894, 2: 123–9, 269–70, 408–9, 1881c, 17–19; Darwin 1888, 3: 205–8). Darwin, for his part, thought it only "fair to bear [his] share of the abuse poured in so atrocious a manner on all physiologists" by Miss Cobbe. He helped initiate the Science Defence Association in 1881, gave it financial support, and even briefly considered accepting its Presidency (Darwin 1888, 3: 206–10, 1903, 2: 437–41). This became the highly

—"ANIMIS CŒLESTIBUS IRÆ!"

A MODERN SCIENTIFIC DISCUSSION.

Miss Fanny (a gentle and most veracious Child). "YAH! YOU CRUEL COWARD! YOU AND YOUR FRIENDS SKINNED A LIVE FROG!"

Master Victor (an industrious but very touchy little Boy). "YOU'RE A LIAR! THE FROG WAS DEAD, AND YOU KNOW IT!"

Miss Fanny. "BOOHOO! WHETHER IT WAS DEAD OR NOT, YOU'VE GOT NO RIGHT TO CALL NAMES; 'COS I'M A GIRL, AND CAN'T PUNCH YOUR HEAD!"

Master Victor. "IT'S JUST BECAUSE YOU'RE A GIRL THAT *I* CAN'T PUNCH YOURS! YOU SHOULD HAVE THOUGHT OF THAT BEFORE YOU CALLED ME A COWARD!"

Figure 7.1 *Punch's* comment on the sexual and scientific politics of Cobbe's antivivi-section crusade. *Punch* 103 (5 November 1892): 205. University of Sydney Library, Rare Books and Special Collections.

influential Association for the Advancement of Medicine by Research, a powerful coalition of leading British biologists and medical men who successfully lobbied behind the scenes on behalf of experimental medicine against the interventions of the antivivisectionists (French 1975, 200–19).

In the end, Cobbe was brought down by that very emphasis on feminine moral superiority that underpinned her feminism and antivivisectionism. In 1892, she was made personally responsible for the wilful distortions of a major publication compiled by her Victoria Street Society that failed to acknowledge the routine use of anaesthetics in the "brutal" animal experiments it otherwise quoted directly from the research literature. The medical and lay press gloated over this evidence of moral fallibility from one who had "assumed a superior morality, a higher scientific knowledge, and a pontifical right to anathematise medicine and all her most honoured followers throughout the world." Was Cobbe's perversion of truth now to be discounted as the "privileges of womanhood"? Was this where the vaunted superior moral sense of women led (Hart 1892, 710–11)?

The media assault seriously damaged Cobbe's credibility and that of her movement, and had more general repercussions for the participation of women in science and public affairs (French 1975, 249–50; Cobbe 1894, 2: 306–11). *Punch*, that arbiter of establishment opinion, summed up the sexual and scientific politics of Cobbe's public humiliation with the adjoining cartoon, which says it all.

III. Conclusion

Lynn Linton, Cobbe and the Darwinians all drew the qualities they attributed to women from the same model of femininity, an illustration of the extent to which evolutionary science and feminism were both bound by the ideology of their time and place. This inevitably set the parameters of their debate on women's condition and future prospects. But what is striking is the variety of ways in which the individuals concerned negotiated or reworked the assumption of sexual difference for different ends: the Darwinians to assert their social and professional hegemony, Lynn Linton in biological determinist opposition to feminist aims but also to argue the necessity of female participation in scientific societies, and Cobbe to promote female agency through moral superiority. Cobbe rejected evolutionary justification of women's subordination, while the Darwinian Lynn Linton endorsed it in contradiction to her own situation as a woman intellectual in Victorian England. Cobbe's theistic ideology was more compatible with the leadership role she assumed for women in the antivivisection campaign, but could not be sustained in a context of religious decline and the growing authority and prestige of a science geared to the needs of a capitalist economy and the gendered nature of the public sphere. In the late Victorian period it was the Darwinians who articulated the dominant constructions of femininity, sexuality, and science and naturalized the barriers against feminine intellectual and social equality in order to protect Darwinian institutional and social interests against the threat posed by the burgeoning women's movement.

The Darwinian redrawing of traditional boundaries was made all the more dev-astating by the problem that many feminists themselves were deeply committed to naturalistic scientific explanations and to the new Darwinism. In the face of the evolutionary onslaught, a number of them retreated from the egalitarian ideal to claim for woman a biologically based "complementary genius" to man's – a "genius" that was rooted in her innate maternal and womanly qualities (Alaya 1977). The American feminist Antoinette Brown Blackwell, for instance, did not dispute Darwin's view that the mental differences between men and women were biologically based and the product of evolution; rather she disputed whether woman's innate mental differences could properly be called inferior to man's. She balanced man's greater strength, reasoning powers and sexual love against wom-en's greater endurance, insightfulness and parental love, and argued that social progress was dependent upon the evolution and perpetuation of these sexually divergent traits (Tedesco 1984). Such argumentation had a dangerous tendency to reinforce traditional stereotypes and cater to the drawing of biological limits to feminine potentiality.[1]

Eliza Lynn Linton, the contradictions notwithstanding, is best located amongst such advanced women, whose confidence in the liberating powers of science, and whose opposition of naturalistic interpretations of human nature and society to conventional theological wisdom and authority, ultimately betrayed them when science, especially Darwinism, gave a naturalistic basis to the class and sexual divisions of Victorian society (Richards 1989, 279–80).

Thus Lynn Linton, in old age (still campaigning indefatigably against the "shrieking sisterhood"), came to endorse her own Huxley-engineered expulsion from organized science on Darwinian grounds. In 1885 she published a bizarre novel, *The Autobiography of Christopher Kirkland*, which was a dramatization of her own life in a male persona. Here, she travestied William Linton as a radi-cal feminist whom Kirkland, against his better judgement, marries and tries to reform to the Victorian ideal of womanhood. His attempt to wean his wife from the "platform" to the "fireside" fails, and Kirkland finds solace in Darwinism. He associates with leading Darwinian scientists and frequents the meetings of the learned societies. But here we find no hint of the arguments for women's right to knowledge and membership of scientific societies with which Eliza petitioned Huxley all those years ago. Kirkland endorses Eliza's claims of women's rights to property, divorce, and to an education different but equal to men's, but on the whole, he subscribes to the doggerel that "Women's Rights are Men's Lefts." He

---

1 It should be noted that there was potential within Darwinism for female agency, for females as sexual selectors and the main agents of social progress, and this form of Darwinism was promoted by some socialist feminists, notably the American visionary Eliza Burt Gamble, the English secular-ists and socialists Annie Besant and Edward Aveling, and Alfred Russel Wallace, co-founder with Darwin of the theory of natural selection. But such attempts to radicalize sexual selection and give women an active and central role in evolutionary theorizing received little attention or support from mainstream feminists and Darwinians (Richards 1995; Gamble 1916; Wallace 1890, 1913, 125–49).

is a rigid biological determinist who defends his anti-feminism on the Darwinian grounds that "unless we accept the creed . . . that the moral sense is as much a matter of evolution as is the intellectual – we are lost in a sea of contradictions." For Kirkland, the stereotypically Victorian intellectual and moral differences between men and women are the products of evolution and are grounded in "the material fact of sex." They are the foundations of society and morality, "the division of labour and function, against which women revolt [in vain], and men must fare forth while they bide within" (Linton 1976, 3: 5, 12, 167–72).

Here, in her masculine *alter ego*, Eliza finally won the entry into science that was denied her in real life. But her fictional victory is achieved only at the cost of the acceptance into a science from which her *alter ego* Kirkland, logically, would also have excluded her as a woman. Kirkland is the ultimate Darwinian fellow traveller, a Spencerian, from the likes of whom the more subtle Huxley was soon carefully to dissociate himself. But Eliza/Kirkland has simply taken Huxley's position on the woman question to its logical extreme. It was Huxley, after all, who both excluded women from science in the name of science and redefined that science to ratify their exclusion.

Cobbe, by contrast, offered her followers a radical critique of Victorian science, based on her opposition of spiritual and feminine values to the materialism and masculine tyranny that constituted that "exquisite kind of vice" that found expression in vivisection and the maltreatment of women (Caine 1992, 145–9). However, while her rejection of the Darwinian naturalization of mind may have provided ideological support for those opposing the imposition of naturalistic limits on women's aspirations, Cobbe's doctrine of theistic ethics could not provide a real political alternative for those confronting the antifeminist applications of Darwinism. The only solution she offered was the unrealizable goal of the individual moral and religious reform of those scientists like Huxley who had constituted themselves the new secular priesthood of Victorian society. Furthermore, in certain significant respects, Cobbe's theism and antivivisectionist stance conduced to the Victorian feminization of feeling and the masculinization of reason, to the legitimation of Huxley's exclusion of women from science and the Darwinian definition of feminine nature as essentially incommensurate with the masculine pursuit of science.

Bibliographical note

The significant role played by leading Darwinians, including Darwin himself, in imposing naturalistic, scientific limits to the claims by nineteenth century feminists for political and social equality has been well-documented by feminist scholars. The more notable of these, Conway (1970), Fee (1974), Rosenberg (1975), Alaya (1977), Mosedale (1978), Rosser and Hogsett (1984), Russett (1989), and Jann (1994) have all contributed to our understanding of the concerted Darwinian refutation of the natural equality of men and women. As well, Love (1983), Tedesco (1984) and Egan (1989), and Erskine (1995), among others, have undertaken studies of the response to Darwinism by some prominent nineteenth century feminists, notably Charlotte Perkins Gilman, Olive Schreiner and Antoinette Brown Blackwell.

However, many of these studies, especially the earlier ones, useful as they are, fall into the category of "feminist empiricism," i.e., they maintain the integrity of the standard view of science as objective and value-free, and argue that androcentrism in science is socially caused and is, therefore, the result of "bad" science. The solution to such sexist science is to be found in a closer adherence to the proper methodologies of scientific enquiry, and nineteenth century evolutionists and anthropologists are censured for their failure to conform to these standards. There is a disjuncture between such studies and the contextual or constructivist evolutionary historiography pioneered by Robert Young in the 1960s, which was, in its turn, regrettably gender-blind (Young 1985), and more recent contextual feminist analyses (e.g., Outram 1987; Haraway 1989; Hall 1992). My own studies of Darwin, Huxley and Victorian feminism (Richards 1983, 1989, 1995) are attempts to integrate feminist insights into Victorian science and society with contextual Darwin historiography. Readers should also consult the recent biographies and studies by Desmond, Moore and others (Desmond 1989, 1994; Desmond and Moore 1991; Moore 1989).

There is a sympathetic portrait of Eliza Lynn Linton by Nancy Fix Anderson (1987) and an excellent study of Frances Power Cobbe by Barbara Caine (1992, 103–49), both of which have enriched our understanding of the individual struggles of these important but neglected Victorians to give meaning to their lives and circumstances within the confines of the Victorian sphere of femininity.

Bibliography

Alaya, Flavia. 1977. "Victorian Science and the 'Genius' of Woman." *Journal of the History of Ideas* 38: 261–80.

Anderson, Nancy Fix. 1987. *Woman Against Women in Victorian England: A Life of Eliza Lynn Linton.* Bloomington: Indiana University Press.

Caine, Barbara. 1992. *Victorian Feminists.* Oxford: Oxford University Press.

Cobbe, Frances Power. 1862. "What Shall We Do with Our Old Maids?" *Fraser's Magazine* 66: 594–610.

———. 1863. "Celibacy v. Marriage." In *Essays on the Pursuits of Women*, 38–57. London: Emily Faithfull.

———. 1872. "Darwinism in Morals." In *Darwinism in Morals, and Other Essays*, 1–33. London: Williams and Norgate.

———. 1878a. "Wife Torture in England." *Contemporary Review* 32: 56–87.

———. 1878b. "The Little Health of Ladies." *Contemporary Review* 31: 276–96.

———. 1881a. *The Duties of Women. A Course of Lectures.* London: Williams and Norgate.

———. 1881b. "The Medical Profession and its Morality." *Modern Review* 2: 630–50.

———. 1881c. "Mr. Darwin on Vivisection." *The Zoophilist*, Special Supplement, 1: 17–19.

———. 1894. *Life of Frances Power Cobbe. By Herself.* 2 vols. London: Richard Bentley & Son.

Conway, Jill. 1970. "Stereotypes of Femininity in a Theory of Sexual Evolution." *Victorian Studies* 14: 47–62.

Darwin, Charles. [1838] 1986. "Darwin's Notes on Marriage." In *The Correspondence of Charles Darwin*, edited by Frederick Burkhardt and Sydney Smith, vol. 2, 443–5. Cambridge: Cambridge University Press.

———. 1871. *The Descent of Man, and Selection in Relation to Sex*. 2 vols. London: John Murray.

———. 1888. *Life and Letters of Charles Darwin*. Edited by Francis Darwin. 3 vols. London: John Murray.

———. 1903. *More Letters of Charles Darwin*. Edited by Francis Darwin. 2 vols. New York: D. Appleton and Company.

———. 1994. *A Calendar of the Correspondence of Charles Darwin, 1821–1882*. Edited by Frederick Burckhardt and Sydney Smith. Cambridge: Cambridge University Press.

———. Archive. Cambridge University Library.

Desmond, Adrian. 1989. *The Politics of Evolution*. Chicago and London: University of Chicago Press.

———. 1994. *Huxley: The Devil's Disciple*. London: Michael Joseph.

Desmond, Adrian, and James Moore. 1991. *Darwin*. London: Michael Joseph.

Egan, Maureen L. 1989. "Evolutionary Theory in the Social Philosophy of Charlotte Perkins Gilman." *Hypatia* 4: 102–19.

Elston, Mary Ann. 1987. "Women and Anti-Vivisection in Victorian England, 1870–1900." In *Vivisection in Historical Perspective*, edited by Nicolaas A. Rupke, 259–94. London: Croom Helm.

Erskine, Fiona. 1995. "*The Origin of Species* and the Science of Female Inferiority." In *Charles Darwin's "The Origin of Species": New Interdisciplinary Essays*, edited by David Amigoni and Jeff Wallace, 95–121. Manchester and New York: Manchester University Press.

Fee, Elizabeth. 1974. "The Sexual Politics of Victorian Social Anthropology." In *Clio's Consciousness Raised: New Perspectives on the History of Women*, edited by Mary Hartman and Lois Banner, 86–102. New York: Harper.

French, Richard D. 1975. *Antivivisection and Medical Science in Victorian Society*. Princeton, NJ and London: Princeton University Press.

Gamble, Eliza Burt. 1916. *The Sexes in Science and History: An Inquiry into the Dogma of Woman's Inferiority to Man*. New York: G. P. Putnam's. (Revised edition of *The Evolution of Woman*, 1894.)

Hall, Catherine. 1992. "Feminism and Feminist History." In *White, Male and Middle Class; Explorations in Feminism and History*, 1–40. Cambridge: Polity Press.

Haraway, Donna. 1989. *Primate Visions*. New York and London: Routledge.

Hart, Ernest. 1892. "Women, Clergymen, and Doctors." *New Review* 7: 708–18.

Huxley, Thomas Henry. [1865] 1968. "Emancipation – Black and White." In *Collected Essays*, 3: 66–75. New York: Greenwood Press.

———. 1900. *Life and Letters of Thomas Henry Huxley*, edited by Leonard Huxley. 2 vols. London: Macmillan.

———. Papers. Huxley Archives. Imperial College.

Jann, Rosemary. 1994. "Darwin and the Anthropologists: Sexual Selection and its Discontents." *Victorian Studies* 37: 287–306.

Lansbury, Coral. 1985. "Gynaecology, Pornography, and the Antivivisection Movement." *Victorian Studies* 28: 413–37.

Linton, Eliza Lynn. 1858. "The Unities of Nature." *The National Magazine* 5: 52–7.

————. 1870. *Ourselves: A Series of Essays on Women*. London and New York: G. Routledge and Sons.

————. 1883. *The Girl of the Period and Other Social Essays*. London: Richard Bentley
and Son.

————. [1885] 1976. *The Autobiography of Christopher Kirkland*. 3 vols. New York: Garland.

Love, Rosaleen. 1983. "Darwinism and Feminism: The 'Woman Question' in the Life and
Work of Olive Shreiner and Charlotte Perkins Gilman." In *The Wider Domain of Evolutionary Thought*, edited by David Oldroyd and Ian Langham, 113–31. Dordrecht: Riedel.

Moore, James R., ed. 1989. *History, Humanity and Evolution*. Cambridge: Cambridge University Press.

Moscucci, Ornella. 1990. *The Science of Woman; Gynaecology and Gender in England,
1800–1929*. Cambridge: Cambridge University Press.

Mosedale, Susan Sleeth. 1978. "Science Corrupted: Victorian Biologists Consider 'The
Woman Question'." *Journal of the History of Biology* 11: 1–55.

Outram, Dorinda. 1987. "The Most Difficult Career: Women's History in Science." *International Journal of Science Education* 9: 409–16.

Owen, Alex. 1990. *The Darkened Room: Women, Power, and Spiritualism in Late Victorian England*. Philadelphia: University of Pennsylvania Press.

Richards, Evelleen. 1983. "Darwin and the Descent of Woman." In *The Wider Domain of Evolutionary Thought*, edited by David Oldroyd and Ian Langham, 57–111. Dordrecht: Riedel.

————. 1989. "Huxley and Woman's Place in Science: The 'Woman Question' and the
Control of Victorian Anthropology." In *History, Humanity and Evolution*, edited by
James R. Moore, 253–84. Cambridge: Cambridge University Press.

————. 1995. " 'The New Woman Worship': T. H. Huxley and Victorian Feminism."
Paper presented at the International Conference "T. H. Huxley: Victorian Science and
Culture." Imperial College of Science, Technology and Medicine, London.

Ritvo, Harriet. 1987. *The Animal Estate*. Cambridge, MA: Harvard University Press.

————. 1988. "Sex and the Single Animal." *Grand Street* 1, Spring: 124–39.

Rosenberg, Rosalind. 1975. "In Search of Woman's Nature: 1850–1920." *Feminist Studies* 3: 141–54.

Rosser, Sue V., and A. Charlotte Hogsett. 1984. "Darwin and Sexism: Victorian Causes,
Contemporary Effects." In *Feminist Visions: Toward a Transformation of the Liberal
Arts Curriculum*, edited by Diane L. Fowlkes and Charlotte S. McClure, 42–52. Univer-
sity, AL: Alabama University Press.

Russett, Cynthia Eagle. 1989. *Sexual Science: The Victorian Construction of Womanhood*.
Cambridge, MA: Harvard University Press.

Smith, F. B. 1973. *Radical Artisan: William James Linton*. Manchester: Manchester University Press.

Taylor, Barbara. 1983. *Eve and the New Jerusalem: Socialism and Feminism in the Nineteenth Century*. New York: Pantheon Books.

Tedesco, Marie. 1984. "A Feminist Challenge to Darwinism: Antoinette L. B. Blackwell on
the Relations of the Sexes in Society and Nature." In *Feminist Visions: Toward a Transformation of the Liberal Arts Curriculum*, edited by Diane L. Fowlkes and Charlotte S.
McClure, 53–65. University, AL: Alabama University Press.

Wallace, Alfred Russell. 1890. "Human Selection." *Fortnightly Review* 48: 325–37.

————. 1913. *Social Environment and Moral Progress*. London and New York: Cassell
and Company.

Young, Robert M. 1985. *Darwin's Metaphor*. Cambridge: Cambridge University Press.

8

"THE GREATEST OF ALL POSSIBLE EVILS TO MANKIND"

Annie Besant vs. Darwin at the Knowlton trial and beyond

The Knowlton Trial of 1877 was initiated by the leading Secularists, Annie Besant and Charles Bradlaugh, to challenge the law that defined contraceptive literature as obscene and blocked its distribution beyond a privileged minority. Their intention, in keeping with their Neo-Malthusian freethought philosophy, was to get clear and straightforward contraceptive advice to working class women at a price they could afford. As they saw it, birth control would not only emancipate women from the fear and bondage of child-bearing, but it also represented the only rational "scientific" check to the otherwise inevitable overcrowding, hunger and poverty of the lower classes. Accordingly, they republished and openly distributed a venerable work on contraception by a Massachusetts physician, Charles Knowlton, with the innocent seeming title, *The Fruits of Philosophy*, volunteered themselves for arrest, and were duly charged and arraigned for trial under the Obscene Publications Act.

The famous trial that followed hinged on the ostensible question of whether the object of the "Knowlton pamphlet" (described as a "dirty, filthy book", by the prosecution) was the "legitimate one of promoting knowledge in a matter of human interest" or whether "science and philosophy are merely made the pretence of publishing a book . . . calculated to arouse the passions of those who peruse it".[1] Behind these euphemisms lurked the mid-Victorian obsession with female chastity and the dire consequences for the family and society should it be corrupted by readily available contraceptive information. It was her rightful fear of illegitimate pregnancy and social ruin that kept Victorian woman to the straight and narrow path of virtue and submission to the patriarchal control of family, church and state. To remove this fear, to cut the tie between sex and reproduction, was to subvert the laws of God and nature, to licence female promiscuity, to threaten the very foundations of society.

---

1 Quoted in Bonner 1895, 2: 21; Langer 1975, 685.

In the weeks leading up to their trial the accused subpoenaed a number of prominent medical and social theorists as witnesses in their defence that the "doctrine of the limitation of the family" was to be found in other works in general circulation. They thereby severely tested the liberal principles of those whose writings were associated with Neo-Malthusianism, but who were aghast at the prospect of being forced into open defence of two militant atheists – one of them a woman living apart from her clergyman husband and suspected of more than mere friendship with Bradlaugh – in a sensational trial that was being splashed all over the press. In the event, only the faithful few, less prestigious and more radical, rallied to the cause.[2] Most notables went to ground. Henry Fawcett, M.P., professor of political economy and advocate of family limitation, reacted to his proposed testimony for the defence by refusing to accept his subpoena. As for the monstrous suggestion that the leading campaigner for women's rights, Millicent Garrett Fawcett, might support the defence, Fawcett threatened to send his wife out of the country should she be called to the witness stand.[3] In the case of the ageing and ultra-respectable Charles Darwin, now well on his way to earthly sanction by church and state, the defendants made an even more disastrous miscalculation.

On 5 June 1877, Bradlaugh wrote courteously to Darwin, informing him of his pending subpoena and in the confident expectation that the famous and enlightened author of the *Origin of Species* would support their case for a scientific solution to the Malthusian law of population growth and its attendant poverty and misery.[4] An appalled Darwin thoroughly quashed this unwarranted optimism by return of post.

Darwin's letter to Bradlaugh cannot be traced, but the original draft in the Cambridge collection, with its many deletions and interlineations, the more vividly recreates his obvious anxiety and the urgency of his need to dissociate himself from any implied support for such a disreputable cause. It begins with the familiar litany of his chronic ill health, his forced retirement for many years from "all society and public meetings", and the "great suffering" it would be for him to be a witness in court. He would probably be unable to attend. Even were he able to sustain this major threat to his health, he would be forced to testify *against* the

---

2 One of them was the trained midwife and qualified chemist and druggist, Alice Vickery, the only woman to testify for the defendants. Her life-long partner, Charles Drysdale (brother of George Drysdale, the author of the notorious *Elements of Social Science*, see text), was the only qualified doctor to appear for the defence; A. Taylor 1992, 117–18; Bland 2002, 202, 207. Vickery subsequently took out her medical degree and became a leading advocate for birth control and women's suffrage. For details on Vickery and her involvement with the Malthusian League, see Benn 1992. Alison Bashford argues that it was Vickery, rather than Besant, who articulated the feminist rationale for birth control, who "grafted a late-Victorian feminism onto British economic thought on population, arguing for a sexual politics based on the rights to bodily integrity of an individual woman"; Bashford 2014, 42.

3 Bonner 1895, 2: 23; Banks and Banks 1964, 92–3.

4 CUL-DAR 160, Darwin Archives, Cambridge University Library.

defendants, not on their behalf as they assumed. He did not doubt their good intentions, but they should be aware that his evolutionary views had long caused him to hold a "very decided" opinion in opposition to theirs on the supposed benefits of contraception for social progress: "I have long held an opposite opinion, as you will see in the enclosed extract, & this I shd. think it my duty to state in court".[5]

The extract was a well-known passage in the conclusion to his *Descent of Man* where Darwin had stated his conviction that man's advance was dependent on his continuing "rapid multiplication" and consequent severe struggle for existence. "Hence, our natural rate of increase, though leading to many and obvious evils, must not be greatly diminished by any means" (Darwin 1871, 2: 403). So that Besant and Bradlaugh should be left in no possible doubt of his meaning, Darwin had recourse to some uncharacteristic plain speaking: the "any means," he now spelt out, meant "artificial means of preventing conception". But this evolutionary deduction did not represent the full extent of the horrors of birth control. In an agitated welter of crossings out, interpolations, and rewrites, Darwin pressed manfully on:

> But besides the evils here alluded to I believe that any such practices would in time spread to unmarried women & would destroy chastity, on which the family bond depends; & the weakening of this bond would be the greatest of all possible evils to mankind; & [it] must be my duty to state in court. So that my judgment would be in the strongest opposition to yours.[6]

The feelings of the recipients may be imagined. With their trial less than two weeks away, the great Darwin, scientific hero of radicals and freethinkers, stood revealed as a typical middle-class conservative on this pressing social issue. At the very least it must have jolted their conviction that progressive science was on their side in the struggle for a better, secular world.

In her father's biography, Hypatia Bradlaugh, ever the dutiful daughter, explained their subsequent decision to "manage" without Mr. Darwin's evidence as motivated by the concern of the defendants for his fragile state of health. She could not resist publishing those portions of the famous man's letter but suppressed the rest, leaving the reader with the impression that Darwin, had his health permitted, would have testified for the defence (Bonner 1895, 2: 24, 28).

The redoubtable Annie Besant, however, was more truthful, and certainly more resourceful, than Hypatia – no great admirer of her father's former protégé – implied. As the transcript of the trial reveals, Besant, who must have undertaken a crash course in Darwinism, devoted a large part of her eloquent defence

---

5 CUL-DAR 202, Darwin Archives. A copy is reproduced in Richards 2017, 495.
6 CUL-DAR 202, Darwin Archives, Cambridge, transcribed with the kind assistance of Stephen Pocock; see also Peart and Levy 2008, 347–8. This transcription differs from that offered by James Moore 1988, 306; Browne 2002, 443–4; Dawson 2007, 141.

to contesting Darwin's arguments on the need for sustained population growth and arguing the indefensibility of his advocacy of the "natural checks" of poverty, overcrowding and starvation in preference to the "scientific check" of contraception. She referred to the letter that Mr. Darwin had written "a few days since" and, after reading the relevant passage from the *Descent* to the judge and jury, proceeded to dispute it on the great man's own territory: "That is Mr. Darwin's position, and putting aside for a moment the awful amount of human misery which he accepts as the necessary condition of progress, let us see if the position be defensible."[7]

Darwin's panicky letter may have ensured that he was not physically present at the notorious Knowlton trial, but, as in any major confrontation over serious social questions in the late Victorian period, his name and theoretical presence were indispensable to the proceedings. Over the next few years Besant was to become a leading re-interpreter of Darwinism for the birth control and secular and socialist causes. Her trial testimony became the basis of her best-selling pamphlet, *The Law of Population: Its Bearing Upon Human Conduct and Morals* (1877), and her revisionist views of natural and sexual selection were incorporated into her prolific writings and lectures on women and marriage, and the evolution of society and socialism. With typical thoroughness and enthusiasm, Besant went even further into Darwinism than this. With the help of the young anatomy lecturer Edward Bibbins Aveling, fervent Darwinian and incipient Marxist, she formally qualified and enrolled for a science degree at London University. Together she and Aveling established, and for many years taught, the highly successful classes in science at the Hall of Science, the London headquarters of the National Secular Society. These popular classes, designed to educate young women and men in the scientific method that purportedly underpinned the secularist platform and to prepare those interested for entry to University, featured the Aveling-Besant version of Darwinism.

Annie Besant

Seen from today, Annie Besant is a fascinating, but also a most puzzling woman. Her life was an extraordinary pilgrimage, leaping from one great cause to another, until, with a final spectacular backflip that stunned and outraged her secular and socialist admirers, the hard-headed atheist "went head over heels" into a mystical and dogmatic Theosophy and repudiated her former belief in the reforming powers of birth-control, materialism and socialism. Each cause was embraced with passionate conviction and illumined by the inexhaustible energy and oratorical brilliance that dazzled Besant's contemporaries and made her one of the most remarkable women of her day. And each, as her detractors pointed out, was embodied by a

---

7 "Special Trial Number," *National Reformer*, XXIX (25), 23 June 1877, 401–14, 412–13. The trial transcription was published in full in the *National Reformer* (hereafter referred to as *N.R.*); see also Chandrasekhar 1981, 26–54; Peart and Levy 2008.

Figure 8.1 The evolution of Annie Besant, satirized. Clockwise from top left: clergyman's wife; birth control crusader at the Knowlton Trial with Charles Bradlaugh; militant socialist under the red flag, being pelted with vegetables; Theosophist; and, finally, reincarnated as spiritually enlightened Mahatma. © The British Library Board: "How to become a Mahatma!" *St. Stephen's Review* Presentation Cartoon, 12 September 1891.

different man. They included, besides the stalwart and high-minded Bradlaugh who offered her secularism but, it seems, no sex, the insidious, womanizing Aveling, who coached her in sexual selection (supposedly in more ways than one) before he opted for commitment, of a sort, to Eleanor Marx; and the elusive and allusive George Bernard Shaw, the joker in the pack, who teased and enticed Besant into love and Fabian Socialism, but who would only laugh when she allegedly presented him with a contract setting forth the terms under which she proposed they should live together as man and wife: "Good God! This is worse than all the vows of all the churches on earth. I had rather be legally married to you ten times over".[8]

It is all too easy for the historian to play up Shaw's mock-heroic view of Besant as a "player of genius", a "tragedian," a "born actress" with a "different leading man every time". Shaw was to exploit Besant's comedic possibilities as the model for the beautiful but impossibly high-minded Raina ("the noble attitude and the thrilling voice") in his *Arms and the Man* (Nethercot 1961, 241, 287). The much-publicized exploits and contradictions of her life make Besant an easy mark for ridicule. But if we are to make sense of this gifted and fearless woman, we must shield her from the jocular Shavian gaze and restore to view the serious choices she perceived confronting her.

At bottom was the choice between the "arbitrary, meaningless universe, without mercy, morality or benevolence" of Besant's secular Darwinian period, and the universe of her adolescent Christianity and her later Theosophy in which "individual effort mattered, pain served a purpose, and justice regulated all creation". The historian Janet Oppenheim has argued that the whole of Besant's life may be better understood as dominated by her highly personal quest for spiritual truth. Her public crusades for social causes and even her zealous atheism are to be viewed as alternative outlets for her repressed religious fervour. Throughout the 1870's and 1880's Besant was engaged in a process of continuing re-evaluation until she finally became convinced that the secular Darwinian cosmos led nowhere and that the other set of beliefs, previously rejected in their Christian guise, would, in the form of *The Secret Doctrine* of Madame Helena Blavatsky, lead her to the Good and the True.[9] In 1889, when she was forty-two years old, Besant espoused Theosophy, and thereafter until her death in 1933, in her own words, she "never wavered" (cited in Nethercot 1961, 307).

It would be a mistake, therefore, to interpret Besant's conversion to Theosophy as entailing a complete break with her previous concerns and beliefs. Nor should Besant's spiritual quest be overly intellectualized. The bare bones of this interpretation do not disclose the emotionality of her pilgrimage, nor that it was bound up with her pursuit of earthly love. Besant came to Theosophy when she was disillusioned not only with a cold and sterile atheism and a faction-ridden socialism, but

8 Cited in Nethercot 1961, 240. Anne Taylor, Besant's most recent biographer, disputes Shaw's claim, arguing that it was William Thomas Stead – the messianic editor of the *Pall Mall Gazette* – whom Besant pursued in an unrequited love affair; Taylor 1992, 184–7, 197–202.
9 See Oppenheim 1989.

also with a series of unsatisfactory relationships. In certain respects, Theosophy, in its celebration of celibacy and its promise of ultimate freedom from the physical body and earthly desires, was as much a refuge for Besant from sexual rejection and the politics of late-Victorian sexuality as it was a source of enduring spiritual comfort and inspiration. The crisis of faith that propelled Besant into materialism and from thence into Theosophy was also a crisis of desire.

The young Annie Wood was a pretty, pious, middle-class girl with bluestocking tendencies. At the age of twenty, her religious fervour seduced her into a loveless marriage. Frank Besant was an ambitious and authoritarian young Anglican clergyman with whom Annie naively envisaged a life of prayer and good works among the poor. The wedding night was a rude awakening. Annie was later scathingly to condemn the Victorian convention that kept young women in "such infantile ignorance" of sex before marriage. So "scared and outraged at heart" was she at this unprepared for violation of her person, that she could not, then or later, control her aversion to her husband's sexual advances (Besant 1908, 44, 49). Nor could the serious-minded, rebellious girl accept the containment entirely within the domestic sphere that her husband considered proper. Nevertheless, in quick succession, she dutifully bore a son and a daughter to Frank Besant, who became increasingly harsh in his treatment of his "unsatisfactory" wife. They quarrelled over Annie's wish to limit their family and over her attempts at writing moral tales and tracts. Her deepening marital misery led her to question her religion, now personified by her domineering clergyman husband. She began the critical scrutiny of Christian dogma that was to lead her out of the Church and her miserable marriage. After a final confrontation over her refusal to participate in Holy Communion, Annie took the terrifying step that would put her outside the conventions of Victorian society. Divorce was out of the question, but a legal separation was agreed in 1873. Frank Besant insisted on keeping their son, and Annie settled in London with the infant Mabel on a barely adequate allowance.[10]

These events laid the foundations of Besant's feminism. She was determined to find significant work that would assuage her frustrated literary and intellectual ambitions. She put in many hours of study at the British Museum, reading Mill, Darwin, and Comte, on a trajectory that took her from theism to atheism. "The difference between a theist and an atheist lies only in the addition of a prefix," she was later to proclaim (Besant 1877, Preface). Within a year, Besant had joined the National Secular Society, and, at the age of twenty-six, had thrown in her lot with Bradlaugh and the notorious Bible-smashing, and (it was said) free-loving, freethinkers.

Sex, secularism and birth control

The Secularists were a body of women and men from the working and lower middle classes who attributed the contemporary political and social injustices to the

---

10 Biographical detail on Besant is taken from Nethercot 1961; Dinnage 1986; Taylor 1992.

evil effects of organized religion. They were committed to the radical restructuring of society by rational and peaceful means, primarily through the discrediting of Christianity and its institutions. They were republicans and atheists in Queen Victoria's England, in an age that, outwardly at any rate, was deeply religious (Royle 1980, x).

Their tradition of freethought, with its emphasis on the twin evils of "kingcraft" and "priestcraft" may be traced back to the earlier radicals, Thomas Paine and Richard Carlile. Late-Victorian freethinkers also drew upon the anti-clerical and ethical teachings of Robert Owen, but not his socialism. The aged George Jacob Holyoake was the connecting link between the Owenite tradition and the Secularists. He founded the movement and coined the term "secularism" in 1851 as a "constructive" alternative to the mere disbelief implied by the "infidelity" or "atheism" of the earlier radicals. Like Owen, Holyoake did not attack Christianity as such, except in so far as it impeded "rational progress", and was even prepared to work with Christians in promoting the secular good. He was a clever writer and a committed radical, but was too cautious and lacking in leadership to make secularism into a national movement. He "never moved on from being the pilot to being the captain of the ship". By the late 1850s, the ship of Secularism had acquired a captain who was not afraid to steer full tilt against the tide of respectable religious and social opinion of the prosperous 50s and 60s, and to run before the winds of change in the 70s. Charles Bradlaugh appropriated and transformed both Holyoake's movement and his word. He made the term "secularism" synonymous with atheism and militant anti-Christianity, and, as a movement, he brought it to the forefront of British radical politics (Royle 1971, 51–4, 1980, 109–21).

A charismatic personality and born orator, motivated by his passionate hatred of religion and oppression of all kinds, Bradlaugh hammered out a gospel of self-help, individual liberty and science, and travelled the country preaching atheism, freedom of the press, and radical political reform. This was the "golden age" of Secularism, dominated by Bradlaugh's attempts to enter Parliament on the Radical ticket, when his public reputation was at its highest. Radicals of all opinions united around his campaign and his newspaper, the *National Reformer*. By 1880, membership of the National Secular Society had reached six thousand, and by 1885 there were over a hundred branches around the country. At the height of his fame, Bradlaugh could draw audiences of more than three thousand to his outdoor meetings (Royle 1971, 57, 1980, 88–92). It was a heady and hopeful time for radical Secularism, and Annie Besant was in the thick of it at Bradlaugh's side.

She took up his offer of a staff job on the *National Reformer*, and was soon ranging with great verve across the boundaries of economics, science, and literature. She also became a familiar figure on the Secularist lecture circuit, touring the country in a barn storming, passionate denunciation of religious bigotry and obscurantism, rivalling her mentor with her eloquence and powers of crowd control. Her very first public lecture, "On the Political Status of Women", drew upon her own bitter experiences to castigate religious influences for the inferior position in which women were placed (Besant 1874). The well bred Mrs. Besant travelled on trains and in farm carts, ate with miners and weavers, was abused and spat at, but was cheered as well. And she thrived on it all. Her many converts were charmed by her youth and beauty, her extraordinary fluency and "bell-like voice", and, above all, by the piquant contrast between her chaste, ladylike deportment and the subversive content of her speeches and writings. She was an icon of advanced political opinion. Young men purchased her portrait (on sale at her lectures) and carried it around with them. In 1875, while still a relative newcomer to the movement, she was elected Vice-President of the Society. She was the first lady of the Secularists, and her name and shockingly radical views were becoming known to the general public. Together she and Bradlaugh were a striking pair, facing a hostile world on behalf of liberty and truth.[11]

Just how far their partnership actually went is debatable. They lived separately, a few minutes apart, in St. John's Wood and so preserved the proprieties. Still, when they were not out on the lecture circuit, they spent their days together, reading and writing. Like Annie's, Bradlaugh's marriage was over in all but name. He lived apart from his wife (allegedly a hopeless alcoholic) with his two devoted daughters, Alice and Hypatia. The repeated joint platform appearances of the High Priest and Priestess of Secularism inevitably caused a deal of scandal and tongue wagging. That their relationship had all the mutual regard and companionship of a successful marriage seems beyond doubt. That they were ever lovers seems unlikely.

In spite of, or because of, his scandalous reputation and the supposed association of freethought with free love, Bradlaugh was known as something of a sexual puritan. He held very conventional views on sex and marriage, going so far as to state (perhaps with reference to his and Besant's situation) that while divorce might be appropriate in some circumstances, he did not regard remarriage as proper while a divorced wife was still alive (Arnstein 1984, 111). He had earlier publicly allied himself with the campaigns of Josephine Butler, who, against the double standards of the day, upheld the ideal of "purity" for men as well as women. In this Bradlaugh was typical of most freethinkers, who held, in opposition to their religious critics, that the point of Secularism was that it provided a secular basis for morality that enabled them to maintain conventional standards of decent behaviour without the need for divine sanction. It was only in his advocacy of birth control that Bradlaugh was prepared to challenge the Victorian sexual conventions (Royle 1980, 250–4). In any case, Besant had her own professed moral scruples. She and Bradlaugh were the "teachers of a lofty morality", and she prized her "pure reputation" and her "good name". It is hard to believe that she and Bradlaugh would have provoked their prosecution for obscenity and outfaced what Besant called the "horrible misconceptions" and "odious imputations on honour and purity" that the Knowlton trial brought upon them, had they not been secure in the conviction that their own standards and behaviour would withstand the closest scrutiny (Besant 1908, 207–8; Banks and Banks 1964, 88).

11 See A. Taylor 1992, 86–100; Dinnage 1986, 29–37.

Annie stood to lose more than her good name by casting all discretion to the winds. Frank Besant, spurred on by the prurient press association of the Besant name with atheism and worse, had already made one attempt to remove their daughter Mabel from this alleged hotbed of vice and moral contagion. When this failed, he had Annie shadowed by detectives. That they never found any evidence of immoral conduct that he might use against her speaks for itself. Annie's attitude towards Bradlaugh was more that of worshipping acolyte than lover. During their ten-year partnership, she was powerfully influenced by him, accepting his views as her own, even, according to one contemporary account, copying his manner-isms and methods of argument. Bradlaugh "was not so much her friend as her idol" (Adams 1903, 2: 405; Nethercot 1961, 114–15; A. Taylor 1992, 125). It was not the saintly Bradlaugh who tempted Annie to bed, but the philandering Aveling. But by then she had lost Mabel to a vindictive Frank Besant in the wake of the Knowlton trial.

From 1874 to around 1885, Secularism, as defined and personified by Brad-laugh, absorbed all the passionate devotion Annie once had given to the Chris-tianity she had revolted against. In the Bradlaugh-Besant version of the secular religion of humanity, men and women looked to themselves for help, not to a God for whose existence there was no scientific evidence (Arnstein 1984, 11; Royle 1980, 115). Theirs was an atheistic, materialist creed, whereby the human race could progress towards perfection by its own efforts. Besant followed Bradlaugh in combining anti-religious propaganda with the advocacy of a succession of radi-cal causes including the abolition of royalty and the House of Lords, Home Rule for Ireland, and (a cause nearer to Besant's heart) women's rights to the vote, education, property, divorce, the custody of their children, and, most notorious of all, birth control.

Bradlaugh himself had been an advocate of birth control or (in contemporary usage) Neo-Malthusianism since around 1861, when he and Dr. George Drysdale (of whom more below), had formed the Malthusian League. The birth control movement had a long association with earlier freethought or "infidelity". Utili-tarians had adopted the original highly influential Malthusian creed that the root cause of pauperism was the excessive procreation of the lower classes; but, in opposition to Malthus, they urged the logical necessity of contraception rather than the unworkable remedy of "moral restraint" that was all that Malthus could bring himself to offer. Their advocacy of "artificial" practices, such as the use of the "spunge," which Malthus condemned as immoral, unnatural and un-Christian, amounted to no more than the hints and allusions that may be inferred from the writings of Bentham or James Mill. Middleclass Utilitarians were not prepared to compromise their reputations by speaking more openly. The need for discretion was brought home to John Stuart Mill, a precocious convert to the birth control cause, when, as a youth of seventeen, he was briefly imprisoned for handing out the "diabolical handbills" of the Benthamite artisan Francis Place. It was Place and, following him, the unstoppable Richard Carlile, outspoken freethinker and champion of the "pauper press" and female emancipation, who first printed and

distributed explicit contraceptive advice to the purportedly overpopulated working classes (Fryer 1965, 46–86; B. Taylor 1983, 22–3).

Carlile was the first person in England to put his name to a birth control pamphlet, which he embarked upon while serving a long term in Dorchester jail for
defying the press laws. It was Carlile who linked freethought, contraception and
free love inextricably together in the public mind. His *Every Woman's Book: or,
What is Love?* (1826) was not so much an argument for population control as for
greater freedom in sexual relations. Carlile offended the respectable and radical
alike with his claim that the pleasures of sexual love without fear of unwanted
pregnancy were the right of every healthy woman after the age of puberty; that
female disorder "in nine cases out of ten" was the result of unsatisfied sexual
desire; and by exhorting young women to take the initiative in choosing their
lovers.[12]

Carlile brought down on his unrepentant head not only the charge of propagat-
ing immoral doctrines and promoting female prostitution, but also the polemics
of avowedly anti-Malthusian radicals, such as William Cobbett, who alleged that
birth control served the interests of the propertied by seeking to eliminate the
"redundant" labouring population. In response, Carlile protested that he was as
much opposed to the "monstrous" propositions of "PARSON Malthus" as Cobbett, and did not simplistically connect poverty and unemployment with a super-
abundance of workers. Against the continuing hostility of his anti-Malthusian
critics he argued that there was no contradiction between defending contraception
and struggling for a reformed society.[13]

In fact, Carlile's advocacy of birth control had more to do with his crusade
against priestcraft than his conviction of the threat of a redundant population.
Every Woman's Book interspersed its examination of contraceptive practices with
suggestive asides on the sexual proclivities of nuns and the moral hegemony of
the Church. It was necessarily an appeal to a morality other than that sanctioned
by a Christian God, and in this lay much of the attraction of the birth control
movement for materialistic freethinkers. To teach workers that they could control
the very creation of life was to undermine the Christian faith in the most intimate
and effective way. The sequence of babies was not an act of God, not inevitable,
and might be avoided by a little self-help and the knowledge of some simple contraceptive practices (McLaren 1976, 244–5).

Charles Knowlton's *Fruits of Philosophy*, which appeared in New York in
1832, and was pirated and reprinted in London in 1833, sprang from this same tradition of freethought inspired birth control. The American connection was forged
by Robert Dale Owen, son of Robert Owen, at that stage optimistically seeking

12 Carlile 1838, 8. *Every Woman's Book* first appeared in Carlile's *Republican* of 1825, was published
as a pamphlet in 1826, and circulated in numerous subsequent editions. He claimed sales of more
than 10,000. See also McLaren 1976; Langer 1975.

13 McLaren 1976, 245. For the reaction of the "pauper press" to the discussion of contraception, see
also Fryer 1965, 79–86; Thompson 1975, 853–5.

to remake the New Moral World in America. The younger Owen, a convinced Neo-Malthusian, wrote his *Moral Physiology; or A Brief and Plain Treatise on the Population Question* (1831) in an effort to dissociate Neo-Malthusianism from the charges of gross indecency and free love with which Carlile had saddled it. He spelt out the economic advantages of contraception for the labouring classes, and dismissed the claim that the "power of preventing conception" would undermine chastity and turn women into prostitutes as a "libel on the sex". He made the radical counter claim that women should have the right to exercise control of their reproduction. Birth control, Owen argued, would not compromise more fundamental forms of social change, but advance it by improving the standard of living of working-class families, giving them more time and the means for self-improvement, and by increasing the value of their labour in the market. The real causes of poverty were "iniquitous laws, false education, and a vicious order of things"; but not even the "most perfect system of political or social economy in the world" could "of itself, prevent the *ultimate* evils of superabundant population".[14]

Knowlton's *Fruits of Philosophy*, which appeared the following year, owed a good deal to Owen's book. Knowlton was a practising physician, committed materialist and would-be author. His earlier self-published polemic, *Elements of Modern Materialism* (1829), had failed to find an audience and sent him badly into debt. His next publishing venture, *Fruits of Philosophy, or, the Private Companion of Young Married People*, grew out of a slim manuscript on methods of contraception that he had circulated amongst his patients. Knowlton had become convinced that too frequent births impaired the health of the women he attended, as well as imposing excessive financial burdens on parents. The "reproductive instinct" was a natural, healthy desire that should be gratified so long as it brought pleasure to women and men rather than misery. It was as much the physician's duty to provide the knowledge of the means of preventing the evils that were liable to arise from its gratification, as it was to inform his patients how they might keep clear of the gout or dyspepsia. Prudery should not compel his silence. Those practices that increased the sum of human happiness could not be called sinful or wicked.[15]

The Fruits of Philosophy was essentially a compilation of anatomical description, physiological information, contraceptive advice, and Knowlton's materialist philosophy, glossed with Owen inspired, Neo-Malthusian, economic arguments. Knowlton's originality lay in his unblinking detailing of human reproductive parts and processes, including the signs of pregnancy, and in his extended discussion of contraceptive "checks" in a work designed for public distribution. The fundamental message was that conception was the result of natural causes and was consonant with certain laws of the animal economy. There was nothing wonderful or

14 Owen 1832, 24, 37–9. On the connections between Owen's *Moral Physiology* and Knowlton's *Fruits of Philosophy*, see Fryer 1965, Ch. 10.
15 C. Knowlton, "Philosophical Proem" to *The Fruits of Philosophy*, reprinted in Chandrasekhar 1981, 95–7. Biographical detail on Knowlton from Fryer 1965, Ch. 10; Chandrasekhar 1981, 21–5.

inexplicable about it, and it could be prevented by a number of simple remedies.[16] Unlike Knowlton's previous publication, *Fruits of Philosophy* sold briskly, both in its American and pirated English editions. Less rewarding for its author were the prosecutions launched against him under the puritanical "Blue Laws" of Massachusetts and the fines and three months' hard labour in the house of correction that ensued. Knowlton was accused of providing a *"Complete Recipe* [for] the trade of a Strumpet" and of promoting "infidel" views (Fryer 1965, 104–5).

In Massachusetts, as across the Atlantic, atheism and immorality went hand in glove. Birth control, whether advocated by libertarian freethinkers like Carlile, or by earnest Neo-Malthusians like Bradlaugh and Besant, was necessarily implicated in this widespread attribution. The advocates of birth control were as responsible as their opponents for confusing the issue. Throughout the nineteenth century, "birth control was never presented simply as a means of limiting family size; it was portrayed as having as much if not more to do with poverty, politics and promiscuity" (McLaren 1990, 183).

In England the printing and circulation of *Fruits of Philosophy* was carried out entirely under the auspices of various leading freethinkers.[17] But it was not the identification of freethought with Knowlton's book that was the origin of what they referred to as their "worst difficulties" for Bradlaugh and Besant (Besant 1908, 197–8), but freethought's association with another work, the explosive *Elements of Social Science*. This so-called "Bible of the Brothel" had been written by George Drysdale (co-founder with Bradlaugh of the Malthusian League) while still a medical student at Edinburgh University. It was first published anonymously in 1854 by the prominent freethinker Edward Truelove, and went through many more editions. It was a "trumpet-blast against sexual abstinence, prostitution, and poverty" (Fryer 1965, 111) that echoed back to Richard Carlile and forward to scientific naturalism. In place of an oppressive and obstructive Christianity, Drysdale advocated a "physical, sexual and natural religion" based on the scientific knowledge of natural laws. Chief among these was the law of population, which was as true and demonstrable as the law of gravitation. Upon it depended the "grand problems which are at present convulsing society; the wages of labour, poverty and wealth, &c." The poor must be taught that the means of improving their condition lay in their own hands. The sexual abstinence advocated by

---

16 Knowlton discussed withdrawal, the *baudruche*, and the sponge, but recommended syringing the vagina immediately after "connection" with a solution of sulphate of zinc, alum, pearl-ash, or any other suitable salt. This "chemical check", according to Knowlton, had the advantages of greater certainty and of health and cleanliness for women (Knowlton, *Fruits of Philosophy*, reprinted in Chandrasekhar 1981, 87–147: 120–21, 137–40).

17 In their preface to the edition that sparked their spectacular trial, Bradlaugh and Besant claimed that for the last forty years Knowlton's book had been "identified with Freethought, advertised by leading Freethinkers, published under the sanction of their names, and sold in the head-quarters of Freethought literature. If during this long period the party has thus – without one word of protest – circulated an indecent work, the less we talk about Freethought morality the better . . ." See "Publishers' Preface", Knowlton, in Chandrasekhar 1981, 89–90.

Malthus was physically and socially injurious. Medical science could prove that it caused "more real disease and misery in one year . . . than sexual excesses in a century". Women and men should "exercise fully [their] sexual organs" and satisfy the needs of their bodies, especially in the years immediately after puberty. If society permitted the "natural" expression of the sexual appetite, there would then be less masturbation, prostitution, venereal disease and female hysteria. The only scientific solution to the natural problems of poverty and the potency of the sex instinct was contraception or "preventive intercourse". The main opposition, Drysdale claimed, came from men who thought that if women no longer feared becoming pregnant, they "would indulge their sexual desires, just as men do".

> Hence the vehement prejudices in favour of our present code of sexual morality, and of the institution of marriage, together with the determined hostility to anything in the shape of unmarried intercourse, at least on the part of women.
>
> (Drysdale 1861 [1854], 347–52)[18]

It was all extremely shocking, the sort of thing that gave medicine and science a bad name, let alone atheism. It exacerbated the deep-seated hostility of the medical profession, the religious and the respectable to Neo-Malthusianism, and it split the ranks of the Secularists themselves. Bradlaugh repudiated free love, but he could never escape the odium of having advertised and even commended the *Elements of Social Science* in the *National Reformer*. Much of the opposition to the Knowlton pamphlet was really opposition to Drysdale's work. It did not help matters that the Bradlaugh-Besant edition of the *Fruits of Philosophy* was updated with medical notes by "G. R." – whom they identified in their Preface as "the author of the 'Elements of Social Science'". It was this association that the prosecution seized upon at the Knowlton trial by alleging that Knowlton advocated pre-marital sex. He did not, but Drysdale most certainly did.[19]

In England, then, atheism implied the doctrine of free love, and the Knowlton trial reinforced this widespread popular impression. Like it or not (and they did not), Bradlaugh and Besant were irrevocably tied to this promiscuous image by a variously titillated or horrified public. Even Bradlaugh's rival secularist critics capitalized on this image by dubbing his organization the "Erotic School of Freethought" (Arnstein 1984, 22).

When Besant and Aveling began touting the Secularist variant of Darwinism initiated by Besant during the Knowlton trial, the reaction of the Darwinians was as much inspired by their fear of moral taint as of the expropriation of their creed to radical political ends. Another term was necessary to dissociate respectable Darwinian freethought from those who flaunted the tag of "atheist" with all its

18 See also the discussions by Ledbetter 1976, 8–19; Benn 1992, 3–11.
19 Chandrasekhar 1981, 90–1; Fryer 1965, 123–31; Royle 1980, 251.

suspect moral and political connotations. The leading Darwinian Thomas Henry Huxley had coined the gentlemanly and neutral sounding "agnostic" in 1867. The definition he proposed was remarkably similar to that given by Bradlaugh for "atheist", as the Secularists lost no time in pointing out. Agnosticism, protested Bradlaugh, was a "mere society form of Atheism". But, as that arch social strategist Huxley knew, a "society form" was crucial to the social and professional hegemony of the Darwinians. There was a lot in a name, and Huxley was very good at names. In 1879, Darwin, recoiling from the bellicose flag-waving atheism and Neo-Malthusianism of the Bradlaughites, took cautious shelter under Huxley's proffered standard: "I think that generally (and more and more as I grow older), but not always, that an Agnostic would be the more correct description of my state of mind", he told a correspondent.[20]

Besant vs. Darwin at the Knowlton trial

It was against this background that Besant and Bradlaugh decided to challenge the law when a leading English freethought publisher, Charles Watts, was prosecuted for publishing *Fruits of Philosophy* in 1876.[21] When Watts declined to be martyred for the cause and destroyed his plates and stock, Bradlaugh and Besant formed the Freethought Publishing Company, brought out a new edition of Knowlton's book, delivered the first copy to the chief clerk at the Guildhall, and informed the police of where and when they intended to sell it. The Secularists, meanwhile, were riven by fierce controversy over their actions, which temporarily abated when Bradlaugh and Besant were arrested and committed for trial. What rallied the Secularists was not the defence of Neo-Malthusianism, about which they remained bitterly divided, but their commitment to freedom of the press. While sales of Knowlton's book and the *National Reformer* (which exhaustively reported the preliminary hearing at the Guildhall and the subsequent trial) boomed,[22] the defendants prepared their case and set up a defence fund. The press gave excited coverage, and public opinion was mobilized for and against Bradlaugh and Besant. It was mostly against.

Hypatia Bradlaugh was to claim that it was Besant rather than Bradlaugh who had initiated the crusade, and that "she hardly realised all the gravity of her

20 Royle 1980, Bradlaugh quote at 115; Darwin cited in Moore 1988, 308–9; but see Lightman 2002 on the failure of Huxley's strategy.
21 The prosecution of Watts followed on that of a Bristol bookseller who, unknown to Bradlaugh and Besant at the time they decided to make a test case of the matter, had added "obscene" illustrations (possibly illustrations of the male and female genital systems) to the copies he sold. The bookseller, Henry Cook, was sentenced to two years hard labour. See Fryer, 160–4; Nethercot, 119–22; Ledbetter, 29–32.
22 Between the 1830s and 1877, *Fruits of Philosophy* sold about seven hundred copies a year. Five hundred copies were sold in the first twenty minutes of its sale by Bradlaugh and Besant, and, in the three months between their arrest and trial, some 125,000 copies were sold; Fryer 1965, 162; Royle 1980, 15–16.

situation" in the "atmosphere of excitement and admiration in which she was living" (Bonner 1895, 2: 23). Besant, who at not quite thirty was still a youthful and beautiful woman, was undoubtedly the star of the show, even if her radical views and activities made her a rather questionable luminary. Against the counsel of the more conservative Secularists that if she persisted it would mean ruin to her "as a lady", Besant insisted on appearing in the dock beside Bradlaugh and conducting her own defence.[23] This was a remarkable enough action for a woman, but it was made doubly so by the subject matter of the trial, and it added to the frenzy of the press.

Earlier, at their committal hearing, an attempt had been made by the bench to exclude women from the court on the ground that the case would necessitate "reference to the private parts of women," to the "various stages of generation," and other gross indecencies not fit for female ears. Bradlaugh and Besant had upheld "the right and the duty" of women to be present at the trial "where the question for the discussion of which we are struggling is a question of vital importance to the female sex".[24] Few women other than the one in the dock, however, dared to brave public condemnation and attend what the *Saturday Review*, with unusual restraint, called "the nasty case". Even Bradlaugh, in view of their youth and unmarried status, allowed himself to be persuaded not to insist upon the Misses Bradlaugh maintaining their "right of presence" in the court.[25]

On 18 June 1877, the case of the Queen v. Bradlaugh and Besant came on at the Queen's Bench before the Lord Chief Justice, Sir Alexander Cockburn. The trial lasted four days, and Bradlaugh was very much the supporting act to Besant's leading testimony of some 40,000 words, which took up almost two full days. Cockburn was courteous, attentive, even admiring, as the erudite and earnest Mrs. Besant tutored the jury in Malthusian economics and its terrible consequences: "our object in publishing this pamphlet was not in any way to arouse the passions of the unmarried, but simply to prevent the curse of pauperism". This "great social question" could only be solved by medical science, by knowledge of human physiology, and this knowledge should be available to all, but especially to the poor and oppressed who needed it most. It would be better to have no science at all, argued Besant, if that science was only to drive people to despair by giving them the knowledge of an approaching misery, while it was forbidden to discuss the means by which it might be averted:

> [Y]ou can no more discuss the population question without physiology than you can solve an arithmetical problem without figures. . . . The economical law teaches the danger of over-production; nature teaches us

23 Nethercot 1961, 126; A. Taylor 1992, 109.
24 *N.R.*, Vol. XXIX (16), 22 April 1877, 241; Besant, "Notes on the Trial and Committal," *N.R.*, XXIX (17), 29 April 1877, 267.
25 See note 2. Banks and Banks 1964, 90. Besant received little support, if any, from feminists, who in general opposed contraception; see Bland 1995, Ch. 5; Benn 1992.

that men and women will marry, and the object of medical science is to teach us how the law of political economy, and the law of nature, may be shown so to dovetail into each other, that they may be both obeyed by human beings without wrong or harm being done to anyone.[26]

Sexual desire, declared the fearless freethinker, was a natural human instinct, and its "lawful" satisfaction by the married should be a pleasure, not a source of anxiety, ill health and poverty:

> There is no harm in feeling thirsty because people get drunk . . . and there is no harm in gratifying the sexual instinct if it can be exercised without injury to anyone else, and without harm to the morals of society. . . . [T]here is a false and spurious kind of modesty, which sees harm in the gratification of one of the highest instincts of human nature – an instinct that goes through all the world, not only in the animal but in the vegetable kingdom – if you are to blame Dr. Knowlton because he recognises a great natural fact, then it is your duty to blame the constitution of the world, and the arrangements of nature, because you find that the reproductive instinct is attended with pleasure in its due gratification.[27]

In any case, and in spite of the views of "ascetics" intent on enforcing the unenforceable Malthusian remedies of late marriage and sexual abstinence, men and women ("but more especially men") would not remain celibate. Either they indulged in profligacy and prostitution, or they married and produced too many children, which led to great suffering and want. "Early marriage with restraint upon the numbers of the family" was the only moral and rational solution. Indeed, given the dreadful rate of infant mortality among the poor which was the only present check to the Malthusian population explosion, Besant put it to the jury that it was "more moral to prevent the birth of children than it is after they are born to murder them as you do today, by want of food, and air, and clothing, and sustenance".[28]

Of all the Malthusian authorities that Besant brandished before the court (and they included Mill, Martineau and the two Fawcetts)[29] Darwin (also in absentia of course) was her leading witness. She read to the jury a lengthy extract from "Darwin's great work on the 'Origin of Species'" in demonstration of the famous naturalist's "strongest" endorsement of the Malthusian doctrine and the subsequent universal struggle for existence. She went on to structure lower-class life in these same Darwinian terms, and gave a harrowing description of the "severe struggle for existence" that prevailed in the overcrowded tenements and slums

26 *N.R.*, XXIX (25), "Special Trial Number," 23 June 1877, 403, quotation at 405.
27 Ibid., 404.
28 Ibid., 407.
29 See note 3.

of Victorian England, where up to fourteen persons of both sexes and all ages might occupy one small, poorly ventilated, chamber. The physical and moral consequences – starvation, disease, infanticide, incest, illegitimacy, prostitution, criminality – were "fearful to contemplate". These were the real obscenities that were the "justification of this pamphlet". Surely, "the birth-restricting check", which did not compromise human happiness or national morality, was preferable to these "terrible checks" attributed to "nature and providence".[30]

The problem was, as she admitted, that Besant's authority on the universal operation of the Malthusian laws throughout the animal and vegetable kingdoms had himself, a "few days hence" and in the *Descent of Man*, strongly asserted the exact opposite opinion. If the case for the defence was not to be undermined it was necessary to undermine the master. Besant was equal to the task. With the aid of the relevant passage from the *Descent*, she explained to the court that Mr. Darwin had put forward the argument that these same terrible checks, the so-called "natural checks", are "good for the human species, and in this he is supported to a certain extent by Mr. Herbert Spencer". But Mr. Darwin, who was more used to dealing with "brutes" than with humans, had got it wrong:

> I have no doubt that if natural checks were allowed to operate right through the human, as they do in the animal world, this result would follow. But I may be allowed to direct attention to the point that Mr. Darwin has over-looked, the fact that these natural checks are not so allowed to operate among men and women. . . . Among the brutes the weaker are driven to the wall, the diseased fall out in the race of life, and the old brutes, when feeble or sickly, are killed. . . . If that were the case among men – if the drunken and improvident were over-ridden in the struggle for existence by those who were careful and temperate – the result might be to improve those who survived, and Mr. Darwin's position might be true.[31]

But this would mean that those who were sickly should be allowed to die without the help of medicine or science, those who were old and useless put to death, and the improvident allowed to starve. Were the Gentlemen of the Jury willing to do this, or allow it to be done? If they were not, then they must face the consequences of removing the natural checks and putting nothing in their place.

Besant had previously claimed that their appalling conditions "unhumanized" the poor and demoralized, and that the children born of such parents were "literally a lower race".[32] She now pressed home to the jury the "deterioration of the

---

30 *N.R.*, 23 June 1877, 407, 409–12.

31 Ibid., 413.

32 "All these causes operate in a double fashion, causing premature death among the adults, and in an even more terrible fashion upon the unhealthy children who are born of drunken and dissolute parents, amidst such awful associations, growing up only half human in many ways. It has often been remarked that when you get to the lowest grade of the criminal classes you observe a kind of

race" that must follow from the protection and fostering of the improvident, the criminal and the diseased, while the more provident and thrifty refrained from marriage. This objection, Besant thought, was "fatal to the ground that [Mr. Darwin] has taken up".[33]

The Lord Chief Justice was inclined to agree. "I think that is a point very well worthy the consideration of Mr. Darwin," he observed:

> Whether there may result, as a consequence of the struggle for existence among mankind, the survival of a smaller number of the strongest, or a larger number of the weaker, and whether, should it be found that the weaker survive, the race is not by that means in process of deterioration. The process might result in a few of a higher race, but the effect on the masses would be an increase of suffering and of misery.[34]

"That, my lord", concurred Besant triumphantly, "is just the point that I have been endeavouring to make".

> My contention is that these natural checks cannot have free operation among men. . . . [I]t is for that reason that we seek to bring in an artificial check. Nature balances herself, but if we remove her checks by civilization, and cure those whom she would kill, we must put some others in their place.[35]

Besant read again from her copy of the *Descent* to hoist Darwin with his own petard. Darwin had himself agreed that the "civilizing checks" of the poor laws and modern medicine were preserving the weak members of society, and that this was, in the long term, "highly injurious to the race of man."

> But Mr. Darwin does not meet his own argument. He states his case, he puts the fact upon record, but he does not suggest any positive remedy. . . . The checks we propose might eliminate the sickly, just as does the struggle for existence.[36]

---

marked type. This is because the lives of their parents have so unhumanized them, that the children born of such parents are literally a lower race than those of parents whose happier circumstances have raised them above that condition" Besant, *N.R.*, 23 June 1877, 411.

33 Ibid., 413.

34 Ibid.

35 Ibid.

36 Ibid. Besant may have adopted this line of argument through her contacts and discussions with one of the writers for the *National Reformer*, a mysterious "D", who contributed articles on evolution and Darwin, and who was possibly Dr. John Drysdale (see note 2), who wrote a number of works on scientific materialism (Nethercot 1961, 156). See for instance, "D", "Lower Animals and Higher," *N.R.*, 23 December 1877, 850–1, which advocates that "rational selection must take the place of natural selection" for humans.

Having out-Darwined Darwin by anticipating the rhetoric of racial degeneration and eugenics of later birth-controllers, Besant wound up her testimony with a rousing patriotic call upon the jury and to the applause of the court. Unless they could honestly believe that she was deliberately intent on corrupting the morals of the young under the pretence of trying to find a solution to "that terrible poverty and misery which is around us on every hand", unless they were prepared to "brand" her with "malicious meaning", "I ask you, as an Englishwoman, for that justice which it is not impossible to expect at the hands of Englishmen–I ask you to give me a verdict of 'Not Guilty', and to send me home unstained".[37]

The prosecuting Solicitor General was not interested in the finer points of Malthusian or Darwinian theory, nor in any other specious justification of such patently "immoral and wicked" practices. He was insistent that the only patriotic course for "any English jury, having any reverence for the married state, for the chastity and purity of their own wives and daughters" was to find the defendants guilty of publishing an obscene work "calculated to degrade and destroy public morals". As he notoriously declaimed,

> I say that this is a dirty, filthy book, and the test of it is that no human being would allow that book to lie on his table; no decently educated English husband would allow even his wife to have it. . . . It does not require any abstruse or recondite arguments. The object of it is to enable persons to have sexual intercourse, and not to have that which in the order of Providence is the natural result of that sexual intercourse.[38]

The Lord Chief Justice was not so sure. The Malthusian theory, he cautioned the jury, was accepted as an "irrefragable truth" by economist after economist. Witnesses in court had testified to the overcrowding of our cities and country towns, to the misery, disease and immorality it caused, to its destructive effects on women and children: "That the evils of over-population are real, and not imaginary, no one acquainted with the state of society in the present day can possibly deny". The defendants honestly believed that the checks they proposed would remedy these evils and improve society. The jury had to decide whether the good intentions of these two "enthusiasts" had led them to do wrong – whether the Knowlton pamphlet and the contraceptive information it contained tended "to corrupt the morals of society, and especially the morals and purity of women", and was therefore "an offence against the law".[39]

It was a fair and favourable summing up, but a bemused and confused jury, having retired for an hour and a half, returned to hand the Lord Chief Justice a no-win verdict: "We are unanimously of opinion that the book in question is calculated to inflame public morals, but at the same time we entirely exonerate the defendants

37 *N.R.*, XXIX (28), "Second Special Trial Number," 30 June 1877, 433–48, 435.
38 *N.R.*, XXIX (30), "Third Special Trial Number," 7 July 1877, 465–78, 474.
39 Ibid., 475–7.

from any corrupt motive in publishing it". This, decided a taken-aback Cockburn, had to be interpreted as a verdict of guilty.[40] He had no option but to fine Bradlaugh and Besant and sentence them to six months jail. They appealed and finally, after a long struggle in which Bradlaugh showed his great command of the law, in February 1878, the indictment was quashed on a technicality.

While Bradlaugh was fighting the case through the Court of Appeal, Besant, who had never been happy with the "coarse" style of Knowlton's pamphlet, rewrote her trial testament as a birth-control tract. *The Law of Population* was first serialized in the *National Reformer*, and then, with the addition of the crucial, more up-to-date medical information, published as a sixpenny pamphlet under Besant's name by the Freethought Publishing Company towards the end of 1877. Besant's pamphlet, which was dedicated to "the poor" and to "British mothers" in particular, superseded Knowlton, Drysdale and the rest, was translated into several languages, and sold hundreds of thousands of copies. It carried her rebuttal of Darwin's objections to birth control and her eugenic and Neo-Malthusian reinterpretation of his evolutionary arguments to all classes of people all over the world until she withdrew it herself in 1890.[41]

Then the woman who had shocked all of England with her public discussion of the unmentionable, who had declared at the Knowlton trial that sexual intercourse was a natural and necessary pleasure for women and men, proclaimed in accordance with her new-found Theosophical principles:

> Now the sexual instinct that he has in common with the brute is one of the most fruitful sources of human misery, and the satisfaction of its imperious cravings is at the root of most of the trouble of the world. . . .
> By no other road than by that of self control and self-denial can men and women set going the causes which on their future return to earth life shall build for them bodies and brains of a higher type.[42]

Much was to happen, however, before a middle-aged Besant joined the ranks of the "ascetics" that her younger self had scorned at the Knowlton trial.

A terrible price had first to be paid for the Knowlton "victory". Before the verdict was quashed Annie had received notice that Frank Besant had applied to the High Court of Justice for custody of their daughter Mabel on the grounds that her conviction for publishing an "indecent and obscene" pamphlet disqualified her from being a fit guardian for her child. The actual petition was not filed in Chancery until April 1878, by which time the grounds had shifted more towards Annie's atheism and her association with Bradlaugh, and all that this supposedly

---

40 Ibid., 478.
41 Besant [1877] 1884; reprinted in Chandrasekhar 1981, 149–201. The Darwin references occur on pp. 163–4, 196–7; see also *N.R.*, XXXIX (4 November 1877), 737–9.
42 *Theosophy and the Law of Population* (1896); reprinted in Chandrasekhar 1981, 203–12, quotation at 211; A. Taylor 1992, 251–2.

implied for Mabel's morals and happiness in this life and the next. The Knowlton pamphlet and Annie's *Law of Population*, however, featured prominently as exhibits, and one of the allegations her estranged husband filed against her was that Annie had stated at the Knowlton trial that children, including Mabel, should be taught the physiological facts contained in these works.

At the trial, Besant and Bradlaugh had claimed in their defence that many physiology texts in wide circulation (including Huxley's *Lessons in Elementary Physiology* and the illustrated *Principles of Human Physiology* by the eminent physiologist William Carpenter, that were used in science classes for young boys and girls), dealt with human reproduction "in a manner beside which our book is the utmost extreme of delicacy". Carpenter's *Physiology* was, on the basis of the same legal definition of obscenity, "immeasurably more obscene" than Knowlton's more circumspect physiological descriptions. This was, however, a two-edged defence: on the one hand, it exploited contemporary ambiguities about the legal status of scientific and medical representations of sexuality and reproduction; while on the other, Besant and Bradlaugh were concerned to argue that such works, including Knowlton's, should be readily and cheaply available to all without the implication of arousing indecent passions and being subject to the charge of obscenity. Indeed, Annie went on explicitly to argue that physiological knowledge of itself could not be considered obscene or corrupt the young when, on the contrary, it inspired "a feeling for Nature of reverence so deep and intense that there is no room left for a coarse or an impure thought". If less mystery were made about the matter, if all children (including her own dearly loved daughter) were given the necessary reproductive knowledge, "you would do away with the whole suggestion of indecency which hangs around it now," and tend to their future health and happiness.[43] These words were now to be turned against her.

Again Annie insisted on fighting her case in person, this time before an openly hostile judge intent on upholding all the Victorian values and virtues she had so blatantly assaulted. Witnesses testified in vain to the loving care and irreproachable education that Mabel was receiving. Her atheism apart, Annie's real crime was to have promulgated a doctrine "so repugnant and so abhorrent to the feelings of the great majority of Englishmen and Englishwomen", a doctrine that was such a violation of their notions of "morality, decency, and womanly propriety", that the "future of a girl brought up in association with such a propaganda would be incalculably prejudiced". If she were permitted to remain with her mother, Mabel

---

43 *N.R.*, XXIX (29), 1 July 1877, 426–7; see also Nethercot 1961, 143; A. Taylor 1992, 117; Dawson 2007, 127, 132, 135. Carpenter was an early promoter of progressive development and recognised as an important advocate of Darwinism. He was subpoenaed and cross-examined by Bradlaugh at the preliminary hearing of their case at the Magistrate's Court, where he appeared as a recalcitrant witness. Dawson argues that his mere association with such a notorious obscenity trial was damaging for Carpenter (his work already under suspicion for its allegedly suggestive content), and exacerbated the public hostility of the Darwinians to Secularism and birth control; ibid., Ch. 4.

would probably "grow up to be the writer and publisher of such works".[44] The double charge of atheism and Neo-Malthusianism was sufficient to decide the issue, and the eight-year old Mabel was taken from her distraught mother.

If anything, the public opprobrium that the Knowlton trial brought upon her and the unforgivable loss of her child hardened Besant's anti-Christianity. One of the weapons she deployed against it was her militant advocacy of the secularist Darwinian view of the world with which she had begun to engage during the trial. Before the trial, Darwin's had been more of a name for Secularists to conjure with. Much of what they wanted from him they already had from other sources, from that same tradition of radical materialism, progressive science and evolutionism that had produced Secularism.[45] Darwinism, however, was on the ascendency in the 1870s and 80s. It symbolized "progressive" modern science, and in the "warfare" between science and religion, Besant and the Secularists knew that Darwin, despite himself, was at the heart of the matter.

The Knowlton trial had forced a closer reading of Darwin on Besant. The opportunity to deepen these biological understandings of men and women and give them a more radical twist came with the advent of Edward Aveling onto the Secularist platform and into Besant's life. Hypatia Bradlaugh later attributed Annie's attraction to science to her attraction to the person of Dr. Aveling (Dinnage 1986, 48–51). Besant's interest in science was long-standing and genuine, but there can be little doubt, as the catty Hypatia observed, that she was powerfully drawn to the fascinating Aveling and he to the celebrated Mrs. Besant. Within a remarkably short time, Besant had decided to put her science studies on a formal footing. Under Aveling's personal tutelage, she qualified and enrolled for a science degree at London University, which, against heavy opposition, had just opened its doors to women. The victory of the progressives was applauded by the *National Reformer*. Science being necessarily antithetic to religion, its study would not simply improve the political situation of women, but bring them "rapidly into the ranks of Freethought".[46]

Secular science and radical sexual selection

Besant and Aveling were soon inseparable. While Aveling coached her in science and mathematics, with poetry readings on the side, Besant encouraged him to write for the *National Reformer*. His very first published contribution, early in 1879, was a pseudonymous article on "Darwin and his Views," in the course of which Aveling expounded Darwin's principles of sexual selection and transmission, and,

44 "Judgment of Lord Justice James," *N.R.*, 9 April 1879, 270–1.
45 Royle 1980, 170–2; Desmond 1989 is the authoritative text on the radical appeal of evolutionary ideas in pre-Darwinian Britain.
46 "What has Freethought to gain by the Higher Education of Women?" *N.R.*, 17 February 1878, 1031.

with his star pupil in mind, defended the mental abilities of women against their male detractors.[47]

Towards the end of 1879, the alliance of science and radical politics forged by Besant and Aveling at a personal level was consolidated institutionally with the establishment of the Science School at the Hall of Science, the London headquarters of the National Secular Society. These classes, initially taught by Aveling with the amateur but enthusiastic assistance of Besant, were affiliated with the South Kensington branch of London University and recognized as formal preparation for the South Kensington matriculation examinations.[48] They meant that Secularism could now lay claim to "science" in more than a merely nominal sense.

Simultaneously, Aveling cast aside his pseudonym and came out with a flourish under his own name in the *Reformer*. He renounced his ambition for any professional scientific position that could not be filled by an atheist, and joined the shock troops of the National Secular Society on the public lecture circuit. In short course, this seemingly dazzling new recruit, with his valuable scientific credentials, belligerent atheism, and outstanding lecturing skills, was taken onto the staff of the *National Reformer* and elected a Vice-President of the Society. Bradlaugh generously called in and paid off his accumulated debts – a substantial sum of £480.16.0. The Bradlaugh-Besant duo now became a threesome, irreverently dubbed "the Trinity" by fellow Secularists.[49]

What are we to make of the Dr. Edward Aveling who made such an impact on Besant and the Secularists? Four years younger than Besant, Aveling was the son of a Congregationalist minister, a brilliant scholar and would-be dramatist, Fellow of University College and lecturer in comparative anatomy at the London Hospital.[50] So much is fairly clear. The man himself remains, at best, ambiguous. Aveling has suffered such a bad press at the hands of vengeful Marxists that it is difficult to separate the young and promising science lecturer that Besant knew from the womanizing, black-mailing, cold-hearted has-been who provoked (and allegedly abetted) the suicide of Marx's favourite daughter (Feuer 1962). According to Shaw, who knew him well, "as a borrower of money and a swindler and seducer of women his record was unimpeachable. . . . He had the art of coaching [girl students] for science examinations. . . . The more fortunate ones got nothing worse for their money than letters of apology for breaking the lesson engagements. The others were seduced and had their microscopes appropriated." Shaw, never one to waste an opportunity, was to set Aveling up for Marxists and Fabians

---

47 [Aveling], "Darwin and his Views," *N.R.*, XXXIV, 9 February 1879, 83. See also [Aveling], "On Educated Women," ibid., 1 June 1879, 356–7; 15 June 1879, 394–5.

48 *N.R.*, 31 August 1879, 569; 2 November 1879, 714; 16 May 1880, 310–11.

49 Nethercot 1961, 163–4; Tribe 1971, 228–9; Aveling to Bradlaugh, 28 January 1880, Bradlaugh Papers, 585.

50 The fullest account of Aveling's early career is given in Standring 1881. See also Kapp 1972, 1: 258–72; Nethercot 1961, 158–67.

to recognize as the charming but unscrupulous Louis Dubedat in *The Doctor's Dilemma*.[51]

Aveling did indeed make off with the microscope of Mrs. Mary Reed, or so this promising pupil of the Science School alleged. Whether he seduced her as well is not known. When Aveling was finally drummed out of the Secularist movement in late 1884 after he had notified Bradlaugh of his intention to live with Eleanor Marx, Bradlaugh charged him with a long and squalid history of unpaid debts, fraud and deceit.[52] But all this lay in the future, and even Shaw, in a rare serious moment, testified to Aveling's redeeming feature: he was "morbidly scrupulous as to his religious and political convictions". He may have had "absolutely no conscience in his private life", but Aveling "would have gone to the stake for Socialism or Atheism".[53] Besant's attraction to such a man becomes more understandable.

Together, she and Aveling built up the flourishing Science School classes. By the end of 1881, Besant (along with Hypatia and Alice Bradlaugh) was formally certified as a science teacher by the Science and Arts Department of the Government, and became eligible for the government grants awarded by the Department to authorized teachers. These were paid on the basis of the number of students who successfully passed the state examination. In 1880–81 some sixty-two Science School students sat for the examinations. By 1882 the number of candidates had almost doubled and their success rate was well above the national average, as a jubilant Aveling reported in the *Reformer*.[54]

In 1880, Bradlaugh, after many years of campaigning, was finally elected to the Commons as the member for Northampton, the first declared atheist ever elected to Parliament. For the next six years Bradlaugh fought to take his seat, a struggle that at times descended into farce over the issue of whether he would be permitted to affirm, and when that was rejected, over the refusal of the Commons to allow an atheist to take the required Bible oath. He was twice taken into custody, once violently ejected from the House, and battled his way through an exhausting series of legal actions and re-elections, as the moderate Liberal government under Gladstone, besieged by Irish nationalists on one side and Bradlaugh and assorted radicals on the other, wrestled with the dilemma of how to deal with such a notorious atheist and convicted purveyor of obscenity. It was a *cause célèbre* that brought the dread triad of atheism, revolution and immorality to the very doors of Parliament. All the most offensive allegations of the Knowlton trial resurfaced in the face of this threat to the citadel of respectability and statesmanship.[55]

---

51 See Nethercot 1961, 160.
52 Notes for Bradlaugh's speech against Aveling. Bradlaugh Papers, 1156. See also Royle 1980, 105–6.
53 Cited in Kapp 1972, 1: 270–1.
54 *N.R.*, 16 May 1880, 310–11; 23 January 1881, 59–60; 31 July 1881, 135; Vol. 40, 1882, 54–5. See also Royle 1980, 317–19.
55 The best and most detailed account of Bradlaugh's parliamentary struggle is by Arnstein 1984.

Besant and Aveling threw themselves headlong into the fray, hammering on the doors of Parliament alongside Bradlaugh, and eloquently denouncing Christian bigotry and the denial of civil liberties to crowded audiences around the country. This brought their efforts at the scientific reformation of society and the Hall of Science classes under parliamentary scrutiny. Besant's conviction for obscenity and the legal loss of her daughter were invoked to allege her unfitness as a teacher of the young and to make political capital out of connecting Gladstone's ministry with the authorization of public moneys for the dissemination of atheism, republicanism, Neo-Malthusianism and worse. As evidence of the subversive intent of the science classes, Aveling's recently published evolutionary account ("in direct opposition to Paley") of certain properties of the frog was targeted as "condemnatory of God". The press had a field day. The *Standard* wondered whether the frog was a Conservative or a Radical, while the *Evening News* pointed out that "men like Professor Huxley, who are in receipt of large Government salaries, hold and teach the same doctrines on the evolution theory as Dr. Aveling" – a connection that may have gratified Aveling but hardly Huxley who was, among other things, Director of the Science and Arts Department that had certified Besant's teaching qualification and eligibility for a government grant.[56] While Huxley maintained a discreet silence, Anthony Mundella, the Liberal minister accountable for the Department, was compelled into exasperated defence of the Hall of Science classes as "very excellently" taught and the teachers as "lady-like and well-educated women". What they taught was simply "true science" without any reference to religion. If the Education Minister were to begin enquiring into the religious views of all science teachers, perhaps he had better begin at the top with Professor Huxley himself![57]

These interchanges were gleefully reported in the *National Reformer*, where Aveling gave elaborate thanks for this "gratuitous advertisement" for the science classes, claiming that enrolments had shot up as a result of the free publicity. Less amusing was the successful eviction of Aveling from his lecturing post at London Hospital for alleged insubordination and incompetence. Perhaps avoiding any closer investigation of his activities, Aveling did not contest his dismissal and transferred his comparative anatomy class to the Hall of Science.[58]

These skirmishes and setbacks simply reinforced the enthusiasm of the Secularists for science. Aveling repudiated the accusation that the Hall of Science teachers brought their politics and unbelief into the classroom: "We strive to teach science pure and simple". But "not for one moment" did he doubt that "ultimately the teachings of science must tell against Conservatism and old orders of thought, replacing these by new".[59] In keeping with their conviction of the radicalizing pow-

---

56 Arnstein 1984, 55–6, 250–1, 262–3; *N.R.*, 28 August 1881, 212; 4 September 1881, 234–5.
57 *N.R.*, 20 August 1882, 132–3.
58 Aveling, *N.R.*, 28 August 1881; 4 September 1881, 234–5; Besant, *N.R.*, 18 December 1881, 474–7; Nethercot 1961, 188–9.
59 Aveling, *N.R.*, 4 September 1881, 235.

ers of science, the "Trinity" embarked on an ambitious new publishing venture of a series of scientific works guaranteed to spread "heresy" among the masses. The first of these was Besant's translation of *Mind in Animals* by the German materialist, Darwinian, and freethinker, Ludwig Büchner. The second volume in the series was to be *The Student's Darwin*, which was based on Aveling's fast-accumulating articles on all aspects of Darwinian biology in the *National Reformer*.[60] Aveling now had the bold notion of securing the patronage of the very man before whose "genius" he had so often genuflected in the pages of the *Reformer*.

Before his emergence as a gazetted atheist, Aveling had sent Darwin copies of some articles he had published in a student magazine, and Darwin had written to thank this promising young popularizer of evolution and asked to see future instalments.[61] He got more than he bargained for when the newly prominent Secularist proposed dedicating his forthcoming book to his mentor. *The Student's Darwin*, Aveling informed Darwin, was to appear in the "International Library of Science and Freethought" under the editorship of "my friends Mrs. Annie Besant and Charles Bradlaugh, M. P."[62]

Darwin, chary of any public connection with the notorious duo, made even more alarmed by the furore over Bradlaugh's parliamentary campaign, was forced to some rapid back-pedalling from his earlier encouragement of Aveling as he sought to shake off these utterly undesirable fellow travellers. He could not permit the dedication, he told Aveling in a four-page letter conspicuously marked "Private":

> . . . though I am a strong advocate for free thought on all subjects, yet it appears to me (whether rightly or wrongly) that direct arguments against christianity & theism produce hardly any effect on the public; & freedom of thought is best promoted by the gradual illumination of men's minds, which follow[s] from the advance of science. It has, therefore, been always my object to avoid writing on religion, & I have confined myself to science.

Perhaps, added Darwin, shielding himself as an added precaution behind wife Emma's protective skirts, he had been "unduly biased by the pain" it would give "some members" of his family, if he "aided in any way direct attacks on religion".

---

60 *The Student's Darwin* was a revised version of the series "Darwin and his Works" by Aveling which appeared in 28 instalments in the *National Reformer* from 16 November 1879 until 19 September 1880.

61 Aveling to Darwin, 23 September 1878, and Darwin to Aveling, both CUL-DAR 202.

62 Aveling to Darwin, 12 October 1880, CUL-DAR 159. With his letter, Aveling sent Darwin a copy of Besant's translation of another of Buchner's pamphlets, *The Influence of Heredity on Freewill* (London: Freethought Publishing Co., 1880), which contained an advertisement explaining that the aim of the series was to "spread heresy" among the "reading masses". This, even were he unaware of Aveling's recent activities, would have been more than enough to alert Darwin to the dangers of identifying himself with such an enterprise.

Besides he was old, and had very little strength, and looking over Aveling's proofs would fatigue him very much.[63]

Repulsed but not repressed, Aveling sent Darwin a copy of *The Student's Darwin* on publication, begging forgiveness if its contents went beyond what Darwin might allow. When Darwin replied that he had "no objection" to people carrying his arguments further than he himself would consider "safe", Aveling was emboldened to telegraph the great man, requesting an interview for himself and Büchner who was in London attending the Congress of the International Federation of Freethinkers. Surprisingly, Darwin agreed, inviting them to Down on the following day. There, supported by a loyal but apprehensive Emma (who was buffered from too close contact with the dangerous atheists by the reassuring Tory presence of the Reverend Brodie Innes, the Down village vicar), by children and grandchildren, a patriarchal Darwin sat down to lunch between Büchner and Aveling.[64]

Lunch over, Darwin, his son Francis, and the two guests withdrew to the privacy of his study for some serious scientific discussion. There, unconstrained by wife, daughters, or clergy, the great man might indulge his curiosity and air his own version of freethought: "Why do you call yourselves Atheists?" In a masculine pall of cigarette smoke, they debated Aveling's rejoinder that "'Agnostic' was but 'Atheist' writ respectable, and 'Atheist' was only 'Agnostic' writ aggressive." Darwin, comfortably ensconced amidst his rural wealth, reproductive success, scientific fame, and upper-class privilege, challenged his radical, heavily indebted, socially and professionally marginalized disciple: "Why should you be so aggressive?" What was to be gained by trying to force these new ideas upon the masses? It was all very well for educated, cultured, thoughtful people, but were the masses yet ripe for it? Aveling turned his questions back at Darwin: What if the "revolutionary truths of Natural and Sexual Selection" had been enunciated only to the "judicious few", instead of being openly discussed and brought into the marketplace and home? Darwin's "own illustrious example" was an encouragement, a command, to every freethinker to proclaim his truths "abroad from the housetops", not keep them "only for the study". It was a well-aimed shaft, but it made no impression on Darwin who had done just that for the twenty years before 1859, and still refused to compromise his respectability by joining the assault on religion. Unperturbed, he agreed with his guests that Christianity was "not supported by evidence", but he had not forced this conclusion on himself, let alone the masses: "I never gave up Christianity until I was forty years of age."[65]

---

63 Darwin to [Aveling], 13 October 1880, in Feuer 1975, 2–3. Darwin's letter to Aveling was for a long time thought to have been addressed to Marx, and gave rise to the myth that Marx had wished to dedicate *Capital* to Darwin: Colp 1982; Desmond and Moore 1991, 643–5.

64 Aveling to Darwin, 27 September 1881. DAR 159. See Moore 1988, 310–12; Desmond and Moore 1991, 656–8.

65 Aveling, "A Visit to Charles Darwin," *N.R.*, 22 October 1881, 273–4; 29 October, 291–3.

The rigidly excluded, unmentionable but unavoidable, presence on this occasion was, of course, the notorious Besant – leading Secularist, dedicated evolutionist and birth controller, threat to the family and the future of "mankind", who had dared to contest Darwin's authority in court, who had put her interpretation of the Darwinian case for "the absolute right of women" to space their pregnancies and limit their families on sale at sixpence a copy (more than 40,000 copies of the *Law of Population* sold by 1881, and still rolling off the press),[66] but who might only be present by proxy through Aveling.

If they could not claim Darwin in life, the Secularists fought for him in death. When six months later Darwin died and was piously entombed in Westminster Abbey, Aveling shouted his "righteous indignation" from the rooftops (or at least from the *Reformer*). Not content with burying the man they had "reviled and maligned" with full religious honours, the hypocritical clergy now had the "audacity" to say that the teaching of evolution was "wholly in accord with that of the Church and the Bible". This "evil" attempt to resurrect the "brother and ally" of all freethinkers as a Christian, was combatted by Aveling with the published revelations of his "sacred" conversation at Down with his "master". The great Darwin was "on our side, fighting our great fight", as his own words to Aveling testified. There were, to be sure, a few regrettable phrases in his writings, but these were probably the result of "inheritance or even of reversion" to his class background: "Not wholly consciously to himself, he was, and ever will be, working with our cause." Aveling was proud to number himself amongst Darwin's "scientific children", and his "duty", his "joy", would be to "work out yet further the noble ideas" of his "great teacher".[67] These, as it happened, included particularly the "revolutionary truth" of sexual selection, with which both Aveling and Besant were much concerned around this time.

The earliest intimations of this are to be found in *The Student's Darwin* (which sat unmarked on Darwin's bookshelves along with his famously uncut presentation copy of Marx's *Das Kapital* – later translated by Aveling). Here was to be found a good deal that Darwin could hardly have considered "safe". In discussing the "Secondary Sexual Characteristics of Man", Aveling made a radical departure from the teachings of his master, and sketched out a dominant and responsible evolutionary role for female choice – that aspect of sexual selection that

66 By the time Besant withdrew her pamphlet in 1890, some 175,000 copies had sold; Chandrasekhar 1981, 57. Of the various birth control checks, Besant advocated the use of a "fine piece of sponge" which women should "pass to the end of the vagina" to prevent impregnation; thus giving women the "absolute right" to space their pregnancies and limit the number of children; ibid., 177, 188; McLaren 1990, 185–6. It might be noted here that Darwin's ideological objections to contraception evidently extended to his personal practices, that Emma Darwin had been almost continuously pregnant for the first eighteen years of their marriage, had borne ten children and suffered a number of miscarriages, giving birth to their last child, a baby with Down syndrome, at the age of forty-eight (Richards 2017, 51–2).

67 Aveling, "Charles Darwin," *N.R.*, 30 April 1882, 338–9. See also *N.R.*, 1 October 1882, 235; 22 October 1882, 273.

predominated among birds, but that Darwin had largely denied to human females in the *Descent of Man*, where he had asserted the predominance of male choice in human evolution and racial divergence.[68] At first, claimed Aveling, when brute force was all in all, it was not to be expected that the female would choose the "intellectual or artistic man". Bodily strength would be the main attractant to the female, and the law of battle would decide which men would breed and transmit their qualities to their descendants. Against Darwin and other commentators, Aveling argued that there was evidence amongst even the "very lowest tribes" that women exercised "a very considerable amount of choice". As women evolved increasing complexity of brain structure and function, they would begin to admire qualities other than mere bodily size and strength, colour and hairiness, and to choose between men on the basis of their mental and moral characters, not forgetting "the artistic nature of those from whom selection was to be made". Sexual selection for the advanced woman had become "far more complex" and its results "far more momentous" for society.

> To-day, woman has to consider beauty of face, beauty of form, social position, strength of mind, strength of moral character. She has to make her choice between men differing in infinite degrees in all particulars, and she has to make that choice conscious that her selection will influence not alone her life and his, but the lives which will spring from them. When . . . each man and woman in the search for the partner that is to make existence life, seeks after one that possesses both body and mind of the highest possible order, then will spring "the crowning race of human kind".
>
> (Aveling 1881a, 2: 321–4)

Aveling elaborated further on the eugenic potentialities of the intelligent female selection of just such a man as he fancied himself to be in his *Darwinism and Small Families* of 1882. Besant's influence is evident in his argument that birth control was compatible with natural selection and racial improvement, but now Aveling contributed the vital added factor of an informed choice of mate. The important point that the Darwinian critics of Neo-Malthusianism had missed was that it was not the *quantity* of the progeny that was crucial for social progress, but their *quality*. Women could play a powerful part in the "development of our race" through the "infinite care" they might exercise in their choice of partner. The socially aware and intelligent man might also contribute by not seizing upon the first woman he desired, as in the semi-brute days, but by deliberating, observing, and choosing a "true help-meet for her fairness and for the mental qualities that mark her as after his own heart" (Aveling 1882, 7–8).

---

68 In effect, Darwin naturalized female choice in animals, but, for what I have argued were primarily social and cultural reasons, normalized male choice in humans. See Richards 2017.

Besant's part in these theoretical constructions may be inferred from her pamphlet, *Marriage, As It Was, As It Is, and As It Should Be: A Plea for Reform*, which she published around the same time. This was based on the articles she had written in 1878 while she was preparing for her court battle for custody of Mabel. Annie, with her penchant for transmuting her personal experiences into public causes, had then written a revealing indictment of the institution of marriage.[69] The pamphlet version of 1882 gives some interesting insights into her state of mind at the time when her relationship with Aveling was at its most intense. It was a scathing denunciation of the Victorian marriage laws that made women the property and prisoners of their husbands and denied them their "rights" of control over their own bodies, health, reputation, personal liberty, property and children. Besant the Darwinian took an evolutionary view of the institution, but her account of woman's continuing "inferior" situation, shaped by her own experiences and her radical ideology, owed more to "Mrs. and Mr. Mill" whom she approvingly cited,[70] than to Darwin.

In the "savage state", Besant wrote, women were "naturally" kept in submission because of their smaller and less muscular stature and the demands of childbirth and childcare. Marriage was a "matter either of force, fraud, or purchase". Woman then had suffered from the "right of the strongest" even more than man. Woman's situation in late Victorian England was due not to her physical disadvantages, but to the "survival of the old system" which excluded women from all political life and "prevented the redressal that man has wrought for himself"; men had freed themselves from tyranny while their "own tyranny in the home" was maintained. The existing English marriage laws perpetuated the sexual inequalities of the savage state: "one would imagine that matrimony was a crime for which a woman deserved punishment, and that confiscation and outlawry were the fit rewards for her misdeeds". It is "impossible not to recognise the fact" that marriage is a "direct disadvantage" to women. The unmarried woman was much better off. For one thing, she had the "absolute right" to custody of her children: "none can step in and deprive her of her little ones". "Unlegalised unions", however, had the major disadvantages of illegitimacy for the children and "social disapprobation" for the woman, although "of late years, as women have been coming more to the front", this latter "difficulty has been very much decreased" among men and women "of advanced views". "As the law now is", there is only a choice between evils, but the evils are "overwhelmingly greater" on the side of legal unions.

> So great are these that a wise and self-respecting woman may well hesitate to enter into a contract of marriage while the laws remain as they are, and a man who really honours a woman must reluctantly subject her to

---

69 *N.R.*, vol. XXXI, beginning 17 February 1878, 1027–8.
70 Mill claimed that although *The Subjection of Women* (1869) was written after the death of his wife and long term companion, Harriet Taylor, it was their "joint production"; see Richards 2017, 449.

the disadvantages imposed on the English wife, when he asks her to take him as literally her master and owner.[71]

In place of the present legal form of marriage, Besant advocated a "simple declaration publicly made" like the Quakers, or, better still, and perhaps with Mill's famous declaration to Harriet Taylor in mind, a contract in writing – a "deed of partnership". Women would have a "fairer chance of happiness and comfort" in such partnerships. Besant urged those who were happily married but wanted reform to "support and strengthen by their open countenance and friendship those who enter into the unlegalised public unions here advocated". What was needed was "courage and quiet resolution" to change the laws, and, above all, equal and monogamous love between men and women. Amongst "savages", crude sexual selection operates:

> it is the female, not the woman, who is loved, although the savage rises higher than the lower brutes, and is attracted by individual beauty. The civilized man and woman need more than sex-difference and beauty of form; they seek satisfaction of mind, heart, and tastes as well as for body; each portion of the complex nature requires its answer in its mate. Hence it arises that true marriage is exclusive, and that prostitution is revolting to the noble of both sexes. . . . The fleeting connections supposed by some Free Love theorists are steps backward and not forward . . .

In the future, Besant concluded optimistically, men and women would be "bound by love instead of by law".[72] While Besant made it clear that she was advocating "equal and monogamous" love, not free love, her plea for the recognition of unlegalized public unions between men and women laid her open to wilful misconstruction. It confirmed the worst prejudices of those who, like Darwin, equated Besant's views and activities with the collapse of the nuclear family and "decent" civilized society (Bland 1995, 149–56).

If Besant had indeed reached the stage of contemplating a publicly contracted union with Aveling, as her writings at this stage might seem to imply, she may have had second thoughts with the discovery around this time that he was already married, though, like herself, separated from his wife and unable to obtain a divorce. Besant, or possibly Bradlaugh on her behalf, must have called him to account. Preserved among Bradlaugh's papers is a written declaration by Aveling of his marital situation, dramatically headed "Written Thursday Jan. 26. 1882. 6.45 a.m." This artful suggestion of a night of turmoil is followed by Aveling's revelations of incompatible marriage in his youth, religious conflict with his "High Church" wife, cruel betrayal when she committed adultery with a clergyman, and

---

71 Besant 1882, 4–5, 23, 32–3.
72 Ibid., 34–7, 58–60.

manly determination to forget and live a celibate and lonely life dedicated to his work – until now. "I thought then", wrote Aveling significantly, "I'd never need to do anything but try to forget the whole matter in work. I did not believe that the occasion would ever come for recalling the circumstances. I told her when we parted that I should be alone all my time & I thought thus then."[73]

Aveling told other accounts of his marriage at other times. This was his most histrionic, a version of events carefully calculated to win all Besant's Secularist sympathy and override the awkwardness of his amnesia on the matter. Besant may have been swayed, but Bradlaugh evidently had his suspicions. Perhaps his well-known antipathy to irregular unions was enough in itself to keep Besant from publicly taking a step that in Bradlaugh's eyes would have undermined his movement and ongoing Parliamentary campaign, still bedevilled by the far-reaching reverberations of the Knowlton trial. Whatever the reasons, the fervour of the romance cooled. Possibly Besant would not commit herself to the sexual relationship that Aveling clearly expected without the public declaration that she felt unable to make. Perhaps the inveterate philanderer Aveling had given her other reason to doubt his commitment to the "exclusive true marriage" she desired. Or Aveling may simply have tired of trying to live up to Besant's demandingly high moral standards. By late 1882, he began to move from Secularism and Besant towards the new socialism and its personification in the dark-eyed and excitingly bohemian daughter of Karl Marx, who had, it was rumoured, "peculiar views on love, etc . . . and somewhat 'natural' relations with men!"[74] It was with Eleanor Marx, not Besant, that Aveling was to work through and write his mature version of radical sexual selection and its relation to the woman question (Richards 2017, 500–2).

Besant lost not only Aveling (which was not perhaps such a great loss) but also her science degree. The presence of the notorious Mrs. Besant in the classroom had been a continual embarrassment to the liberal gentlemen who governed London University. Matters came to a head when Besant was refused admission to the botany classes at University College. The matter was widely publicized by the Secularists and Aveling persuaded a reluctant Huxley, that leading agnostic and enlightened advocate of female education, to sign a petition in protest against her unwarranted exclusion. But when it came to the crunch, Huxley, who had just been elected President of the Royal Society, and was rapidly putting as much distance as possible between himself and these persistent and undesirable fellow travellers, abstained from voting on the issue at the meeting of the College Council. The motion was lost, and Besant had to find another venue for her botany

---

73 Aveling, 26 January 1882, Bradlaugh Papers, 939; Tribe 1971, 227–9. Anne Taylor represents Aveling's letter as a "confession exacted under duress" after Bradlaugh and Besant became aware that Aveling had concealed the fact that he was married from Eleanor Marx, with whom Aveling already had begun a relationship (Taylor 1992, 167).

74 According to Beatrice Potter, later Mrs. Sydney Webb; quoted in Kapp 1972, 1: 284; see also A. Taylor 1992, 166–7.

instruction.[75] Huxley, it emerged, had a "strong feeling that freedom of thought should be carefully distinguished from laxity of morals." If Besant, as he had reason to think, was responsible for advocating "safeguards of sexual intercourse among unmarried people . . . we are out of the range of speculation and into that of practice – and I have no objection to her exclusion." "Freethinking," said Huxley primly, "does not mean Free Love."[76]

Huxley's need to dissociate "decent" Darwinism and his respectable self from such disreputables with their scandalous birth control, free love, and subversive evolutionary interpretations was compounded by the continuing success of the Hall of Science classes. These were competing for just that freethinking work-ing-class constituency that Huxley regarded as particularly his own. In addition, under the imprimatur of the Freethought Publishing Company, Aveling was busily producing a flood of ultra-cheap publications and pamphlets, with inclusionary, eye-catching titles (such as *The People's Darwin; or, Darwin Made Easy* and *The Gospel of Evolution*). These offered accessible introductions to evolutionary science to those with little education, who were hungry for new knowledge and impatient of old orthodoxies. The radical secular brand of Darwinism and sexual selection, connoting atheism and suspect morality, was reaching into territory unplumbed by professional Darwinians and middle-class popularizers (Paylor 2006; Richards 2017, 498–9).

With Darwin dead, Huxley was the self-anointed "High Priest" of evolution, defender of the faith, preacher of a puritanical "moralizing naturalism", intent on dissociating respectable Darwinian science from any taint of sensualism and immorality. Although Huxley recognised the pressing problem of over-population and its attendant "hordes of vice and pauperism" that would "destroy modern civilization", he eschewed any consideration of contraceptive methods that might morally compromise reputable Darwinism and his credo of scientific naturalism. Indeed, he was soon to argue, social struggle engendered by the "Malthusian ser-pent" of overpopulation was necessary and ultimately beneficial, ensuring the dominance of that "exceptionally endowed minority" with whom Huxley identi-fied. Increasingly, from the 1880s, Huxley tied Darwinian scientific naturalism to middle-class values and political conservatism.[77] His rhetoric of neutrality and balance upheld the social conventions, culminating in his celebrated "Evolution and Ethics" of 1893, which promoted an ethical code based firmly in traditional Victorian domestic ideology and conventional family pieties – one in which a patriarchal Huxley defended the social virtues of self-restraint, duty, and propriety against the incursions of those, like Besant, whose activities threatened to weaken

75 Aveling 1883b; Nethercot 1961, 188–9; A. Taylor 1992, 157–60; the *National Reformer* gave extensive coverage to the whole affair; *N.R.*, 1883, XLI: 361–2, 380–2, 409–10.
76 Letter from Huxley to Crompton, 16 July 1883, in Tribe 1971, 227; see Desmond 1997, 146–7.
77 Dawson 2007, Ch. 6; Desmond 1997, 147–51; Hale 2014, 209–21.

the bonds of the family and therefore the fabric of society, to bring about what Darwin had so greatly feared – "the greatest of all possible evils to mankind."[78]

"I am a believer in evolution"

Historians are agreed that the Knowlton Trial, by giving unprecedented publicity to the advocates of birth control, played a significant part in the long and complex history of contraception and the transition to the modern, conception limiting family. Besant's role in opening up public discussion of contraception in the face of the great opprobrium it brought down on her is well acknowledged, in spite of her subsequent renunciation.[79] Less recognised is the part she played (along with Aveling) in the dissemination and promotion of evolutionary ideas relevant to late-Victorian debates on the woman question and eugenics, particularly in relation to sexual selection and the Darwinian principle of female choice. Besant was among those assorted writers, feminists, social purists, neo-Malthusians, utopians, political radicals, socialists and sex reformers who kept the notion of female choice in play during a period when it was without serious support among professional naturalists and biologists (Richards 2017, 491–3).[80]

Ironically, it was the spiritualist and socialist Alfred Russel Wallace, co-founder with Darwin of the theory of natural selection but long-term opponent of the biological notion of sexual selection, who in a dramatic about-turn in 1890 put forward one of the better known theories of human progress based in free and informed female choice. The accepted inspiration for Wallace's *volt face* is the American socialist Edward Bellamy's best-selling social utopian novel *Looking Backward* (1888); but Wallace's adoption of the idea had a longer inception in the history of earlier deployments, such as those by Besant and Aveling, of radical versions of sexual selection. Female choice was commensurate with the continuous tradition of sexual radicalism and feminist utopianism that that ran from the time of the Owenite socialists to its re-emergence among the more radically minded toward the close of the Victorian period.[81] Both Wallace and Besant (whom he knew) were prominent exponents of this late-Victorian amalgam of evolutionary science, feminism, social reform and eugenics.

Wallace, seen by many after Darwin's death as the "greatest living champion" of evolutionary science and with his international renown at its height, might withstand the odium generated by orthodox Darwinians like Huxley, intent on

78 For a gendered reinterpretation of Huxley's "Evolution and Ethics", see Richards 1995, 2017, 511–15.

79 Chandrasekhar 1981, 46 and passim; McLaren 1990, 180–83; Cook 2004, 60–1; Sreenivas 2015, 510.

80 Kimberly Hamlin argues for a direct connection between the Darwinian concerns of Besant and the London Neo-Malthusians and the origins of the organized American birth control movement, primarily through Margaret Sanger who studied in London in 1915. She sees the Darwinian concept of female choice adopted by the Neo-Malthusians as a "major intellectual impetus" for Sanger's promotion of women's control of reproduction (Hamlin 2014, 149–65).

81 Richards 2005, 241, 2017, 492–3; B. Taylor 1983, 275ff. and passim; Hale 2010, 2014; Sreenivas 2015, 524–9.

excluding the political and supernatural from scientific discourse in the cause of an avowedly objective, "ideologically pure" professional science (Fichman 2004, 104–34, 211–57, 2015; Desmond 1997, 192–3, passim); but Besant had no such credentialing and little defence against the continuing prejudice and misrepresentation she encountered. Wallace, for all his "other" status as spiritualist and socialist and his lack of formal scientific qualifications, was no disreputable sexual reformer or birth control advocate, and was almost as chary as Darwin and Huxley of any hint of moral laxity.[82] Where Besant gave women an active (though monogamous) sexuality in the new society, Wallace's version of sexual selection retained a fairly traditional rendering of femininity. Wallace's feminism did not escape Victorian assumptions of an essentially passive female sexuality and of the central significance of the institution of marriage in any society, including the society of the future. His "woman of the future" of the post-socialist society, she of the "cultivated mind and pure instincts", was not the seeker but the sought. Her sexual and social power lay in her economically and socially secured ability to say no to those suitors who did not come up to her exacting moral standards. Wallace's promotion of free female choice thus had less to do with sexuality than with the exercise of woman's spiritually endowed, superior moral judgment (Wallace 1890, 329–30; Richards 2017, 400–2, 506–11).

Wallace's imprimatur on the notion of biologically and socially improving free female choice was given enhanced validity in the close integration of late nineteenth century feminism with emerging eugenic concerns. Women's rights activists (often in association with the social purity movement) recast female virtue as eugenic virtue: women's marriage choices, in contrast with the inherently "unhealthy tendency of men to promiscuity and vice", were not only moral, but also the sign of rationally exercised and socially and biologically improving female will. For the most part, they wanted as little to do with Besant, birth control, and the taint of atheistic free love as Huxley.[83] Her radical advocacy of birth control in the 1870s placed Besant almost as far outside the political and sexual mainstream of the feminist movement as of Victorian society at large.[84] In any case, by the time Wallace enunciated his version of female choice, Besant too had converted to spiritualism, denounced "brutish" sexual passion, and publicly endorsed celibacy. "Gone to Theosophy", declared Shaw, striking her name from the roll call of Fabians (A. Taylor 1992, 251–5).

---

82 Wallace argued that under socialism the Malthusian ratio of population growth to food production would be subverted by appropriation of production and better distribution and exchange: hence the artificial competition engendered by capitalism would be ameliorated, the subsistence needs of a growing population would be accommodated, women would be better educated and freed from financial dependence upon men, early marriage and its consequences would be discouraged, and fertility rates would drop without the need for birth control (Wallace 1890; Hale 2010, 23–4).

83 Richardson 2003, 48–9; Bland 1995, 229–35; Phillips 2004, 258–90; Sreenivas 2015, 524–9.

84 Western birth control campaigns were primarily eugenic, rather than feminist; see Carey 2012.

After Aveling's defection, although she had continued to teach in the School of Science for a few years, Besant gradually lost heart in the enterprise. She abandoned her science degree, and then her Secularism, for Fabianism. But she retained her faith in evolution. Looking to biology, rather than economics, to inform her politics, Besant now proclaimed that she was a socialist first and foremost because "*I am a believer in evolution*". The "great tree of life" linked all organisms and ensured both organic and social progress. Society could not but evolve from the "lowest savagery" through increasing understanding and united social effort to the realization of the fabled "Golden Age" of brotherhood, equality and liberty, here "in reality on earth".[85]

Fabian socialism famously lived by the slogan "Evolution, not Revolution", upholding a policy of patient economic reform, of gradual, incremental social change. Besant's prodigious energy, militancy and feminism were decidedly mismatched to the political reformism, the middle-class quietism, of the entrenched inner male coterie of Fabians. A "fifth wheel to the Fabian coach" (the description is Shaw's, as is his characterization of Besant during this period as a one-woman "expeditionary force"), Besant tried unsuccessfully to propel these armchair intellectuals into more direct political action, and herself led the famous matchgirls' strike at Bryant & May's London factory in 1888.[86]

After further failure at putting radical sexual selection into personal practice with the relentlessly facetious and evasive Shaw (who later claimed she had no sex appeal),[87] and, following this, a desperate, unrequited passion for the crusading editor of the *Pall Mall Gazette*, William T. Stead, Besant put on the ring of Madame Helena Blavatsky and espoused Theosophy (A. Taylor 1992, 188–202). Theosophy, like spiritualism at large, gave women a voice that Victorian sexual politics mostly denied them. The "darkened room" was one of the few arenas in which women could shine, where women might play influential, even dominant, roles (A. Owen 1990). It has to be said, though, that Besant had never permitted her particular blend of feminist aspiration, self-assertion and political evangelism to be outshone, nor her right to a voice to be denied. With characteristic verve, single-minded determination and organizational prowess, she rocketed up the ranks of the Theosophists, and after Blavatsky's death in 1891 Besant assumed leadership of the worldwide Theosophy Society. She made her headquarters in India for the last forty years of her life, awaiting the "Coming of the World Teacher", while becoming energetically involved in new educational projects (featuring the integration of information from the astral plane with practical experimental science lessons[88])

---

85 Besant 1886a, 2–3, 1886b, 24. On Fabian socialism and its evolutionary model, see MacKenzie and Mackenzie 1977, 191–206; Hale 2014, 175–6, 189–202.

86 A. Taylor 1992, 182; Oppenheim 1989, 15; Raw 2009.

87 Nethercot 1961, 241; Taylor represents this as "Shavian self-regard", and casts doubt on Shaw's version of events; A. Taylor 1992, 184–5.

88 Conlin suggests that Besant's Indian based work with Charles Ledbetter is credited in some circles with pioneering research into nuclear physics, and thus makes it hard to dismiss her as a lonely crank; Conlin 2014, 211.

and playing a prominent (and typically controversial) political role in the nascent Indian Nationalism movement, confronting racism and British imperialism (A. Taylor 1992, 259–326; Sreenivas 2015).

Removed from the contradictions of a radical program of sexual selection in late Victorian England, the self-styled Dr. Besant propounded a kind of spiritual evolutionary progress via successive reincarnations. This process, unlike the Darwinian and Secular versions, was not random, not dependent on a cruel, capricious and ultimately humiliating natural and sexual selection, but was guided by the benevolent Masters who had themselves attained Nirvana and freedom from the necessity of rebirth in the flesh (Oppenheim 1989, 15–18).

It was both a denial of desire and a vision of an alternative spiritualized evolutionary science that sustained Besant for the rest of her long and active life.

Acknowledgements

Earlier versions of this paper were read at the Annual Conference of the American Society for History of Science, Santa Fe, 1993; at the T. H. Huxley: Victorian Science and Culture Conference, Imperial College, London, 1995; and at the Annual Conference of the Australasian Association for History, Philosophy and Social Studies of Science, University of Sydney, 2002. It benefited from comments and discussions on these occasions. I should also like to thank the libraries of Cambridge University, the British Library, and the Bishopsgate Institute for permission to study correspondence and manuscript material.

Bibliography

Newspaper articles and manuscript sources are cited fully in the notes, and are not included in this list of references.

Adams, William E. 1903. *Memoirs of a Social Atom*. 2 vols. London: Hutchinson & Company.
Arnstein, Walter L. 1984. *The Bradlaugh Case: Atheism, Sex, and Politics among the Late Victorians*. Columbia: University of Missouri Press.
Aveling, Edward Bibbins. 1881a. *The Student's Darwin*. 2 vols. London: Freethought Publishing.
———. 1881b. "A Visit to Charles Darwin." *National Reformer*, October 22, 273–4, October 29, 291–93.
———. 1882. *Darwinism and Small Families*. London: Freethought Publishing.
———. 1883a. "At University College." *National Reformer*, July 29, 67–9.
———. 1883b. *The Religious Views of Charles Darwin*. London: Freethought Publishing.
———. 1884. *The Gospel of Evolution*. London: Freethought Publishing.
———. n.d. *The People's Darwin; or, Darwin Made Easy*. London: R. Forder.
Banks, J. A., and Olive Banks. 1964. *Feminism and Family Planning in Victorian England*. Liverpool: Liverpool University Press.
Bashford, Alison. 2014. *Global Population: History, Geopolitics and Life on Earth*. New York: Columbia University Press.

Benn, J. Miriam. 1992. *The Predicaments of Love*. London: Pluto Press.

Besant, Annie. 1874. *The Political Status of Women*. 3rd edn. London: Freethought Publishing.

———. [1877] 1884. *The Law of Population: Its Consequences, and Its Bearing upon Human Conduct and Morals*. London: Freethought Publishing.

———. [1877] 1885. *My Path to Atheism*. 3rd edn. London: Freethought Publishing.

———. 1882. *Marriage, as It Was, as It Is, and as It Should Be*. London: Freethought Publishing.

———. 1886a. *Why I Am a Socialist*. London: Printed by Annie Besant and Charles Bradlaugh.

———. 1886b. *The Evolution of Society*. London: Freethought Publishing.

———. [1893] 1908. *An Autobiography*. 2nd edn. London: Theosophical Publishing House.

———. 1896. *Theosophy and the Law of Population*. London: Theosophical Publishing Society, reprinted in Chandrasekhar 1981, 149–201.

Bland, Lucy. 1995. *Banishing the Beast: English Feminism and Sexual Morality, 1885–1914*. Harmondsworth: Penguin.

Bonner, Hypatia Bradlaugh. 1895. *Charles Bradlaugh: A Record of his Life and Work*. 2 vols. London: T. Fisher Unwin.

Browne, Janet. 2002. *Charles Darwin: The Power of Place*. London: Jonathan Cape.

Carey, Jane. 2012. "The Racial Imperatives of Sex: Birth Control and Eugenics in Britain, the United States and Australia in the Interwar Years." *Women's History Review* 21: 733–52.

Carlile, Richard. [1818] 1838. *Every Woman's Book; or, What is Love?* London: A. Carlile.

Chandrasekhar, S. 1981. *"A Dirty, Filthy Book": The Writings of Charles Knowlton and Annie Besant on Reproductive Physiology*. Berkeley: University of California Press.

Colp, Ralph. 1982. "The Myth of the Darwin-Marx Letter." *History of Political Economy* 14: 461–82.

Cook, Hera. 2004. *The Long Sexual Revolution: English Women, Sex, and Contraception 1800–1975*. Oxford: Oxford University Press.

Darwin, Charles. 1871. *The Descent of Man, And Selection in Relation to Sex*. 2 vols. London: John Murray.

Dawson, Gowan. 2007. *Darwin, Literature and Victorian Respectability*. Cambridge: Cambridge University Press.

Desmond, Adrian. 1989. *The Politics of Evolution; Morphology, Medicine and Reform in Radical London*. Chicago: University of Chicago Press.

———. 1997. *Huxley: Evolution's High Priest*. London: Michael Joseph.

Desmond, Adrian, and James Moore. 1991. *Darwin*. London: Michael Joseph.

Dinnage, Rosemary. 1986. *Annie Besant*. Harmondsworth: Penguin.

Drysdale, George. [1854] 1861. *Elements of Social Science; or, Physical, Sexual, and Natural Religion*. London: E. Truelove.

Feuer, Lewis S. 1962. "Marxian Tragedians: A Death in the Family." *Encounter* 19: 23–32.

———. 1975. "Is the 'Darwin-Marx Correspondence' Authentic?" *Annals of Science* 32: 1–12.

Fichman, Martin. 2004. *An Elusive Victorian: The Evolution of Alfred Russel Wallace*. Chicago: University of Chicago Press.

———. 2015. "Alfred Russel Wallace on Science and the Problems of Progress." *Victorian Review* 41 (2): 71–89.

Fryer, Peter. 1965. *The Birth Controllers*. London: Secker & Warburg.

Hale, Piers J. 2010. "Of Mice and Men: Evolution and the Socialist Utopia. William Morris, H. G. Wells, and George Bernard Shaw." *Journal of the History of Biology* 43: 17–66.

———. 2014. *Political Descent: Malthus, Mutualism, and the Politics of Evolution in Victorian England*. Chicago: University of Chicago Press.

Hamlin, Kimberly A. 2014. *From Eve to Evolution: Darwinian Science and Women's Rights in Gilded Age America*. Chicago: University of Chicago Press.

Kapp, Yvonne. 1972. *Eleanor Marx*. 2 vols. London: Lawrence and Wishart.

Knowlton, Charles. n.d. *Fruits of Philosophy: An Essay on the Population Question*. 2nd edn. London: Freethought Publishing Co., reprinted in Chandrasakar 1981, 87–147.

Langer, William L. 1975. "The Origins of the Birth Control Movement in England in the Early Nineteenth Century." *Journal of Interdisciplinary History* 14: 669–86.

Ledbetter, Rosanna. 1976. *A History of the Malthusian League, 1877–1927*. Columbus: Ohio State University Press.

Lightman, Bernard. 2002. "Huxley and Scientific Agnosticism: The Strange History of a Failed Historical Strategy." *British Journal for the History of Science* 35: 271–89.

Mackenzie, Norman I., and Jeanne Mackenzie. 1977. *The Fabians*. New York: Simon and Schuster.

McLaren, Angus. 1976. "Contraception and the Working Class: The Social Ideology of the English Birth Control Movement in its Early Years." *Comparative Studies in Society and History* 18: 236–51.

———. 1990. *A History of Contraception, From Antiquity to the Present Day*. Oxford and Cambridge: Blackwell.

Moore, James R. 1988. "Freethought, Secularism, Agnosticism: The Case of Charles Darwin." In *Religion in Victorian Britain*, edited by Gerald Parsons, 274–319. Manchester: Manchester University Press.

Nethercot, Arthur. 1961. *The First Five Lives of Annie Besant*. London: Rupert Hart-Davis.

Oppenheim, Janet. 1989. "The Odyssey of Annie Besant." *History Today* 39 (9): 12–18.

Owen, Alex. 1990. *The Darkened Room: Women, Power and Spiritualism in Late Victorian England*. Philadelphia: University of Pennsylvania Press.

Owen, Robert Dale. 1832. *Moral Physiology; or, A Brief and Plain Treatise on the Population Question*. 8th edn. London: E. Truelove.

Paylor, Suzanne. 2006. "Edward B. Aveling: The People's Darwin." *Endeavour* 29: 66–71.

Peart, Sandra J., and David M. Levy. 2008. "Darwin's Unpublished Letter at the Bradlaugh-Besant Trial: A Question of Divided Expert Judgment." *European Journal of Political Economy* 24: 343–53.

Phillips, Melanie. 2004. *The Ascent of Woman: A History of the Suffragette Movement and the Ideas Behind It*. London: Abacus.

Raw, Louise. 2009. *Striking a Light: The Bryant and May Matchwomen and their Place in Labour History*. London: Continuum.

Richards, Evelleen. 1995. "Gendering the Romanes Lecture: The Sexual Politics of T. H. Huxley's *Evolution and Ethics*." Paper presented at T. H. Huxley: Victorian Science and Culture Conference, Imperial College, London.

———. 2005. "The Whole Wallace: Mapping the Multi-dimensional Man." *Metascience* 14: 237–41.

———. 2017. *Darwin and the Making of Sexual Selection*. Chicago: University of Chicago Press.

Richardson, Angelique. 2003. *Love and Eugenics in the Late Nineteenth Century: Rational Reproduction and the New Woman*. Oxford: Oxford University Press.

Rover, Constance. 1970. *Love, Morals and the Feminists*. London: Routledge and Kegan Paul.

Royle, Edward. 1971. *Radical Politics, 1790–1900: Religion and Unbelief.* London: Longman.
———. 1980. *Radicals, Secularists and Republicans: Popular Freethought in Britain, 1866–1915.* Manchester: Manchester University Press.
Sreenivas, Mytheli. 2015. "Birth Control in the Shadow of Empire: The Trials of Annie Besant, 1877–1878." *Feminist Studies* 41 (3): 509–37.
Standring, George. 1881. "Biography of Edward Aveling." *The Republican* 7: 353–4.
Taylor, Anne. 1992. *Annie Besant, A Biography.* Oxford and New York: Oxford University Press.
Taylor, Barbara. 1983. *Eve and the New Jerusalem: Socialism and Feminism in the Nineteenth Century.* New York: Pantheon Books.
Thompson, Edward P. 1975. *The Making of the English Working Class.* Harmondsworth: Pelican.
Tribe, David. 1971. *President Charles Bradlaugh, M.P.* London: Elek Books.
Wallace, Alfred Russel. 1890. "Human Selection." *Fortnightly Review* 48: 325–37.

INDEX

Note: Numbers in *italics* indicate a figure.